Curves and Surfaces

Curves and Surfaces

Sebastián Montiel
Antonio Ros

Translated by Sebastián Montiel

Graduate Studies
in Mathematics

Volume 69

American Mathematical Society
Real Sociedad Matemática Española

This work was originally published in Spanish by Proyecto Sur de Ediciones, S. L. under the title *Curvas Y Superficies* © 1998. The present translation was created under license for the American Mathematical Society and is published by permission.

Translated by Sebastián Montiel

2000 *Mathematics Subject Classification.* Primary 53A04, 53A05, 53C40.

For additional information and updates on this book, visit
www.ams.org/bookpages/gsm-69

Library of Congress Cataloging-in-Publication Data

Montiel, Sebastián, 1958–
  [Curvas y superficies. English]
  Curves and surfaces / Sebastián Montiel, Antonio Ros.
    p. cm. — (Graduate studies in mathematics, ISSN 1065-7339 ; v. 69)
  Includes bibliographical references and index.
  ISBN 0-8218-3815-6 (alk. paper)
  1. Curves on surfaces. 2. Geometry, Differential. 3. Submanifolds. I. Ros, A. (Antonio), 1957– II. Title. III. Series.

QA643.M6613   2005
516.3′62—dc22

2005048190

This book is dedicated to Eni and Juana, our wives, and to our sons and daughters. We are greatly indebted to them for their encouragement and support, without which this book would have remained a set of teaching notes and exercises.

Contents

Preface to the English Edition   xi

Preface   xiii

Chapter 1. Plane and Space Curves   1

  §1.1. Historical notes   1

  §1.2. Curves. Arc length   2

  §1.3. Regular curves and curves parametrized by arc length   8

  §1.4. Local theory of plane curves   9

  §1.5. Local theory of space curves   14

  Exercises   20

  Hints for solving the exercises   24

Chapter 2. Surfaces in Euclidean Space   31

  §2.1. Historical notes   31

  §2.2. Definition of surface   32

  §2.3. Change of parameters   38

  §2.4. Differentiable functions   40

  §2.5. The tangent plane   44

  §2.6. Differential of a differentiable map   46

  Exercises   55

  Hints for solving the exercises   58

Chapter 3. The Second Fundamental Form   67

  §3.1. Introduction and historical notes   67

§3.2.   Normal fields. Orientation   69

§3.3.   Gauss map and the second fundamental form   76

§3.4.   Normal sections   81

§3.5.   Height function and the second fundamental form   85

§3.6.   Continuity of the curvatures   88

Exercises   94

Hints for solving the exercises   97

Chapter 4.   Separation and Orientability   107

§4.1.   Introduction   107

§4.2.   Local separation   108

§4.3.   Surfaces, straight lines, and planes   111

§4.4.   The Jordan-Brower separation theorem   116

§4.5.   Tubular neighbourhoods   120

Exercises   125

Hints for solving the exercises   126

§4.6.   Appendix: Proof of Sard's theorem   131

Chapter 5.   Integration on Surfaces   135

§5.1.   Introduction   135

§5.2.   Integrable functions and integration on $S \times \mathbb{R}$ 136

§5.3.   Integrable functions and integration on surfaces   139

§5.4.   Formula for the change of variables   144

§5.5.   The Fubini theorem and other properties   145

§5.6.   Area formula   151

§5.7.   The divergence theorem   157

§5.8.   Brower fixed point theorem   161

Exercises   163

Hints for solving the exercises   165

Chapter 6.   Global Extrinsic Geometry   171

§6.1.   Introduction and historical notes   171

§6.2.   Positively curved surfaces   173

§6.3.   Minkowski formulas and ovaloids   182

§6.4.   The Alexandrov theorem   186

§6.5.   The isoperimetric inequality   188

Exercises   195

Hints for solving the exercises 198

Chapter 7. Intrinsic Geometry of Surfaces 203

§7.1. Introduction 203

§7.2. Rigid motions and isometries 206

§7.3. Gauss's Theorema Egregium 210

§7.4. Rigidity of ovaloids 214

§7.5. Geodesics 219

§7.6. The exponential map 229

Exercises 239

Hints for solving the exercises 242

§7.7. Appendix: Other results of an intrinsic type 252

Chapter 8. The Gauss-Bonnet Theorem 275

§8.1. Introduction 275

§8.2. Degree of maps between compact surfaces 276

§8.3. Degree and surfaces bounding the same domain 284

§8.4. Index of a field at an isolated zero 289

§8.5. The Gauss-Bonnet formula 295

§8.6. Exercise: The Euler characteristic is even 304

Exercises: Steps of the proof 305

Chapter 9. Global Geometry of Curves 309

§9.1. Introduction and historical notes 309

§9.2. Parametrized curves and simple curves 313

§9.3. Results already shown on surfaces 319

§9.4. Rotation index of plane curves 329

§9.5. Periodic space curves 338

§9.6. The four-vertices theorem 346

Exercises 350

Hints for solving the exercises 353

§9.7. Appendix: The one-dimensional degree theory 365

Bibliography 371

Index 373

Preface to the English Edition

This book was begun after a course on the classical differential geometry of surfaces was given by the authors over several years at the University of Granada, Spain. The text benefits largely from comments from students as well as from their reaction to the different topics explained in the lectures.

Since then, different parts of the text have been used as a guide text in several undergraduate and graduate courses. The more that time passes, the more we are persuaded that the study of the geometry of curves and surfaces should be an essential part of the basic training of each mathematician and, at the same time, it could likely be the best way to introduce all the students concerned with differential geometry, both mathematicians and physicists or engineers, to this field.

The book is based on our course, but it has also been completed by including some other alternative subjects, giving on one hand a larger coherence to the text and on the other hand allowing the teacher to focus the course in a variety of ways. Our aim has been to present some of the most relevant global results of classical differential geometry, relative both to the study of curves and that of surfaces.

This text is indeed an improved and updated English version of our earlier Spanish book *Curvas y Superficies* [**12**], published by Proyecto Sur de Ediciones, S. L., Granada, in 1997, and republished in 1998. We are indebted to our colleague Francisco Urbano for many of the improvements and corrections that this English version incorporates. We also recognize a great debt to Joaquín Pérez, another colleague in the Departmento de

Geometría y Topología of our university, for creating the sixty-four figures accompanying the text.

It is also a pleasure to thank Edward Dunne, Editor at the American Mathematical Society, who proposed the possibility of translating our Spanish text, giving us the opportunity of exposing our work to a much wider audience. We also owe a great debt of gratitude to our Production Editor, Arlene O'Sean, for substantially improving the English and the presentation of the text.

The authors gratefully acknowledge that, while this book was being written, they were supported in part by MEC-FEDER Spanish and EU Grants MTM2004-00109 and MTM2004-02746, respectively.

S. Montiel and A. Ros
Granada, 2005

Preface

Over the past decades, there has been an outstanding increase in interest in aspects of global differential geometry, to the neglect of local differential geometry. We have wished, for some years, to present this point of view in a text devoted to classical differential geometry, that is, to the study of curves and surfaces in ordinary Euclidean space. For this, we need to use more sophisticated tools than those usually used in this type of book. Likewise, some topological questions arise that we must unavoidably pay attention to. We intend, at the same time, that our text might serve as an introduction to differential geometry and might be used as a guidebook for a year's study of the differential geometry of curves and surfaces. For this reason, we find additional difficulties of a pedagogical nature and relative to the previous topics that we have to assume a hypothetical reader knows about. These different aspects that we want to consider in our study oppose each other on quite a number of occasions. We have written this text by taking the above problems into account, and we think that it might supply a certain point of equilibrium for all of them.

To read this book, it is necessary to know the basics of linear algebra and to have some ease with the topology of Euclidean space and with the calculus of two and three variables, together with an elementary study of the rudiments of the theory of ordinary differential equations. Besides these standard requirements, the reader should be acquainted with the theory of Lebesgue integration. This novelty is one of the features of our approach. Even though students of our universities usually learn the theories of integration before starting the study of differential geometry, in the past, these theories have been used only in a superficial way in the introductory courses to this subject. Instead, we will use integration as a powerful tool to obtain geometrical results of a global nature.

The second basic difference in our approach, with respect to other books on the same subject, is the particular attention that we will pay, on several occasions throughout the text, to some questions of a topological character. The understanding of such questions will appear sometimes as a goal in itself and often as an essential step to obtaining results of a purely geometrical type.

A third remarkable characteristic of this course on classical differential geometry, this time of a pedagogical nature, although it also has some consequences in the theoretical development of the text, is the authors' determination to use a language free of coordinates whenever possible. We think that, in this way, we get a clearer statement of the results and their proofs.

We would like to thank Manuel Ritoré for his work on drawing the figures accompanying the text. We would also like to thank someone for having typed it and for having decoded the numerous and endless error messages of LATEX when it was processed, but, in truth, we cannot. Instead, it is easy to recognize our indebtedness to the students of the third year of mathematics at the University of Granada, who, for three years, brought about, with their questions and misgivings, new proofs and points of view relative to diverse problems of classical differential geometry. We are also indebted to earlier textbooks on this subject, mainly to those cited in the bibliography and in a very special way to the book *Differential Geometry of Curves and Surfaces* ([2]) of M. P. do Carmo. With it, we passed from ignorance to surprise with respect to the knowledge of differential geometry.

Last, we point out that, for the sake of self-containedness, we have included answers to many of the exercises that we posed within the main text and in the list of exercises that we have included in each chapter. The exercises that we have chosen to answer are not necessarily the most difficult, but those more significant from our point of view. We have marked them with a vertical arrow, like ↑, at the beginning of the exercise.

S. Montiel and A. Ros
Granada, 1997

Chapter 1

Plane and Space Curves

1.1. Historical notes

At the very beginning of calculus, LEIBNIZ and NEWTON started to deal with the first problems of plane curves. In 1684, LEIBNIZ already tried to define the curvature of a plane curve in his *Meditatio nova de natura anguli contactus et osculi*, although, in the 14th century, the bishop of the French city of Lisieux, NICOLAS D'ORESME, had also tried to do it, at least in a qualitative way, in his *Tractatus de configurationibus qualitatum et motum*. In fact, the first great scholar was LEONHARD EULER (1707–1783), who initiated so-called intrinsic geometry in 1736 by introducing the *intrinsic coordinates* of a plane curve: *arc length* and *curvature radius*—instead of curvature. The definition of *curvature* as the rate of variation of the angle that the tangent line makes with a given direction is due to this author himself. The theory of closed curves with constant width was also started by him. Moreover, he made, together with JOHN and DANIEL BERNOUILLI (1667–1748 and 1700–1782, respectively), some contributions to the theory of surfaces. In fact, he was the first to characterize *geodesics* as solutions of certain differential equations, and he showed that a material point which is constrained to move on a surface and is not subjected to any other forces has to move along one of its geodesics.

GASPARD MONGE (1746–1818), the engineer of the French army, began the theory of space curves in 1771, motivated by some problems of a practical nature—military fortification. However his first publications did not appear until 1785. MONGE had a clear precedent: the *Recherches sur les courbes à double courbure* that ALEXIS CLAIRAUT (1713–1765) wrote when he was only sixteen. In this treatise written during his youth, CLAIRAUT already made a notorious attempt to study the geometry of space curves. In 1795 MONGE published his results on space curves in a work that has been looked upon as the first book on differential geometry: *Applications de l'analyse à la géométrie*. The importance that this text attained can be realized by bearing in mind that its fifth edition appeared only thirty years after the death of MONGE. Among his students, we find mathematicians and physicists such as LAPLACE, MEUSNIER, FOURIER, LANCRET, AMPÈRE, MALUS, POISSON, DUPIN, PONCELET, and OLINDE RODRIGUES. The analytical methods to studying curves, already introduced in EULER papers, were not exactly applied by MONGE in the same way as we do it nowadays. In fact, it was AUGUSTIN CAUCHY (1789–1857) who was the first to define notions such as *curvature* and *torsion* in a really modern manner in his *Leçons sur l'application du calcul infinitésimal à la géométrie*.

It was only in 1846 that the first complete compilation of results, written in a more legible manner than usual in this period, was published. It is the treatise on space curves by BARRÉ DE SAINT-VENANT (1791–1886). He was the creator of the term *binormal*. The unification of all the previous efforts is due to F. FRENET and J. SERRET (1816–1868 and 1819–1885, respectively), who independently divulged the so-called *Frenet equations* about 1847 and 1851, respectively. However, these equations had been already written in 1831 by the Esthonian mathematician KARL E. SENFF (1810–1917), a student of M. BARTELS, who, in turn, had been a fellow student of GAUSS at Göttingen. The contributions by SENFF were probably ignored when they appeared because of their lack of the linear algebra language. The last three French authors mentioned above belonged to the group of mathematicians associated to the *Journal de Mathématiques pures et appliquées*, published without interruptions since 1836, the year in which JOSEPH BERNOUILLI (1809–1882) founded it. Finally, we have to mention GASTON DARBOUX (1824–1917), who started the theory of curves as we are used to studying it today, introducing the notion of *moving frame*.

1.2. Curves. Arc length

To start with, there are two ways of thinking about what one may understand intuitively as a *curve*. The first one is as a geometrical locus, that is, as a set of points sharing a certain property. The second one is as the path described by a moving particle. In this case, the coordinates of the curve should be functions of a parameter which usually stands for time. At the moment we will adopt this second approach because it allows us to use more quickly the techniques of calculus to describe the geometrical behaviour of the curve. In spite of this, we are mainly interested in those properties depending only on its image set, that is, on its *geometrical shape*. This is why, in Chapters 2 (Example 2.18 in Section 2.2) and 9, we will again take up the relation between these two ways of thinking of curves.

Definition 1.1. A (differentiable)[1] curve is a C^∞ class map $\alpha : I \to \mathbb{R}^3$ defined on an open, possibly unbounded, interval I of $\mathbb{R}$. Referring to a vector valued function α, the word *differentiable* means that, if

$$\alpha(t) = \big(x(t), y(t), z(t)\big),$$

the three coordinate functions x,y,z are differentiable. The vector

$$\alpha'(t) = \big(x'(t), y'(t), z'(t)\big)$$

whose components are the derivatives of the components of α is called the *tangent vector*—or *velocity vector*—of the curve α at $t \in I$—or at the point $\alpha(t)$, even though this last expression is ambiguous.

We will say that the curve $\alpha : I \to \mathbb{R}^3$ is a *plane curve* if there exists a plane $P \subset \mathbb{R}^3$ such that $\alpha(I) \subset P$. Since we may always dispose of a rigid motion of $\mathbb{R}^3$ (see comments after Proposition 1.6) taking the plane P into the plane $z = 0$ and considering only those properties of the curves which are invariant under rigid motions, we can restrict ourselves to the case $\alpha : I \to \mathbb{R}^3$ where

$$\alpha(t) = \big(x(t), y(t), 0\big),$$

that is, to the case of differentiable maps $\alpha : I \to \mathbb{R}^2$.

Example 1.2. A curve may have *self-intersections* (i.e., it does not necessarily have to be injective), as the plane curve $\alpha : \mathbb{R} \to \mathbb{R}^2$ given by

$$\alpha(t) = (t^3 - 4t, t^2 - 4)$$

shows (see Figure 1.1).

Example 1.3. The image or *trace* of a curve may present some *cusps* even though the curve is differentiable, for instance (see Figure 1.2), the curve $\alpha : \mathbb{R} \to \mathbb{R}^2$ given by

$$\alpha(t) = (t^3, t^2).$$

One can check that $\alpha(\mathbb{R})$ is the graph of the function $f : \mathbb{R} \to \mathbb{R}$ defined by $f(x) = x^{2/3}$, which is not differentiable at the origin.

---

[1]From now on, differentiable will always mean of the C^∞ class.

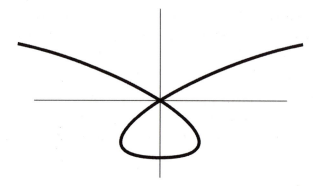

Figure 1.1. *Curve with a self-intersection*

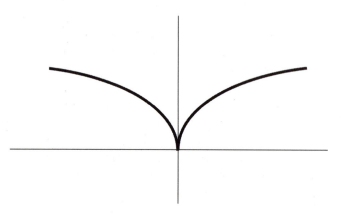

Figure 1.2. *Curve with a cusp*

Example 1.4. Even if a curve is injective, it does not have to be a homeomorphism onto its image. This is the case (see Figure 1.3) for the *Descartes folium* $\alpha : (-1, \infty) \to \mathbb{R}^2$ given by

$$\alpha(t) = \left(\frac{3t}{1 + t^3}, \frac{3t^2}{1 + t^3} \right).$$

Example 1.5. The following curves will be emphasized from now on because of their properties:

$$\alpha(t) = pt + q,$$
$$\beta(t) = c + r \left(\cos \tfrac{t}{r}, \sin \tfrac{t}{r} \right),$$
$$\gamma(t) = \left(a \cos \frac{t}{\sqrt{a^2 + b^2}}, a \sin \frac{t}{\sqrt{a^2 + b^2}}, \frac{bt}{\sqrt{a^2 + b^2}} \right),$$

where $p, q, c \in \mathbb{R}^2$, $r > 0$, and $a, b \neq 0$. They are, respectively, *straight lines*, *circles*, and *circular helices*. (See Figure 1.4.)

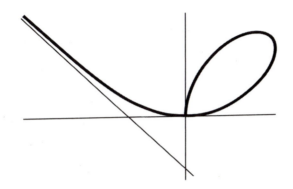

Figure 1.3. *Descartes folium*

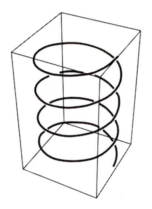

Figure 1.4. *Circular helix*

The first *intrinsic* notion for curves that we will study is *arc length*. Let $\alpha : I \to \mathbb{R}^3$ be a curve and $[a, b] \subset I$ a compact interval. We want to measure the length of the piece $\alpha([a, b])$ of the curve. A first intuitive approach is given by the length of a polygonal line inscribed in the curve. Precisely, let $P = \{t_0 = a < t_1 < \cdots < t_n = b\}$ be a partition of the interval $[a, b]$. We denote by

$$L_a^b(\alpha, P) = \sum_{i=1}^{n} |\alpha(t_i) - \alpha(t_{i-1})|$$

the length of that polygonal line, where $|u|$ is the Euclidean length of any vector $u \in \mathbb{R}^3$. Then we may prove

Proposition 1.6. *If $\alpha : I \to \mathbb{R}^3$ is a curve and $[a, b] \subset I$, then for every $\varepsilon > 0$ there exists $\delta > 0$ such that*

$$|P| < \delta \implies \left| L_a^b(\alpha, P) - \int_a^b |\alpha'(t)| \, dt \right| < \varepsilon$$

where $|P| = \max_{1 \leq i \leq n} |t_i - t_{i-1}|$.

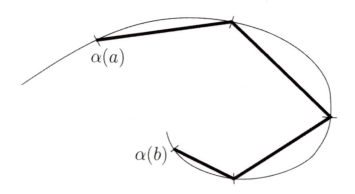

Figure 1.5. *Inscribed polygonal*

Proof. We define the function $f : I^3 \to \mathbb{R}$ by

$$f(t_1, t_2, t_3) = \sqrt{x'(t_1)^2 + y'(t_2)^2 + z'(t_3)^2}$$

which is clearly continuous on I^3 and so uniformly continuous on the compact set $[a, b]^3 \subset I^3$. Thus, for each $\varepsilon > 0$, there is a $\delta > 0$ such that, if $(t_1, t_2, t_3), (t'_1, t'_2, t'_3) \in [a, b]^3$,

$$\left. \begin{array}{l} |t_1 - t'_1| < \delta \\ |t_2 - t'_2| < \delta \\ |t_3 - t'_3| < \delta \end{array} \right\} \implies |f(t_1, t_2, t_3) - f(t'_1, t'_2, t'_3)| < \frac{\varepsilon}{b - a}.$$

On the other hand, by using the Cauchy mean value theorem for the functions x,y,z on the interval $[t_{i-1}, t_i]$, we have

$$|\alpha(t_i) - \alpha(t_{i-1})| = f(\beta_i, \gamma_i, \delta_i)(t_i - t_{i-1})$$

for some $\beta_i, \gamma_i, \delta_i \in [t_{i-1}, t_i]$. Hence

$$L_a^b(\alpha, P) = \sum_{i=1}^n f(\beta_i, \gamma_i, \delta_i)(t_i - t_{i-1}).$$

Moreover

$$\int_a^b |\alpha'(t)| \, dt = \sum_{i=1}^n \int_{t_{i-1}}^{t_i} |\alpha'(t)| \, dt$$

and, by applying the mean value theorem to each of the integrals on the right side,

$$\int_a^b |\alpha'(t)| \, dt = \sum_{i=1}^n |\alpha'(\xi_i)|(t_i - t_{i-1}) = \sum_{i=1}^n f(\xi_i, \xi_i, \xi_i)(t_i - t_{i-1})$$

where $\xi_i \in [t_{i-1}, t_i]$. Now, if $|P| < \delta$, we have $t_i - t_{i-1} < \delta$ for $i = 1, \ldots, n$. As $\beta_i, \gamma_i, \delta_i, \xi_i \in [t_{i-1}, t_i]$, we also have $|\beta_i - \xi_i| < \delta, |\gamma_i - \xi_i| < \delta, |\delta_i - \xi_i| < \delta$

and so

$$\left| L_a^b(\alpha, P) - \int_a^b |\alpha'(t)| \, dt \right| = \left| \sum_{i=1}^n \big(f(\beta_i, \gamma_i, \delta_i) - f(\xi_i, \xi_i, \xi_i) \big)(t_i - t_{i-1}) \right|$$

$$\leq \sum_{i=1}^n \left| f(\beta_i, \gamma_i, \delta_i) - f(\xi_i, \xi_i, \xi_i) \right|(t_i - t_{i-1}) < \frac{\varepsilon}{b-a} \sum_{i=1}^n (t_i - t_{i-1}) = \varepsilon,$$

and this concludes the proof. $\qquad\qquad\qquad\qquad\qquad\qquad\qquad\qquad$ $\square$

This result allows us to call *the length of α from a to b*—or *from $\alpha(a)$ to $\alpha(b)$*—the number

$$L_a^b(\alpha) = \int_a^b |\alpha'(t)| \, dt.$$

An important property of the length of a curve that we have just defined is its invariance through rigid motions of Euclidean geometry. In fact, this invariance property will appear repeatedly within this text and it will be a proof that the concepts that will be defined are really geometrical notions.

Definition 1.7 (Rigid motion). Recall that a rigid motion $\phi : \mathbb{R}^n \to \mathbb{R}^n$ of Euclidean space $\mathbb{R}^n$ is nothing but an isometry of $\mathbb{R}^n$ viewed as a metric space with the usual Euclidean distance. These isometries are necessarily affine maps whose associated linear maps are orthogonal, i.e.,

$$\phi(x) = Ax + b, \quad \forall x \in \mathbb{R}^n,$$

where A is a matrix of the orthogonal group $O(n)$, that is, a matrix of order n such that $AA^t = I_n$, and $b \in \mathbb{R}^n$ is a vector. Depending on either $\det A = \pm 1$, one says that the rigid motion ϕ is either direct or inverse. In the first exercise below we will announce the invariance under rigid motions that we referred to before.

Exercise 1.8. Let $\alpha : I \to \mathbb{R}^3$ be a curve and $M : \mathbb{R}^3 \to \mathbb{R}^3$ a rigid motion. Prove that $L_a^b(\alpha) = L_a^b(M \circ \alpha)$. That is, rigid motions preserve the length of curves.

Exercise 1.9. ↑ Let $\alpha : I \to \mathbb{R}^3$ be a curve and $[a, b] \subset I$. Prove that

$$|\alpha(a) - \alpha(b)| \leq L_a^b(\alpha).$$

In other words, straight lines are the shortest curves joining two given points.

Exercise 1.10. ↑ Let $\alpha : I \to \mathbb{R}^3$ be a curve. Show that

$$L_a^b(\alpha) = \sup\{L_a^b(\alpha, P) \mid P \text{ is a partition of } [a, b]\}.$$

Exercise 1.11. Let $\alpha : I \to \mathbb{R}^3$ be a curve such that $|\alpha'(t)| = 1$ for each $t \in I$. What is the relation between the parameter t and the length $L_a^t(\alpha)$, where $a \in I$? If a curve has this property, we will say that it is *parametrized by the arc length*.

Given two open intervals I and J in $\mathbb{R}$, a diffeomorphism $\phi : J \to I$, and a curve $\alpha : I \to \mathbb{R}^3$, we may consider a new curve $\beta : J \to \mathbb{R}^3$ defined by the composition $\beta = \alpha \circ \phi$. We will say that β is a *reparametrization* of α because it has the same trace as α. The following exercise points out that the arc length is an intrinsic notion, since it only depends on the trace of the curve and not on the precise parametrization being worked with.

Exercise 1.12. ↑ Let $\phi : J \to I$ be a diffeomorphism and $\alpha : I \to \mathbb{R}^3$ a curve. Given $[a, b] \subset J$ with $\phi([a, b]) = [c, d]$, prove that $L_a^b(\alpha \circ \phi) = L_c^d(\alpha)$.

1.3. Regular curves and curves parametrized by arc length

A usual method in differential geometry consists of substituting for a given differentiable object, a curve in our case, the best linear object approaching it near a certain point.

Given a curve $\alpha : I \to \mathbb{R}^3$ and $t \in I$, among all the straight lines in $\mathbb{R}^3$ passing through $\alpha(t)$, the one having direction $\alpha'(t)$ is seemingly the *closest* one to the curve at that point. Of course, this only makes sense when $\alpha'(t) \neq 0$; see Example 1.3 in the previous section. In this case, we will say that this straight line is the *tangent line of α at t*—or at $\alpha(t)$. A curve is said to be *regular* when its tangent vector does not vanish anywhere, i.e., when its tangent line is a well-defined object at any of its points. If this happens, we also may establish the following definition: we say that the *normal line* of a regular plane curve at a given point is the line through this point which is perpendicular to the corresponding tangent line.

Let $\alpha : I \to \mathbb{R}^3$ be any curve and $t_0 \in I$. We define the *arc length function from t_0*, and denote it by $S : I \to \mathbb{R}$, by

$$S(t) = L_{t_0}^t(\alpha) = \int_{t_0}^t |\alpha'(u)| \, du.$$

As the function $u \in I \mapsto |\alpha'(u)|$ is, in general, merely continuous, the function S is only C^1 and

$$S'(t) = |\alpha'(t)|.$$

If we suppose the curve α to be regular, then S is a differentiable increasing open—because of the inverse function theorem—function. Then, if we put $J = S(I)$, the function $S : I \to J$ is a diffeomorphism between two open intervals. Let $\phi : J \to I$ be the inverse diffeomorphism and let $\beta : J \to \mathbb{R}^3$

ɔ reparametrization of α given by $\beta = \alpha \circ \phi$. This new curve satisfies

$$\beta'(s) = \alpha'\big(\phi(s)\big)\phi'(s) = \frac{\alpha'\big(\phi(s)\big)}{|\alpha'\big(\phi(s)\big)|}$$

and so $|\beta'(s)| = 1$ for all $s \in J$; see Exercise 1.11 above. It follows from the above that *any regular curve admits a reparametrization by arc length.* From now on, unless we state explicitly the opposite, we will consider only curves parametrized by arc length, and we will use the abbreviation p.b.a.l.

Exercise 1.13. Consider the *logarithmic spiral* $\alpha : \mathbb{R} \to \mathbb{R}^2$ given by

$$\alpha(t) = (ae^{bt}\cos t, ae^{bt}\sin t)$$

with $a > 0$, $b < 0$. Compute, for $t_0 \in \mathbb{R}$, the arc length function $S : \mathbb{R} \to \mathbb{R}$ corresponding to t_0. Reparametrize this curve by arc length and study its trace.

1.4. Local theory of plane curves

In many geometrical problems the choice of a reference system adapted to the situation that one is dealing with simplifies its resolution remarkably. Next we will associate an orthonormal frame of $\mathbb{R}^2$ to every point along a plane curve p.b.a.l. whose evolution will determine its more important properties.

Let $\alpha : I \to \mathbb{R}^2$ be a curve p.b.a.l. Henceforth we will denote by $T(s)$ its tangent vector $\alpha'(s)$ which, in this case, is a unit vector. So $T : I \to \mathbb{R}^2$ is a differentiable vector valued function with $|T(s)| = 1$ for each $s \in I$. There are only two unit vectors in $\mathbb{R}^2$ perpendicular to $T(s)$. We pick up from them $N(s) = JT(s)$, where $J : \mathbb{R}^2 \to \mathbb{R}^2$ is the ninety degree counterclockwise rotation. We will call $N(s)$ the *normal vector of α at s* —or *at $\alpha(s)$.* With this choice, $N : I \to \mathbb{R}^2$ is differentiable and satisfies $|N(s)| = 1$, $\langle T(s), N(s) \rangle = 0$, and $\det(T(s), N(s)) = 1$ for every $s \in I$, where $\langle\,,\,\rangle$ is the usual scalar product and det is the determinant or mixed product. Then we have built, for each $s \in I$, an orthonormal basis of $\mathbb{R}^2$ which is positively oriented. This basis $\{T(s), N(s)\}$ will be called the *oriented Frenet dihedron* of the curve α at s.

A way of measuring how a curve bends is to observe how the basis $\{T(s), N(s)\}$ associated to each of its points varies when we move along the curve. This change can be controlled by means of the derivatives $T'(s)$ and $N'(s)$ of the two vector functions in that basis. Taking derivatives of the equalities

$$|T(s)|^2 = |N(s)|^2 = 1 \quad \text{and} \quad \langle T(s), N(s) \rangle = 0,$$

we obtain that

$$\langle T'(s), T(s) \rangle = \langle N'(s), N(s) \rangle = 0 \quad \text{and} \quad \langle T'(s), N(s) \rangle + \langle T(s), N'(s) \rangle = 0.$$

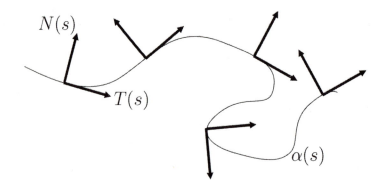

Figure 1.6. *Frenet dihedron*

Hence, the vector $T'(s)$ has the same direction as $N(s)$. Then $T'(s) = k(s)N(s)$ for some $k(s) \in \mathbb{R}$, and this occurs for every $s \in I$. Moreover, by definition,

$$k(s) = \langle T'(s), N(s) \rangle$$

and so the function $k : I \to \mathbb{R}$ is differentiable. The real number $k(s)$ is called the *curvature of the curve* α *at* $s \in I$. Then we have

$$T'(s) = k(s)N(s) \quad \text{and} \quad N'(s) = -k(s)T(s),$$

which are the *Frenet equations* of the curve α.

Remark 1.14. What follows is a first justification for the name of curvature that we have given to the function k. We have $|k(s)| = |T'(s)| = |\alpha''(s)|$, which is the length of the acceleration vector of α. Since $|\alpha'(s)| = 1$ for all $s \in I$, this acceleration is only centripetal and not tangential. Then $|k(s)|$ is *responsible* because the curve changes its direction at $\alpha(s)$. On the other hand, $k(s) = \langle T'(s), N(s) \rangle = \langle \alpha''(s), J\alpha'(s) \rangle = \det(\alpha'(s), \alpha''(s))$, and then the sign of $k(s)$ informs us about the orientation of the basis formed by the velocity and the acceleration of the curve. That is, if $k(s) > 0$, then the curve changes its direction counterclockwise and if $k(s) < 0$, it does so clockwise. This is an important difference with respect to the case of space curves that we will study in the following section, since an everywhere positive curvature will be defined for them.

Remark 1.15. Let $\alpha : I \to \mathbb{R}^2$ be a regular curve, not necessarily p.b.a.l. We know that α can be reparametrized by arc length, that is, there is another interval J and a diffeomorphism $f : J \to I$ such that $\beta = \alpha \circ f$ is a curve p.b.a.l. having the same trace as α. We define the curvature of α at time $t \in I$ by the equality

$$k_\alpha(t) = k_\beta(f^{-1}(t)).$$

Bearing in mind that $f'(s) = 1/|\alpha'(f(s))|$ for each $s \in J$, one easily sees that
$$k_\alpha(t) = \frac{1}{|\alpha'(t)|^3} \det(\alpha'(t), \alpha''(t))$$
or alternatively
$$k_\alpha(t) = \det(T(t), T'(t)), \qquad \forall t \in I,$$
where $T(t) = \alpha'(t)/|\alpha'(t)|$ is the unit vector tangent of α at the instant $t \in I$.

Example 1.16. Straight lines $\alpha : \mathbb{R} \to \mathbb{R}^2$, $\alpha(s) = as + b$, $a, b \in \mathbb{R}^2$, $|a| = 1$, have identically zero curvature, since $\alpha''(s) = 0$ for every $s \in \mathbb{R}$.

Example 1.17. Let $\alpha : \mathbb{R} \to \mathbb{R}^2$, and let $\alpha(s) = c + r \left(\cos \frac{s}{r}, \sin \frac{s}{r}\right)$ be a counterclockwise circle with centre $c \in \mathbb{R}^2$ and radius $r > 0$. We have
$$T'(s) = \alpha''(s) = -\frac{1}{r} \left(\cos \frac{s}{r}, \sin \frac{s}{r}\right),$$
$$N(s) = JT(s) = \left(-\cos \frac{s}{r}, -\sin \frac{s}{r}\right),$$
and so $k(s) = 1/r$ for all $s \in \mathbb{R}$. Analogously, clockwise circles have constant negative curvature $-1/r$.

Exercise 1.18. Let $\alpha : I \to \mathbb{R}^2$ be a curve p.b.a.l. and $M : \mathbb{R}^2 \to \mathbb{R}^2$ a rigid motion. If we denote by β the curve $M \circ \alpha$, prove that
$$k_\beta(s) = \begin{cases} k_\alpha(s) & \text{for all } s \in I, \text{ if } M \text{ is direct,} \\ -k_\alpha(s) & \text{for all } s \in I, \text{ if } M \text{ is inverse.} \end{cases}$$

Exercise 1.19. ↑ Let $\alpha : I \to \mathbb{R}^2$ be a curve p.b.a.l. Prove that α is a segment of a straight line or an arc of a circle if and only if the curvature of α is constant.

Exercise 1.20. ↑ Let $\alpha : I \to \mathbb{R}^2$ be a curve p.b.a.l. If there is a differentiable function $\theta : I \to \mathbb{R}$ such that $\theta(s)$ is the angle which makes the tangent line of α at s with a fixed direction, show that $\theta'(s) = \pm k(s)$.

This last exercise points out that the curvature of a plane curve determines, up to a sign, the angle that its tangent vector at each point makes with a fixed given direction. In this sense, the curvature is a genuine *intrinsic coordinate* for plane curves p.b.a.l. In fact, we will prove, in the following result, that the curvature suffices, up to a direct rigid motion, to determine a curve completely.

Theorem 1.21 (Fundamental theorem of the local theory of plane curves). *Let $k_0 : I \to \mathbb{R}$ be a differentiable function defined on an open interval $I \subset \mathbb{R}$. Then, there exists a plane curve $\alpha : I \to \mathbb{R}^2$ p.b.a.l. such that $k_\alpha(s) = k_0(s)$ for every $s \in I$, where k_α is the curvature function of α. Moreover, if*

$\beta : I \to \mathbb{R}^2$ *is another plane curve p.b.a.l. with* $k_\beta(s) = k_0(s)$, *then there exists a direct rigid motion* $M : \mathbb{R}^2 \to \mathbb{R}^2$ *such that* $\beta = M \circ \alpha$.

Proof. We define a function $\theta : I \to \mathbb{R}$ by

$$\theta(s) = \int_{s_0}^{s} k_0(u)\,du,$$

where $s_0 \in I$ is arbitrary. Then θ is differentiable and, from Exercise 1.20 above, must be, up to a constant, the angle that the tangent line of the curve that we want to construct makes with a fixed direction. This is why we define $\alpha : I \to \mathbb{R}^2$ by

$$\alpha(s) = \Big(\int_{s_1}^{s} \cos\theta(u)\,du, \int_{s_2}^{s} \sin\theta(u)\,du \Big)$$

for any $s_1, s_2 \in I$. The so-defined map α is clearly differentiable and so, for $s \in I$,

$$\alpha'(s) = \big(\cos\theta(s), \sin\theta(s) \big).$$

Thus $|\alpha'(s)| = 1$ for each $s \in I$ and, so, α is a curve p.b.a.l. Its Frenet dihedron is

$$T(s) = \big(\cos\theta(s), \sin\theta(s) \big), \qquad N(s) = JT(s) = \big(-\sin\theta(s), \cos\theta(s) \big).$$

On the other hand, we have

$$T'(s) = \big(-\theta'(s)\sin\theta(s), \theta'(s)\cos\theta(s) \big)$$

and, by the definitions of curvature and of the function θ,

$$k_\alpha(s) = \big\langle T'(s), N(s) \big\rangle = \theta'(s) = k_0(s)$$

as we were expecting.

Now suppose that $\beta : I \to \mathbb{R}^2$ is another plane curve p.b.a.l. with $k_\beta(s) = k_0(s)$ for all $s \in I$. Pick any $s_0 \in I$. Since the Frenet dihedra of α and β at s_0

$$\{ T_\alpha(s_0), N_\alpha(s_0) \} \quad \text{and} \quad \{ T_\beta(s_0), N_\beta(s_0) \}$$

are two positively oriented orthonormal bases of $\mathbb{R}^2$, there exists a unique orthonormal matrix of order two $A \in SO(2)$ such that

$$AT_\alpha(s_0) = T_\beta(s_0) \quad \text{and} \quad AN_\alpha(s_0) = N_\beta(s_0).$$

Let $b = \beta(s_0) - A\alpha(s_0) \in \mathbb{R}^2$ and $M : \mathbb{R}^2 \to \mathbb{R}^2$ be the direct rigid motion determined by A and b, that is,

$$Mx = Ax + b \qquad \text{for every } x \in \mathbb{R}^2.$$

We are going to show that the plane curve $\gamma : I \to \mathbb{R}^2$ p.b.a.l. defined by $\gamma = M \circ \alpha$ coincides with β. We have, in fact, that

$$\begin{aligned}
\gamma(s_0) &= A\alpha(s_0) + b = \beta(s_0), \\
T_\gamma(s_0) &= AT_\alpha(s_0) = T_\beta(s_0), \\
N_\gamma(s_0) &= JT_\gamma(s_0) = JT_\beta(s_0) = N_\beta(s_0).
\end{aligned}$$

Moreover, from Exercise 1.18 above,

$$k_\gamma(s) = k_\alpha(s) = k_0(s) \qquad \text{for each } s \in I.$$

If one defines $f : I \to \mathbb{R}$ by

$$f(s) = \frac{1}{2} \left[|T_\beta(s) - T_\gamma(s)|^2 + |N_\beta(s) - N_\gamma(s)|^2 \right],$$

then one obtains $f(s_0) = 0$ and also

$$f'(s) = \langle T'_\beta(s) - T'_\gamma(s), T_\beta(s) - T_\gamma(s) \rangle + \langle N'_\beta(s) - N'_\gamma(s), N_\beta(s) - N_\gamma(s) \rangle.$$

Using the Frenet equations and the fact that $k_\beta = k_\gamma = k_0$, we have $f'(s) = 0$ for each $s \in I$. Therefore f vanishes everywhere and so

$$T_\beta(s) - T_\gamma(s) = 0 \qquad \text{for each } s \in I,$$

and then

$$\beta(s) - \gamma(s) = \text{constant} \in \mathbb{R}^2.$$

But $\beta(s_0) = \gamma(s_0)$ and so one concludes that $\beta = \gamma = M \circ \alpha$. $\qquad \square$

Exercise 1.22. Let $\alpha, \beta : I \to \mathbb{R}^2$ be two plane curves p.b.a.l. such that $k_\alpha(s) = -k_\beta(s)$ for every $s \in I$. Show that there is an inverse rigid motion $M : \mathbb{R}^2 \to \mathbb{R}^2$ such that $\beta = M \circ \alpha$.

Exercise 1.23. If $\alpha : I \to \mathbb{R}^2$ is a curve p.b.a.l. defined on an open interval of $\mathbb{R}$ containing the origin and symmetric relative to it, we may define another curve $\beta : I \to \mathbb{R}^2$ by the relation $\beta(s) = \alpha(-s)$ for each $s \in I$. Prove that β is p.b.a.l. and that $k_\beta(s) = -k_\alpha(-s)$ for each $s \in I$.

Exercise 1.24. ↑ Let $a \in \mathbb{R}^+$, and let $\alpha : (-a, a) \to \mathbb{R}^2$ be a curve p.b.a.l. with $k_\alpha(s) = k_\alpha(-s)$ for each $s \in (-a, a)$. Prove that the trace of α is symmetric relative to the normal line of α at 0.

Exercise 1.25. ↑ Let $a \in \mathbb{R}^+$, and let $\alpha : (-a, a) \to \mathbb{R}^2$ be a curve p.b.a.l. with $k_\alpha(s) = -k_\alpha(-s)$ for each $s \in (-a, a)$. Prove that the trace of α is symmetric relative to the point $\alpha(0)$.

1.5. Local theory of space curves

Now we intend, as we did for plane curves, to associate a positively oriented orthonormal basis of the three-space to any point of a space curve, whose evolution gives us its geometrical features when we move along the curve.

Let $\alpha : I \to \mathbb{R}^3$ be a curve p.b.a.l. Its tangent vector $\alpha'(s)$ is a unit vector which, as in the case of plane curves, will be denoted from now on by $T(s)$. Then, one has that the vector valued function $T : I \to \mathbb{R}$ is differentiable and satisfies $|T(s)|^2 = 1$ for every $s \in I$. Taking derivatives in this last equality, we obtain

$$\langle T'(s), T(s) \rangle = 0 \qquad \text{for any } s \in I.$$

That is, $T'(s)$ is perpendicular to $T(s)$. We define the *curvature of α at s* by $k(s) = |T'(s)|$. Note that, in this situation, $k(s) \geq 0$ for each $s \in I$, which did not necessarily happen for plane curves. In general, by definition, the curvature function $k : I \to \mathbb{R}$ is only continuous. Assume in what follows that the curve α satisfies $k(s) > 0$ for each $s \in I$. Then the function k will be differentiable and also the vector $T'(s)$ will not vanish anywhere. Thus it makes sense to consider the vector

$$N(s) = \frac{T'(s)}{|T'(s)|} = \frac{1}{k(s)} T'(s).$$

We will call this vector the *normal vector of α at s* and, from previous reasonings, the map $N : I \to \mathbb{R}$ is differentiable and satisfies $|N(s)| = 1$ and $\langle T(s), N(s) \rangle = 0$ for each $s \in I$. So, in order to complete a positively oriented orthonormal basis of $\mathbb{R}^3$, it suffices to add a third vector

$$B(s) = T(s) \wedge N(s),$$

where $\wedge$ represents the vector product of Euclidean three-space, which we will call the *binormal vector of the curve α at s*. Of course, the map $B : I \to \mathbb{R}^3$ is also differentiable.

We have built, at each point $\alpha(s)$ of the curve that we are studying, a positively oriented orthonormal basis $\{T(s), N(s), B(s)\}$ which will be called the *Frenet trihedron* of the curve α at s. It depends differentiably on s. This construction is possible only when the curve $\alpha : I \to \mathbb{R}^3$ p.b.a.l. has curvature $k(s) = |T'(s)|$ strictly positive everywhere.

Now we are interested in the variation of the Frenet trihedron along the curve α. As in the case of plane curves, we will do this by considering the derivatives of the three vector valued functions T, N, and B. It follows, by the definition of curvature, that

$$T'(s) = k(s)N(s)$$

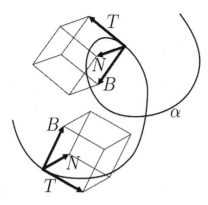

Figure 1.7. *Frenet trihedron*

for each $s \in I$. On the other hand, taking derivatives of the equality $B(s) = T(s) \wedge N(s)$ defining the binormal vector, we have

$$B'(s) = T'(s) \wedge N(s) + T(s) \wedge N'(s) = T(s) \wedge N'(s)$$

since the vectors $T'(s)$ and $N(s)$ have the same direction. Thus

$$\langle B'(s), T(s) \rangle = \det\big(T(s), N'(s), T(s)\big) = 0$$

for each $s \in I$. Moreover, since $|B(s)|^2 = 1$ everywhere, taking derivatives again, we obtain

$$\langle B'(s), B(s) \rangle = 0.$$

That is, the vector $B'(s)$ has no components either in the direction of $T(s)$ or in that of $B(s)$. So, $B'(s)$ must have the same direction as $N(s)$ and we set

$$B'(s) = \tau(s)N(s),$$

where $\tau(s) \in \mathbb{R}$ for each $s \in I$. Since $\tau(s) = \langle B'(s), N(s) \rangle$, the function $\tau : I \to \mathbb{R}$ is differentiable. The real number $\tau(s)$ is called the *torsion of the curve α at s*. Last, it remains to study the derivative $N'(s)$ of the normal vector. Taking derivatives of the equalities $|N(s)|^2 = 1$, $\langle N(s), T(s) \rangle = 0$, and $\langle N(s), B(s) \rangle = 0$, we arrive at

$$\langle N'(s), N(s) \rangle = 0,$$
$$\langle N'(s), T(s) \rangle = - \langle N(s), T'(s) \rangle = -k(s),$$
$$\langle N'(s), B(s) \rangle = - \langle N(s), B'(s) \rangle = -\tau(s),$$

and thus

$$N'(s) = -k(s)T(s) - \tau(s)B(s)$$

for each $s \in I$. The relations that we have just deduced,

$$(1.1) \qquad \begin{cases} T'(s) = k(s)N(s), \\ N'(s) = -k(s)T(s) - \tau(s)B(s), \\ B'(s) = \tau(s)N(s), \end{cases}$$

are the *Frenet equations* of the curve α.

Remark 1.26. A curve $\alpha : I \to \mathbb{R}^3$ p.b.a.l. contained in a plane can be thought of, in order to define its curvature, alternatively as a plane curve or as a space curve. The curvature of α viewed as a space curve is nothing but the absolute value of its curvature when considered as a plane curve.

Example 1.27. A straight line $\alpha : \mathbb{R} \to \mathbb{R}^3$, $\alpha(s) = sa + b$, with $a, b \in \mathbb{R}^3$ and $|a| = 1$, considered as a space curve, has identically zero curvature, since $T(s) = a$ for each $s \in \mathbb{R}$, and so $k(s) = |T'(s)| = 0$.

Example 1.28. Let $\alpha : I \to \mathbb{R}^3$ be a plane curve p.b.a.l. Suppose that the plane where the trace of the curve lies is $P = \{x \in \mathbb{R}^3 | \langle a, x \rangle = b\}$, with $a \in \mathbb{R}^3, |a| = 1$, and $b \in \mathbb{R}$. Suppose also that, as a space curve, it has positive curvature. Then its torsion and its Frenet trihedron are well-defined objects. Since $\langle T(s), a \rangle = 0$ and $\langle T'(s), a \rangle = 0$, the tangent and normal vectors at any point of α are perpendicular to the vector a. Then, $B \equiv \pm a$ and $\tau(s) = 0$ for each $s \in I$; see Exercise 1.32 below.

Example 1.29. As a consequence of the discussions above, a circle with radius $r > 0$ has, as a space curve, constant curvature $1/r$—independently of the rotation sense—and torsion identically zero.

Example 1.30. In Example 1.5 of Section 1.2, we defined the circular helix $\alpha : \mathbb{R} \to \mathbb{R}^3$ by

$$\alpha(s) = \left(a \cos \frac{s}{\sqrt{a^2 + b^2}}, a \sin \frac{s}{\sqrt{a^2 + b^2}}, \frac{bs}{\sqrt{a^2 + b^2}} \right)$$

with $a, b \neq 0$, which is a curve p.b.a.l. Taking derivatives, one has that

$$T(s) = \frac{1}{\sqrt{a^2 + b^2}} \left(-a \sin \frac{s}{\sqrt{a^2 + b^2}}, a \cos \frac{s}{\sqrt{a^2 + b^2}}, b \right),$$

$$T'(s) = \frac{-a}{a^2 + b^2} \left(\cos \frac{s}{\sqrt{a^2 + b^2}}, \sin \frac{s}{\sqrt{a^2 + b^2}}, 0 \right).$$

Hence, the curvature of the circular helix is

$$k(s) = |T'(s)| = \frac{|a|}{a^2 + b^2},$$

that is, a non-null constant. Then, we may consider its Frenet trihedron and its torsion. In fact, its normal vector is

$$N(s) = \frac{1}{k(s)} T'(s) = \frac{-a}{|a|} \left(\cos \frac{s}{\sqrt{a^2 + b^2}}, \sin \frac{s}{\sqrt{a^2 + b^2}}, 0 \right)$$

and its binormal vector will be given by

$$B(s) = T(s) \wedge N(s) = \frac{-a}{|a|\sqrt{a^2 + b^2}} \left(-b \sin \frac{s}{\sqrt{a^2 + b^2}}, b \cos \frac{s}{\sqrt{a^2 + b^2}}, -a \right).$$

Then, by differentiating with respect to s, we have

$$B'(s) = \frac{-a}{|a|(a^2 + b^2)} \left(-b \cos \frac{s}{\sqrt{a^2 + b^2}}, -b \sin \frac{s}{\sqrt{a^2 + b^2}}, 0 \right) = \frac{-b}{a^2 + b^2} N(s).$$

From this equality, one deduces that the torsion of the circular helix is

$$\tau(s) = \frac{-b}{a^2 + b^2};$$

that is, it is also constant.

Exercise 1.31. Let $\alpha : I \to \mathbb{R}^3$ be a curve p.b.a.l. Prove that α is a segment of a straight line if and only if the curvature of α vanishes everywhere.

Exercise 1.32. ↑ Show that a curve $\alpha : I \to \mathbb{R}^3$ p.b.a.l. with $k(s) > 0$ for each $s \in I$ is plane if and only if its torsion vanishes everywhere.

Exercise 1.33. Let $\alpha : I \to \mathbb{R}^3$ be a curve p.b.a.l. and $M : \mathbb{R}^3 \to \mathbb{R}^3$ a rigid motion. If we set $\beta = M \circ \alpha$, show that

$$k_\beta(s) = k_\alpha(s) \text{ and } \tau_\beta(s) = \begin{cases} \tau_\alpha(s) \text{ for each } s \in I, \text{ if } M \text{ is direct}, \\ -\tau_\alpha(s) \text{ for each } s \in I, \text{ if } M \text{ is inverse}. \end{cases}$$

Exercise 1.34. Let $\alpha : I \to \mathbb{R}^3$ be a curve p.b.a.l. with positive curvature defined on an interval $I \subset \mathbb{R}$ symmetric relative to the origin. We define another curve $\beta : I \to \mathbb{R}^3$ by $\beta(s) = \alpha(-s)$, for each $s \in I$. Show that β is a curve p.b.a.l. and that $k_\beta(s) = k_\alpha(-s)$ and $\tau_\beta(s) = -\tau_\alpha(-s)$ for each $s \in I$.

In the same way that a plane curve p.b.a.l. is determined, up to a direct rigid motion, by its curvature, we will be able to recuperate, up to a rigid motion of $\mathbb{R}^3$, a curve $\alpha : I \to \mathbb{R}^3$ p.b.a.l. with positive curvature from its curvature and its torsion. The existence proof that we gave in the fundamental Theorem 1.21 of the local theory of plane curves was a constructive proof. Instead, in the case of space curves, we will use the existence theorem for first order linear ordinary differential equations, applied to a particular equation whose matrix of coefficients is determined by the curvature and the torsion functions of the curve. Before starting with the proof, it is convenient to write in a vectorial form the Frenet equations (1.1) that we got for a space curve p.b.a.l. In fact, if $\{T(s), N(s), B(s)\}$ is the Frenet trihedron of the curve $\alpha : I \to \mathbb{R}^3$ p.b.a.l. with $k(s) > 0$, for each $s \in I$, we define on I a map taking values in $\mathbb{R}^9$ by

$$s \mapsto \begin{pmatrix} T(s) \\ N(s) \\ B(s) \end{pmatrix}, \qquad \forall s \in I,$$

where we are considering $T(s), N(s), B(s)$ as column vectors in $\mathbb{R}^3$. Then

$$\begin{pmatrix} T'(s) \\ N'(s) \\ B'(s) \end{pmatrix} = \begin{pmatrix} k(s)N(s) \\ -k(s)T(s) - \tau(s)B(s) \\ \tau(s)N(s) \end{pmatrix}$$

$$= \begin{pmatrix} 0_3 & k(s)I_3 & 0_3 \\ -k(s)I_3 & 0_3 & -\tau(s)I_3 \\ 0_3 & \tau(s)I_3 & 0_3 \end{pmatrix} \begin{pmatrix} T(s) \\ N(s) \\ B(s) \end{pmatrix}.$$

So, the map

$$s \mapsto \begin{pmatrix} T(s) \\ N(s) \\ B(s) \end{pmatrix}$$

is a solution for a first order linear differential equation. Just this property of the Frenet trihedron of a curve allows us to announce the following result.

Theorem 1.35 (Fundamental theorem of the local theory of space curves). *Let $I \subset \mathbb{R}$ be an open interval and $k_0, \tau_0 : I \to \mathbb{R}$ two differentiable functions with $k_0(s) > 0$, for each $s \in I$. Then there exists a curve $\alpha : I \to \mathbb{R}^3$ p.b.a.l. such that $k_\alpha(s) = k_0(s)$ and $\tau_\alpha(s) = \tau_0(s)$ for each $s \in I$, where k_α and τ_α are the curvature and torsion functions of α. Furthermore, α is unique up to a direct rigid motion of Euclidean space $\mathbb{R}^3$.*

Proof. Consider the following first order linear differential equation

(1.2) $x'(s) = A_0(s)x(s),$

where

$$A_0(s) = \begin{pmatrix} 0_3 & k_0(s)I_3 & 0_3 \\ -k_0(s)I_3 & 0_3 & -\tau_0(s)I_3 \\ 0_3 & \tau_0(s)I_3 & 0_3 \end{pmatrix}.$$

We take $a \in \mathbb{R}^9$ such that the vectors

$$t_0 = (a_1, a_2, a_3), \quad n_0 = (a_4, a_5, a_6), \quad \text{and } b_0 = (a_7, a_8, a_9)$$

form a positively oriented orthonormal basis of $\mathbb{R}^3$. Let $f : I \to \mathbb{R}^9$ be a solution of (1.2) with initial condition $f(s_0) = a$ and $s_0 \in I$ arbitrary. Then, if we define $t, n, b : I \to \mathbb{R}^3$ by

$$t = (f_1, f_2, f_3), \quad n = (f_4, f_5, f_6), \quad \text{and } b = (f_7, f_8, f_9),$$

we have

$$\begin{aligned} t'(s) &= k_0(s)n(s), \\ n'(s) &= -k_0(s)t(s) - \tau_0(s)b(s), \\ b'(s) &= \tau_0(s)n(s). \end{aligned}$$

Let $M(s)$ be the matrix whose entries are the scalar products of the vector valued functions $t(s)$, $n(s)$, and $b(s)$, that is,

$$M(s) = \begin{pmatrix} |t(s)|^2 & \langle t(s), n(s) \rangle & \langle t(s), b(s) \rangle \\ \langle n(s), t(s) \rangle & |n(s)|^2 & \langle n(s), b(s) \rangle \\ \langle b(s), t(s) \rangle & \langle b(s), n(s) \rangle & |b(s)|^2 \end{pmatrix}.$$

From this one can see that $M(s)$ satisfies the differential equation

(1.3) $$M'(s) = A(s)M(s) - M(s)A(s),$$

where

$$A(s) = \begin{pmatrix} 0 & k_0(s) & 0 \\ -k_0(s) & 0 & -\tau_0(s) \\ 0 & \tau_0(s) & 0 \end{pmatrix}.$$

Furthermore M verifies the initial condition $M(s_0) = I_3$. But the matrix function that is identically I_3 also satisfies the same equation (1.3) with the same initial condition. Thus, $M \equiv I_3$ and so $\{t(s), n(s), b(s)\}$ is an orthonormal basis of $\mathbb{R}^3$ for each $s \in I$. As a consequence,

$$\det(t(s), n(s), b(s)) = \pm 1$$

for each $s \in I$. On the other hand, since $\det(t(s_0), n(s_0), b(s_0)) = 1$, by continuity, the basis formed by the vectors $\{t(s), n(s), b(s)\}$ is positively oriented for every $s \in I$.

Now, define $\alpha : I \to \mathbb{R}^3$ by

$$\alpha(s) = \int_{s_0}^{s} t(u)\, du, \quad \forall s \in I.$$

Then α is differentiable and $\alpha'(s) = t(s)$, and so $|\alpha'(s)| = |t(s)| = 1$ and α is a curve p.b.a.l. Moreover $T'(s) = t'(s) = k_0(s)n(s)$, that is, $k_\alpha(s) = |T'(s)| = k_0(s)$ and $N(s) = n(s)$. Hence $B(s) = T(s) \wedge N(s) = t(s) \wedge n(s) = b(s)$ and $B'(s) = b'(s) = \tau_0(s)n(s) = \tau_0(s)N(s)$, and finally $\tau_\alpha(s) = \tau_0(s)$.

The uniqueness can be proved in a way completely analogous to that of the fundamental Theorem 1.21 of the local theory of plane curves. $\qquad \square$

Exercise 1.36. Let $\alpha, \beta : I \to \mathbb{R}^3$ be two plane curves p.b.a.l. with $k_\alpha(s) = k_\beta(s) > 0$ and $\tau_\alpha(s) = -\tau_\beta(s)$ for each $s \in I$. Prove that there exists an inverse rigid motion $M : \mathbb{R}^3 \to \mathbb{R}^3$ such that $\beta = M \circ \alpha$.

Exercise 1.37. Let $\alpha : I \to \mathbb{R}^3$ be a curve p.b.a.l. with $k(s) > 0$ for every $s \in I$. Prove that α is an arc of a circular helix or an arc of a circle if and only if both the curvature and the torsion of α are constant.

Exercises

(1) Let $\alpha : I \to \mathbb{R}^2$ be a regular plane curve and $a \in \mathbb{R}^2 - \alpha(I)$. If there exists $t_0 \in I$ such that $|\alpha(t) - a| \geq |\alpha(t_0) - a|$ for each $t \in I$, prove that the straight line joining the point a with $\alpha(t_0)$ is the normal line of α at t_0. The same happens if we reverse the inequality.

(2) ↑ Let $\alpha : I \to \mathbb{R}^2$ be a regular plane curve and R a straight line in $\mathbb{R}^2$. If one can find a number $t_0 \in I$ such that the distance from $\alpha(t)$ to R is greater than or equal to the distance from $\alpha(t_0)$ to R, for all $t \in I$, and such that $\alpha(t_0) \notin R$, then show that the tangent line of α at t_0 is parallel to R.

(3) ↑ Let $\alpha : I \to \mathbb{R}^2$ be a regular plane curve. Let $[a, b] \subset I$ such that $\alpha(a) \neq \alpha(b)$. Prove that there exists some $t_0 \in (a, b)$ such that the tangent line of α at t_0 is parallel to the segment of the straight line joining $\alpha(a)$ with $\alpha(b)$. (This is a generalization of Rolle's theorem of elementary calculus.)

(4) Prove that a curve $\alpha : I \to \mathbb{R}^2$ p.b.a.l. is a segment of a straight line if and only if all its tangent lines are concurrent.

(5) ↑ Prove that a curve $\alpha : I \to \mathbb{R}^2$ p.b.a.l. is an arc of a circle if and only if all its normal lines pass through a given point.

(6) ↑ Show that a curve $\alpha : I \to \mathbb{R}^2$ p.b.a.l. is a segment of a straight line or an arc of a circle if and only if all its tangent lines are equidistant from a given point.

(7) ↑ Given a curve $\alpha : I \to \mathbb{R}^2$ p.b.a.l. prove that all the normal lines of α are equidistant from a point if and only if there are $a, b \in \mathbb{R}$ such that

$$k(s) = \pm \frac{1}{\sqrt{as + b}}$$

for each $s \in I$. (See Exercise (11) below.)

(8) Let $\alpha : I \to \mathbb{R}^2$ be a curve p.b.a.l., and let $s_0 \in I$. We define the function $f : I \to \mathbb{R}$ by

$$f(s) = \langle \alpha(s) - \alpha(s_0), N(s_0) \rangle ,$$

measuring the oriented distance from the point $\alpha(s)$ to the tangent line of α at s_0. Prove that $f(s_0) = 0, f'(s_0) = 0$, and $f''(s_0) = k(s_0)$. Deduce from this the following statements:

- If $k(s_0) > 0$, then there is a neighbourhood J of s_0 in I such that $\alpha(J)$ belongs to the closed half-plane determined by the tangent line of α at s_0 and the vector $N(s_0)$.

- If there exists a neighbourhood J of s_0 in I such that $\alpha(J)$ is in the closed half-plane determined by the tangent line of α at s_0 and the normal vector $N(s_0)$, then $k(s_0) \geq 0$.

Obtain analogous results for the reverse inequalities.

(9) Let $\alpha : I \to \mathbb{R}^2$ be a curve p.b.a.l. and $s_0 \in I$ with $k(s_0) > 0$. Let $a_\lambda = \alpha(s_0) + \lambda N(s_0)$, $\lambda \in \mathbb{R} - \{0\}$, be a point in the normal line of α at s_0. Define the function $f_\lambda : I \to \mathbb{R}$ by

$$f_\lambda(s) = |\alpha(s) - a_\lambda|^2,$$

measuring the square of the distance from the points of the curve to the point a_λ. Show that $f_\lambda(s_0) = \lambda^2$, $f'_\lambda(s_0) = 0$, and $f''_\lambda(s_0) = 2\big(1 - \lambda k(s_0)\big)$. As a consequence, deduce the following:

- If $\lambda < 1/k(s_0)$, then there exists a neighbourhood J of s_0 in I such that $\alpha(J)$ is out of the circle with centre a_λ and radius $|\lambda|$.
- If $\lambda > 1/k(s_0)$, then there exists a neighbourhood J of s_0 in I such that $\alpha(J)$ is inside of the circle of centre a_λ and radius λ.

We see in this way that the circle centred at $\alpha(s_0) + (1/k(s_0))N(s_0)$ with radius $1/k(s_0)$ is the best circle approaching the curve near $\alpha(s_0)$. This circle will be called the *osculating circle*; its radius, the *curvature radius*; and its centre, *curvature centre*. The curve whose trace consists of all the curvature centres of α, $\beta(s) = \alpha(s) + 1/k(s)N(s)$, provided that $k(s) > 0$, will be called the *evolute* of α.

(10) ↑ Let $\alpha : I \to \mathbb{R}^2$ be a curve p.b.a.l. with $k(s) > 0$ for all $s \in I$. Prove that all the osculating circles of α meet at a given point if and only if α is an arc of a circle.

(11) ↑ Let $\alpha : I \to \mathbb{R}^2$ be a curve p.b.a.l. with $k(s) > 0$ for all $s \in I$. Prove that all the normal lines of α are equidistant from a fix point if and only if the evolute of α is a circle. In such a case we will say that α is an *evolvent* of that circle.

(12) Suppose that a plane curve $\alpha : I \to \mathbb{R}^2$ p.b.a.l. has positive and non-decreasing curvature. Let $\beta : I \to \mathbb{R}^2$ be its evolute. Prove that

$$L_a^s(\beta) = \frac{1}{k(a)} - \frac{1}{k(s)}$$

for any $a \in I$ and $s \in I$ with $s \geq a$.

(13) ↑ Let $\alpha : I \to \mathbb{R}^2$ be a curve p.b.a.l. with positive and non-decreasing curvature and β its evolute. If $a \in I$, prove that

$$|\alpha(s) - \beta(a)| \leq \frac{1}{k(a)}$$

for each $s \in I$ with $s \geq a$, that is, $\alpha([a, \infty) \cap I)$ is inside of the osculating circle of α at a.

(14) ↑ Let $\alpha : I \to \mathbb{R}^2$ be a curve p.b.a.l. and let $a \in \mathbb{R}^2 - \alpha(I)$. If there exists $s_0 \in I$ such that $|\alpha(s) - a| \leq |\alpha(s_0) - a|$ for all $s \in I$, show that

$$|k(s_0)| \geq \frac{1}{|\alpha(s_0) - a|}.$$

(15) Let $\alpha : \mathbb{R} \to \mathbb{R}^2$ a plane curve p.b.a.l. such that $\alpha(\mathbb{R})$ is included in a closed disc of radius $r > 0$ and $|k(s)| \leq 1/r$ for each $s \in \mathbb{R}$. Prove the following.
  - The function $f(s) = |\alpha(s) - a|^2$, where a is the centre of the disc, is bounded from above and satisfies $f''(s) \geq 0$ for all $s \in \mathbb{R}$.
  - Any differentiable function $f : \mathbb{R} \to \mathbb{R}$ bounded from above and such that $f'' \geq 0$ is constant.

Deduce that α is a circle centred at a with radius $r > 0$.

(16) ↑ *Comparison of two curves at a point.* Let α and β two regular plane curves defined, respectively, on two intervals of $\mathbb{R}$ containing the origin. Suppose that $\alpha(0) = \beta(0) = p$ and that $\alpha'(0) = \beta'(0)$, or alternatively $N_\alpha(0) = N_\beta(0)$, that is, α and β are tangent at p and have the same orientation. We say that α is over β at p if there is a neighbourhood of $0 \in \mathbb{R}$ where $\langle \alpha - \alpha(0), N_\alpha(0) \rangle \geq \langle \beta - \beta(0), N_\beta(0) \rangle$. Prove the following.
  - If α is over β at p, then $k_\alpha(0) \geq k_\beta(0)$.
  - If $k_\alpha(0) > k_\beta(0)$, then α is over β at p.

(17) ↑ Two regular plane curves α and β, defined, respectively, on two intervals including the origin, intersect transversely at $p = \alpha(0) = \beta(0)$ if $\alpha'(0)$ and $\beta'(0)$ are not parallel. Show that, if this happens, the traces of α_t and β have non-empty intersection, for t small enough, where α_t is the curve given by $\alpha_t(u) = \alpha(u) + ta$, $a \in \mathbb{R}^2$, $|a| = 1$.

(18) Let $\alpha : I \to \mathbb{R}^3$ a curve p.b.a.l. with positive curvature. Then α is an arc of a circle if and only if it has constant curvature and its trace is contained in a sphere.

(19) ↑ If $\alpha : I \to \mathbb{R}^3$ is a curve p.b.a.l. with positive curvature and $s \in I$, we will call the straight line passing through $\alpha(s)$ with direction $B(s)$ the *binormal line* of α at s. Suppose that $\alpha(I)$ lies in a sphere and that all its binormal lines are tangent to this sphere. Show that α is an arc of a great circle.

(20) ↑ Let $\alpha : I \to \mathbb{R}^3$ be a curve p.b.a.l. with positive curvature. If the trace of α is included in a sphere and it has constant torsion a, prove that there exist $b, c \in \mathbb{R}$ such that

$$k(s) = \frac{1}{b\cos as + c\sin as}.$$

(21) A curve $\alpha : I \to \mathbb{R}^3$ p.b.a.l. and with positive curvature is said to be a *helix* when all its tangent lines are perpendicular to a given direction.

Show that α is a helix if and only if there exists $a \in \mathbb{R}$ such that $\tau(s) = ak(s)$ for each $s \in I$. (This result is known as *Lancret's theorem*.)

(22) ↑ Let $\alpha : I \to \mathbb{R}^3$ be a curve p.b.a.l. with positive curvature, such that $k'(s) \neq 0$ and $\tau(s) \neq 0$ for each $s \in I$. Prove that the trace of α is contained in a sphere of radius $r > 0$ if and only if

$$\frac{1}{k(s)^2} + \frac{k'(s)^2}{k(s)^4 \tau(s)^2} = r^2.$$

(23) ↑ Let $\alpha : I \to \mathbb{R}^3$ be a curve p.b.a.l. with positive curvature and contained in a sphere of radius $r > 0$. If $s_0 \in I$, show that the following assertions are equivalent:
- $k'(s_0) = 0$,
- $k(s_0) = 1/r$ or $\tau(s_0) = 0$.

(24) ↑ We use the term *osculating plane* of a curve $\alpha : I \to \mathbb{R}^3$ p.b.a.l.—with positive curvature—at $s \in I$ for the plane passing through $\alpha(s)$ with normal vector $B(s)$. Let $s_0 \in I$. If there is a plane P passing through $\alpha(s_0)$, including the tangent line of α at s_0 and such that, for any neighbourhood J of s_0 in I, $\alpha(J)$ intersects the two open half-spaces determined by P, show that P coincides with the osculating plane of α at s_0.

(25) Let $\alpha : I \to \mathbb{R}^3$ be a curve p.b.a.l. with positive curvature. Show that the two following assertions are equivalent:
- All the osculating planes of α are concurrent.
- The curve α is a plane curve.

(26) Let $\alpha : I \to \mathbb{R}^3$ be a curve p.b.a.l. with positive curvature. Prove that there exists a curve $\omega : I \to \mathbb{R}^3$ such that the Frenet equations of α can be written as

$$T'(s) = \omega(s) \wedge T(s), \quad N'(s) = \omega(s) \wedge N(s), \quad \text{and } B'(s) = \omega(s) \wedge B(s).$$

The vector $\omega(s)$ is called the *angular velocity* of α at s.

(27) Under the hypotheses of Exercise (26), show that the angular velocity ω is constant if and only if α is an arc of a circular helix.

Hints for solving
the exercises

Exercise 1.9: It is enough to take into account that, if $f : I \to \mathbb{R}^3$ is a continuous function, then

$$\left| \int_a^b f \right| \leq \int_a^b |f|$$

(the Schwarz inequality). Now apply it to the case $f = \alpha'$. After that, one finishes by using the fundamental theorem of calculus.

Exercise 1.10: From Proposition 1.6, it only remains to show that

$$L_a^b(\alpha, P) \leq L_a^b(\alpha) = \int_a^b |\alpha'(t)|\, dt$$

for each partition $P = \{t_0 = a < t_1 < \cdots < t_n = b\}$ of $[a,b]$. This follows by applying Exercise 1.9 to the curve α on each subinterval $[t_{i-1}, t_i]$, $i = 1, \ldots, n$.

Exercise 1.12: By the chain rule,

$$|(\alpha \circ \phi)'(t)| = |\alpha'(\phi(t))||\phi'(t)|.$$

Since ϕ is a diffeomorphism, we have either $\phi' > 0$ or $\phi' < 0$. Now use the theorem for the change of variables for integrals of one variable functions.

Exercise 1.19: The sufficient condition can be shown in this way: if the curvature is identically zero, the Frenet equations imply that α is a segment of a straight line. Instead, if it is a non-null constant k_0, one may prove that the function $f : I \to \mathbb{R}$ given by

$$f(s) = \alpha(s) + \frac{1}{k_0} N(s)$$

is constant, differentiating with respect to s.

Exercise 1.20: Differentiate the equality

$$\cos \theta(s) = \langle T(s), a \rangle,$$

where $a \in \mathbb{R}^2$ is a given unit vector. Notice that, then, $\langle N(s), a \rangle = \pm \sin \theta(s)$.

Exercise 1.24: Let $\phi : \mathbb{R}^2 \to \mathbb{R}^2$ be the plane symmetry whose axis is the normal line of α at $\alpha(0)$. Define another curve $\beta : (-a, a) \to \mathbb{R}^2$ by

$$\beta(s) = \phi(\alpha(-s)) \quad \text{for each } s \in (-a, a).$$

Using Exercises 1.22 and 1.23, check that $k_\beta = k_\alpha$. Also since $\beta(0) = \alpha(0)$ and $\beta'(0) = \alpha'(0)$, one concludes using the fundamental Theorem 1.21.

Exercise 1.25: It can be solved similarly to Exercise 1.24 above, if one considers now that ϕ is the central symmetry of $\mathbb{R}^2$ with respect to the point $\alpha(0)$ and if one bears in mind that this symmetry is a direct rigid motion and not an inverse motion as in the case of axial symmetries.

Exercise 1.32: From Example 1.28 it only remains to study the sufficient condition. If $\tau \equiv 0$, the Frenet equations imply that $B(s) = a$, for all $s \in I$ and some unit vector $a \in \mathbb{R}^3$. Thus

$$\langle \alpha(s), a \rangle' = \langle T(s), a \rangle = 0.$$

Then $\langle \alpha(s), a \rangle = b$, for some $b \in \mathbb{R}$.

Exercise (2): If $a \in R$ and $u \in \mathbb{R}^2$ is a unit vector normal to the straight line R, the function $f : I \to \mathbb{R}$ given by

$$f(t) = \langle \alpha(t) - a, u \rangle$$

measures the (oriented) distance from the point $\alpha(t)$ to the line R. Since $\alpha(t_0) \notin R$, there is a neighbourhood of t_0 in I in which f does not vanish. Hence, f measures, up to a sign, the distance from the points of (the trace of) α to R. Then f has a minimum at $t_0 \in I$ and so $f'(t_0) = 0$.

Exercise (3): It is enough to apply Rolle's theorem of calculus to the function $f : [a, b] \to \mathbb{R}$ given by

$$f(t) = \det(\alpha(t), \alpha(a) - \alpha(b)).$$

Exercise (5): The necessary condition is clear. Suppose that all the normal lines of α pass through $a \in \mathbb{R}^2$. Then the function $f : I \to \mathbb{R}$ given by

$$f(s) = |\alpha(s) - a|^2$$

has derivative $f'(s) = 2\langle \alpha(s) - a, T(s) \rangle$. But $\alpha(s)$ and a are points of the normal line of α at $\alpha(s)$. Then f' vanishes everywhere because $\alpha(s) - a$ is normal.

Exercise (6): Assume that $a \in \mathbb{R}^2$ is the point from which all the tangent lines of α are equidistant. If $s \in I$, then the tangent line of α at s pass through $\alpha(s)$ and is perpendicular to $N(s)$. Then

$$\langle \alpha(s) - a, N(s) \rangle = c$$

for all $s \in I$ and for some $c \in \mathbb{R}$. Taking derivatives and using the Frenet equations, one sees that, for each $s \in I$, either $k(s) = 0$ or $1 + ck(s) = 0$. Since k is continuous, either k is identically zero or identically $-1/c$.

Exercise (7): The normal line of α at $s \in I$ passes through $\alpha(s)$ and the vector $T(s)$ is perpendicular to it. Then, if all the normal lines of α are equidistant from $a \in \mathbb{R}^2$, there exists some $c \in \mathbb{R}$ with

$$\langle \alpha(s) - a, T(s) \rangle = c \quad \text{for each } s \in I.$$

Differentiating with respect to s, we have

$$1 + k(s) \langle \alpha(s) - a, N(s) \rangle = 0.$$

Taking derivatives again, we obtain

$$k'(s) \langle \alpha(s) - a, N(s) \rangle - k(s)^2 \langle \alpha(s) - a, T(s) \rangle = 0.$$

Multiplying by $k(s)$ and using the above equalities, we deduce that

$$k'(s) + ck(s)^3 = 0,$$

or $(1/k^2)'' = 0$ as we are expecting. Conversely, if the equality of Exercise (7) holds, it is obvious that k does not vanish anywhere. Furthermore, $(1/k^2)' = 2(1/k)(1/k)' = 2c$ for some $c \in \mathbb{R}$. Let $f : I \to \mathbb{R}^2$ be the function defined by

$$f(s) = \alpha(s) + \frac{1}{k(s)} N(s) - cT(s), \quad \forall s \in I.$$

Differentiating and using the Frenet equations, we arrive at

$$f'(s) = \left(\frac{1}{k(s)} \right)' N(s) - ck(s)N(s) = 0.$$

Therefore, there is $a \in \mathbb{R}^2$ such that

$$\alpha(s) - a = cT(s) - \frac{1}{k(s)} N(s).$$

Thus $\langle \alpha(s) - a, T(s) \rangle = c$ for each $s \in I$.

Exercise (10): If α is an arc of a circle, its evolute is the curve constantly passing through its centre and, so, the osculating circle at any point is just that one containing α. So α and any of the osculating circles have an infinite number of common points. Conversely, let $a \in \mathbb{R}^2$ be a common point of all the osculating circles of α. We have

$$\left| \alpha(s) + \frac{1}{k(s)} N(s) - a \right|^2 = \frac{1}{k(s)^2}$$

for all $s \in I$. Differentiating, one obtains

$$\left(\frac{1}{k(s)} \right)' \langle \alpha(s) - a, N \rangle \equiv 0.$$

Then, on the open set where $(1/k)' \neq 0$, one has that $\langle \alpha - a, N \rangle = 0$, that is, all the tangent lines pass through a. From Exercise (4), the curvature should vanish on that open set, and this is impossible. Thus $(1/k)' = 0$, and so k is a positive constant.

Exercise (11): Show that the tangent line of the evolute β, at $s \in I$, is just the normal line of α at $s \in I$, and use Exercise (6) and the fact that the curvature of α is positive everywhere. Another way to solve it could be the following: prove that the curvature of the evolute β of α satisfies

$$\frac{1}{k_\beta} = \frac{1}{2} \left(\frac{1}{k^2} \right)',$$

and then apply Exercise (7).

Exercise (13): For each $s \in I$, we have

$$|\alpha(s) - \beta(a)| \leq |\alpha(s) - \beta(s)| + |\beta(s) - \beta(a)|.$$

Using the definition of β and taking into account Exercise 1.9, we deduce that

$$|\beta(s) - \beta(a)| \leq L_a^s(\beta).$$

We finish by considering Exercise (12).

Exercise (14): The function $f : I \to \mathbb{R}$ given by

$$f(s) = |\alpha(s) - a|^2$$

is differentiable and has a local maximum at $s_0 \in I$. It only remains to compute $f''(s)$ and to set $f''(s_0) \leq 0$.

Exercise (16): Define two functions f and g on the interval of common definition of the two curves by

$$f(s) = \langle \alpha(s) - p, u \rangle \quad \text{and} \quad g(s) = \langle \beta(s) - p, u \rangle,$$

where u is the common normal at $s = 0$. From our hypothesis, the function $f - g$ has a local minimum at $s = 0$. To finish, check that $(f - g)''(0) = k_\alpha(0) - k_\beta(0)$.

Exercise (17): Let I be an interval of $\mathbb{R}$ containing 0 on which two curves α and β are defined. We consider the map $F : I \times \mathbb{R} \to \mathbb{R}^2$ given by

$$F(u, t) = \alpha_t(u) - \beta(u) = \alpha(u) - \beta(u) + ta.$$

Then $F(0,0) = 0$ and $F_u(0,0) = \alpha'(0) - \beta'(0) \neq 0$. We may, then, apply the implicit function theorem and deduce that there exists an interval K including the origin and a differentiable function $f : K \to \mathbb{R}$ such that $f(0) = 0$ and $F(f(t), t) = 0$ for all $t \in K$. In other words,

$$\alpha_t(f(t)) = \beta(f(t)) \quad \text{for all } t \in K.$$

Hence, the curves α_t and β meet at each $t \in K$.

Exercise (19): We know that $|\alpha|^2 = 1$ and that the only root of the second degree polynomial $p(t) = |\alpha + tB|^2 - 1$ is $t = 0$. Then $\langle \alpha, B \rangle \equiv 0$. Taking derivatives of $|\alpha|^2 = 1$, one obtains $\langle \alpha, T \rangle \equiv 0$. Differentiating again, we have $1 + k\langle \alpha, N \rangle \equiv 0$. Since α is a unit vector everywhere and $\langle \alpha, T \rangle = \langle \alpha, B \rangle = 0$, we have that $\langle \alpha, N \rangle$ is identically -1. Thus the curvature k is constantly equal to 1. On the other hand, differentiating the expression $\langle \alpha, B \rangle = 0$, we see that the torsion τ vanishes and so the curve is a plane curve.

Exercise (20): Taking derivatives twice of $|\alpha|^2 = 1$, we get

$$\langle \alpha, T \rangle = 0 \qquad \text{and} \qquad \frac{1}{k} = -\langle \alpha, N \rangle.$$

Hence

$$\left(\frac{1}{k} \right)' = k\langle \alpha, T \rangle + \tau\langle \alpha, B \rangle = a\langle \alpha, B \rangle.$$

Differentiating once more, we obtain

$$\left(\frac{1}{k} \right)'' = a\tau\langle \alpha, N \rangle = -a^2 \frac{1}{k},$$

and we are finished.

Exercise (22): If the trace of α lies in a sphere of radius $r > 0$, then $|\alpha|^2 = r^2$. Differentiating this expression three times, we have

$$\langle \alpha, T \rangle = 0, \quad 1 + k\langle \alpha, N \rangle = 0, \quad \text{and } k'\langle \alpha, N \rangle - k\tau\langle \alpha, B \rangle = 0.$$

Therefore

$$r^2 = |\alpha|^2 = \langle \alpha, T \rangle^2 + \langle \alpha, N \rangle^2 + \langle \alpha, B \rangle^2 = \left(\frac{1}{k} \right)^2 + \left(\frac{k'}{\tau k^2} \right)^2.$$

Conversely, if this last equality occurs, we take derivatives and get, using the fact that $k' \neq 0$,

$$\frac{\tau}{k} + \left[\frac{1}{\tau} \left(\frac{1}{k} \right)' \right]' = 0.$$

This equality is just what we need to prove that $\beta' = 0$, where $\beta : I \to \mathbb{R}^3$ is the curve given by

$$\beta = \alpha + \frac{1}{k}N - \frac{1}{\tau}\left(\frac{1}{k}\right)' B.$$

Thus, there exists $a \in \mathbb{R}^3$ such that

$$\alpha - a = -\frac{1}{k}N + \frac{1}{\tau}\left(\frac{1}{k}\right)' B.$$

Taking lengths on this last equality, we have

$$|\alpha - a|^2 = \left(\frac{1}{k}\right)^2 + \left[\frac{1}{\tau}\left(\frac{1}{k}\right)'\right]^2 = r^2.$$

Exercise (23): Some previous exercises should have already convinced us that, if the trace of α lies in a sphere, we have

$$1 + k\langle \alpha, N \rangle = 0 \qquad \text{and} \qquad k'\langle \alpha, N \rangle - \tau k \langle \alpha, B \rangle = 0.$$

From the first equation we deduce that

$$1 = -k\langle \alpha, N \rangle \leq k|\alpha| = rk.$$

Hence, $k(s_0) = 1/r$ if and only if $\alpha(s_0) = -N(s_0)$, and we may finish by using the second equality.

Exercise (24): Assume that the plane P is $\{p \in \mathbb{R}^3 \mid \langle p, a \rangle + b = 0\}$, where $a \in \mathbb{R}^3$ is a unit vector and $b \in \mathbb{R}$. Consider the function $f : I \to \mathbb{R}^3$ defined by $f(s) = \langle \alpha(s), a \rangle + b$. Then

$$f(s_0) = 0, \quad f'(s_0) = \langle T(s_0), a \rangle = 0, \quad \text{and} \quad f''(s_0) = k(s_0)\langle N(s_0), a \rangle.$$

Since f has neither a local maximum nor a local minimum at s_0, we have $\langle N(s_0), a \rangle = 0$, that is, $a = \pm B(s_0)$.

Chapter 2

Surfaces in Euclidean Space

2.1. Historical notes

As happens in many other branchs of mathematics, KARL FIEDRICH GAUSS (1777–1855) was the first author to undertake a systematic study of surfaces in Euclidean space. In this way, he initiated the later so-called *differential geometry* (this name is due to L. BIANCHI). GAUSS was, in the period from 1821 to 1848, a scientific adviser to the governments of Hannover and Denmark with the goal of a complete geodesic study of their territories. The fulfilment of this work led him to think about many problems arising from the precise determination of a piece of Earth's surface. These problems could be translated, in a mathematical way, into some questions about the so-called *skew surfaces*, which were not an entirely new matter since they had already been considered by EULER, CLAIRAUT, and MONGE. The task of these geometers earlier than GAUSS focused mainly on the problem of knowing when a given surface may be *developed*, that is, when it may be mapped, preserving distances or any other geometric characteristic, onto a plane. This problem had its origin, in some sense, in the ancient world: the stereographic projection was already described in the *Geography* of PTOLEMY (c. 150). After this old antecedent, there is nothing new until the appearance of the Mercator—his real name was G. KRÄMER (1512–1594)—projection, whose discoverer knew that it was a conformal projection. Shortly after the beginnings of calculus, EULER started the theory of surfaces in his *Introductio in analysin infinitorum*, published in 1748, and in his *Recherches sur la courbure des surfaces*, published in 1767, where, in addition to studying the problem of the relation between surfaces and planes, he introduced the so-called *normal sections* and a first approach to the notion of the curvature of a surface.

In spite of this precedent and the fact that in 1795 MONGE had also published the first synopsis of the theory of surfaces in *Applications de l'Analyse à la Géométrie*, it was GAUSS who first studied surfaces in a general way in his *Disquisitiones generales circa superficies curvas*, which appeared in 1828. To get an idea of the situation of the study of surfaces, when he began his work, we may have considered that, only two years before, in 1826, CAUCHY had rigorously proved the existence of the tangent plane at each point of a surface. In this text of GAUSS, among other things, the determinant role of the *curvature*—later called *Gauss curvature*—is already clear and glimpses of the distinctions between local/global or intrinsic/extrinsic aspects of the study of surfaces may already be seen. GAUSS was also the first geometer who saw that thinking of surfaces as objects endowed locally with two coordinates is much more convenient for studying some problems than viewing them as subsets of space whose three coordinates satisfy a given relation, or even as boundaries of solid bodies.

One year before the death of GAUSS, in 1854, BERNARD RIEMANN (1826–1866) presented his Habilitation Schrift *Über die Hipothesen welche der Geometrie zu Grunde Liegen*, initiating in this way the second great period for differential geometry and going from the notion of a surface to that of a *Riemann manifold*, which has been so fruitful in several fields of physics and mathematics.

2.2. Definition of surface

In Chapter 1 we applied calculus to study curves. Our explanations were centred on concepts—tangent line, length, curvature,...—arising from some geometrical intuition, but they involved calculus to be formulated in a rigorous manner. Now we propose studying, with the same methods, another important family of figures: *surfaces.* One of the main objectives of this chapter is to transfer the usual analysis on open subsets of the Euclidean plane—differentiability, differential of a function,...—to these figures.

The first problem that we have to handle is that of defining surfaces. In Figures 2.1 and 2.2, one can see some examples of figures that could, in a first approach, deserve the name of surfaces. The definition of a surface that we will choose, and that of course is not the only one possible, is determined mainly by the tool that we will use to study these objects: calculus. Thus our theory will include neither Example 5 of Figure 2.2 nor any of the polygons, so important in geometry, because they have, in a more or less intuitive sense, a lack of differentiability. We will also throw out Example 4 of Figure 2.2 because it has a *strange* point (which one?), Example 6 of

Figure 2.1. *Examples 1, 2, 3*

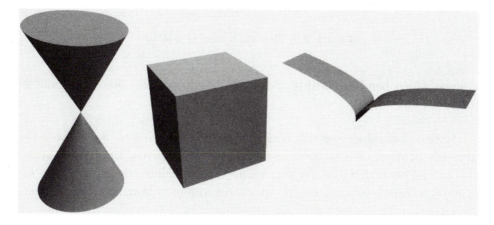

Figure 2.2. *Examples 4, 5, 6*

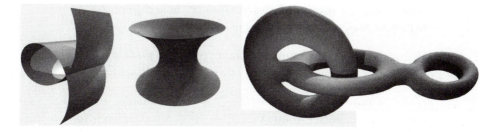

Figure 2.3. *Examples 7, 8, 9*

Figure 2.2, because it has an edge—even though it is defined by means of a differentiable equation—and Example 7 of Figure 2.3, because it intersects itself. The remaining figures which bend smoothly in Euclidean space have no corners or edges and do not intersect themselves are examples of the theory that we want to construct. All of them share a property that we choose as a first approach to the final definition.

Definition 2.1 (Intuitive definition of surface). A surface is a subset of $\mathbb{R}^3$ such that each of its points has a neighbourhood similar to a piece of

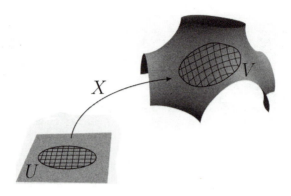

Figure 2.4. *Bending a piece of a plane*

a plane which bends smoothly and without self-intersections when bent in three-space.

Notice that, since we will consider surfaces as subsets, we move away from the point of view that we chose for studying curves. In order to arrive at a formal definition of a *surface*, we need to translate the expression *piece of a plane which bends smoothly* into more rigorous language. To do this, we will argue by analogy with the one-dimensional case. Based on what we learned in Chapter 1, we may replace the expression *piece of a line which bends smoothly and without self-intersections* by the following:

> *Given an open interval I, let $\alpha : I \to \mathbb{R}^3$ be a differentiable injective map with $\alpha'(t) \neq 0$ for all $t \in I$.*

Thus, analogously, we might translate the expression *piece of a plane which bends smoothly and without self-intersections* in the following way:

> *Given a non-empty open set $U \subset \mathbb{R}^2$, we consider a differentiable injective map $X : U \to \mathbb{R}^3$ whose differential $(dX)_q : \mathbb{R}^2 \to \mathbb{R}^3$ is injective for all $q \in U$.*

We will introduce the fundamental notion of this chapter.

Definition 2.2 (Surface). We will say that a non-empty set $S \subset \mathbb{R}^3$ is a surface if, for each $p \in S$, there exist an open set $U \subset \mathbb{R}^2$, an open neighbourhood V of p in S, and a differentiable map $X : U \to \mathbb{R}^3$ such that the following hold.

   i: $X(U) = V$.

   ii: $X : U \longrightarrow V$ is a homeomorphism.

   iii: $(dX)_q : \mathbb{R}^2 \longrightarrow \mathbb{R}^3$ is injective for all $q \in U$.

Remark 2.3. We can write X in terms of its components as $X(u, v) = \big(x(u, v), y(u, v), z(u, v)\big)$. Then, saying that X is differentiable is equivalent to saying that the three functions x, y, and z are differentiable.

Remark 2.4. If X is differentiable and satisfies condition **i**, then $X : U \to V$ is continuous and surjective. Thus, to check **ii**, it is enough to see that X is injective and that X^{-1} is continuous. Condition **ii** is in fact a somewhat technical requirement. For the moment, it suffices to note that this condition forces X to be injective.

Remark 2.5. The fact that $(dX)_q$ is injective for all $q \in U$ is equivalent to requiring that the vectors

$$X_u(q) = \frac{\partial X}{\partial u}(q) \quad \text{and} \quad X_v(q) = \frac{\partial X}{\partial v}(q)$$

be linearly independent.

Remark 2.6. The maps X appearing in Definition 2.2 will be called *parametrizations* of the surface S and their variables u, v *local coordinates* of S. The curves obtained from X by fixing one of its variables are the so-called *coordinates curves*. These parametrizations should be thought of as a way of introducing coordinates on a region of the surface or, equivalently, a way of drawing a map of that region. This is why the term *coordinate neighbourhood* is a common name for the image of such a parametrization.

Example 2.7 (Planes). Set $S = \{(x, y, z) \in \mathbb{R}^3 |\ ax + by + cz = d\}$ with $(a, b, c) \neq (0, 0, 0)$. If $c \neq 0$, we can put $S = \{(x, y, z) \in \mathbb{R}^3 |\ z = Ax + By + C\}$. We define a differentiable map $X : \mathbb{R}^2 \to \mathbb{R}^3$ by $X(u, v) = (u, v, Au + Bv + C)$. Then $X(\mathbb{R}^2) = S$, X is a homeomorphism whose inverse map $X^{-1} : S \to \mathbb{R}^2$ is given by $X^{-1}(x, y, z) = (x, y)$, and its partial derivatives $X_u = (1, 0, A)$ and $X_v = (0, 1, B)$ are linearly independent at each point. Notice that this surface is covered by a unique parametrization. This does not happen, in general, for a given surface.

Example 2.8 (Graphs). Let $U \subset \mathbb{R}^2$ be an open set, and let $f : U \to \mathbb{R}$ be a differentiable map. The graph of f is the following subset of $\mathbb{R}^3$:

$$S = \{(x, y, z) \in \mathbb{R}^3 |\ (x, y) \in U, z = f(x, y)\}.$$

In this situation, the map $X : U \to S$ given by $X(u, v) = \big(u, v, f(u, v)\big)$ is a parametrization with $X(U) = S$. Therefore, S is a surface which is covered again by only one coordinate neighbourhood.

Example 2.9 (Open subsets). If S is a surface and $S_1 \subset S$ is a nonempty open set, then S_1 is also a surface. To check this, it is enough to observe that, if $X : U \to S$ is a parametrization of S with $X(U) \cap S_1 \neq \emptyset$, then $X : X^{-1}(S_1) \to S_1$ is a parametrization of S_1. As a consequence, each connected component of a surface is also a surface because connected components of a locally connected space are open.

Example 2.10. Let $S \subset \mathbb{R}^3$ such that $S = \bigcup_{i \in I} S_i$, where each of the S_i is an open subset of S. If S_i is a surface for all $i \in I$, then S is a surface.

Example 2.11 (Spheres). Let $\mathbb{S}^2 = \{p \in \mathbb{R}^3 | \ |p|^2 = 1\}$ be the sphere centred at the origin with unit radius. Denote by $U = \{(u, v) \in \mathbb{R}^2 | \ u^2 + v^2 < 1\}$ the unit disc of the Euclidean plane and by $f : U \to \mathbb{R}$ the differentiable function defined by

$$f(u, v) = \sqrt{1 - u^2 - v^2}, \qquad \text{for all } (u, v) \in U.$$

For $a \in \mathbb{S}^2$ we consider the open neighbourhood of a in $\mathbb{S}^2$ given by $V_a = \{p \in \mathbb{S}^2 | \ \langle p, a \rangle > 0\}$. Choosing the coordinate system of $\mathbb{R}^3$ so that the vector a has components $(0, 0, 1)$, we have that V_a is the graph of the function f and, from Example 2.8, V_a is a surface. Since $\mathbb{S}^2 = \bigcup_{a \in \mathbb{S}^2} V_a$, using Example 2.10, we see that the sphere $\mathbb{S}^2$ is a surface.

Example 2.12 (Diffeomorphic images). Let $O_1, O_2 \subset \mathbb{R}^3$ be two open sets and $\phi : O_1 \to O_2$ a diffeomorphism. If $S_1 \subset O_1$ is a surface, then $S_2 = \phi(S_1)$ is also a surface; see Exercise 2.45. A proof of this fact consists simply in realizing that, if $X : U \to S_1$ is a parametrization of S_1, the map $\phi \circ X : U \to S_2$ is a parametrization of S_2.

Consequently, any ellipsoid is a surface, because there is an affine transformation of $\mathbb{R}^3$ taking it into $\mathbb{S}^2$. In particular, spheres of arbitrary centre and radius are surfaces.

Example 2.13 (Inverse images of a regular value). Let $O \subset \mathbb{R}^3$ be an open set, $f : O \to \mathbb{R}$ a differentiable map, and $a \in \mathbb{R}$. We will say that a is a *regular value* of f if, for each $p \in O$ with $f(p) = a$, one has $(df)_p \neq 0$ or, equivalently, $(\nabla f)(p) \neq 0$. We will need the following classical result of calculus:

Theorem 2.14 (Implicit function theorem). *Let O be an open set of $\mathbb{R}^3$, $p = (x_0, y_0, z_0) \in O$, $a \in \mathbb{R}$, and let $f : O \to \mathbb{R}$ be a differentiable function. Suppose that $f(p) = a$ and $f_z(p) = (\partial f / \partial z)(p) \neq 0$. Then there exist an open neighbourhood U of (x_0, y_0) in $\mathbb{R}^2$, an open neighbourhood V of z_0 in $\mathbb{R}$, and a differentiable function $g : U \to V$ such that $U \times V \subset O$, $g(x_0, y_0) = z_0$, and*

$$\{p \in U \times V | \ f(p) = a\} = \{(x, y, g(x, y)) \in \mathbb{R}^3 | \ (x, y) \in U\}.$$

In other words, if $f_z(p) \neq 0$, then we may, near p, solve the equation $f(x, y, z) = a$ for the unknown z. A direct application is

Proposition 2.15. *Let $a \in \mathbb{R}$ be a regular value of the differentiable function $f : O \to \mathbb{R}$, where O is an open set of $\mathbb{R}^3$. Then, if $S = f^{-1}(\{a\})$ is a non-empty set, it is a surface.*

Proof. Take $p = (x_0, y_0, z_0) \in S$. Then $f(p) = a$ and $(df)_p \neq 0$. Relabeling the axes, if necessary, one can assume that $f_z(p) \neq 0$. Using the same notation as in the previous implicit function Theorem 2.14, we have that $S \cap (U \times V)$ is a neighbourhood of p in S which is the graph of a differentiable function. We finish by taking Examples 2.8 and 2.10 into account. □

Example 2.16 (Quadrics). Let A be a symmetric matrix of order four. We consider the associated quadric

$$S = \{p \in \mathbb{R}^3 | \ (1, p^t)A \begin{pmatrix} 1 \\ p \end{pmatrix} = 0\}.$$

Suppose that S is non-empty and that, for each $p \in \mathbb{R}^3$, we have $(1, p^t)A \neq 0$; that is, all the singular points of the quadric lie in the plane at infinity. Let us see that, in this case, S is a surface. For this, define a differentiable function $f : \mathbb{R}^3 \to \mathbb{R}$ by

$$f(p) = (1, p^t)A \begin{pmatrix} 1 \\ p \end{pmatrix}.$$

For any $p, v \in \mathbb{R}^3$ we have

$$(df)_p(v) = 2(1, p^t)A \begin{pmatrix} 0 \\ v \end{pmatrix}.$$

If $(df)_p = 0$ for some $p \in S$, we deduce that $(1, p^t)A = (\lambda, 0)$ with $\lambda \in \mathbb{R}$. Also since $f(p) = 0$, we arrive at $\lambda = 0$ and, so, $(1, p^t)A = 0$, which is a contradiction. Hence, S is a surface, as Example 2.13 shows.

Remember that the real quadrics having all their singular points in the plane at infinity are just ellipsoids, hyperboloids, paraboloids, and cylinders.

Example 2.17 (Torus of revolution). Let $\mathbb{S}^1(r)$ be a circle of radius $r > 0$. If we rotate this circle around a straight line in the same plane that it belongs to, at distance a, $a > r$, from the centre of the circle, we generate a geometrical locus that we will call the *torus of revolution*. If we place the coordinate axes of $\mathbb{R}^3$ so that the circle $\mathbb{S}^1(r)$ lies in the plane $x = 0$ with centre at the point $(0, a, 0)$ and if the rotation axis is the line $x = 0, y = 0$, the torus of revolution becomes

$$S = \{(x, y, z) \in \mathbb{R}^3 | \ \left(\sqrt{x^2 + y^2} - a\right)^2 + z^2 = r^2\}.$$

Consider the differentiable function $f : \mathbb{R}^3 - \{z\text{-axis}\} \to \mathbb{R}$ defined by $f(x, y, z) = \left(\sqrt{x^2 + y^2} - a\right)^2 + z^2$. Then $S = f^{-1}(\{r^2\})$. Furthermore

$$\frac{\partial f}{\partial x} = \frac{2x\left(\sqrt{x^2 + y^2} - a\right)}{\sqrt{x^2 + y^2}}, \quad \frac{\partial f}{\partial y} = \frac{2y\left(\sqrt{x^2 + y^2} - a\right)}{\sqrt{x^2 + y^2}}, \quad \text{and} \quad \frac{\partial f}{\partial z} = 2z.$$

Hence, for each $p \in S$ one has $(df)_p \neq 0$, and so S is a surface.

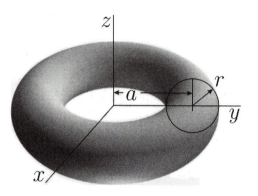

Figure 2.5. *Torus of revolution*

Example 2.18 (Simple curves). We will use the name *simple curve* for a one-dimensional object of the Euclidean space $\mathbb{R}^3$—or of the plane $\mathbb{R}^2$—analogous to a surface. That is, a non-empty subset C of $\mathbb{R}^3$ will be a simple curve when each of its points $p \in C$ possesses an open neighbourhood V in C for which there exists a regular curve $\alpha : I \to \mathbb{R}^3$ defined on an open interval I of the real line mapping I homeomorphically onto $\alpha(I) = V$. Each of these regular curves α will be called, analogously to the case of surfaces, a parametrization of the simple curve C. We may, by following what we have learned in Chapter 1, choose such parametrizations to be curves p.b.a.l. We will prove the following in Theorem 9.10:

> *Any connected component of a simple curve of $\mathbb{R}^3$ or $\mathbb{R}^2$ is homeomorphic either to a circle or to $\mathbb{R}$, depending upon whether it is or is not compact.*

Moreover, the entire curve C is the image of a curve $\alpha : I \subset \mathbb{R} \to \mathbb{R}^3$ p.b.a.l. In the first case α can be chosen to be periodic, and in the second case it can be chosen to be a homeomorphism from I onto C. In fact, straight lines and circles are the simplest examples of simple curves.

Exercise 2.19. ↑ Prove that the cone $C = \{(x, y, z) \in \mathbb{R}^3 \mid z^2 = x^2 + y^2\}$ is not a surface.

Exercise 2.20. Show that compact surfaces and compact simple curves have finitely many connected components.

2.3. Change of parameters

If a point p of a given surface belongs to the ranges of two different parametrizations, we will dispose of two different systems of coordinates near this point. Our purpose now is to find the properties of the map taking coordinates in a system into coordinates in the second one.

Lemma 2.21. *Let S be a surface and $X : U \to S$ a parametrization whose image contains the point p. Let $p_0 \in U$ be such that $X(p_0) = p$. Then there exists an open neighbourhood V of p_0 and an orthogonal projection $\pi : \mathbb{R}^3 \to \mathbb{R}^2$ onto some of the three coordinate planes of $\mathbb{R}^3$ such that $W = (\pi \circ X)(V)$ is open in $\mathbb{R}^2$ and such that $\pi \circ X : V \to W$ is a diffeomorphism.*

Proof. We write explicitly the components of the parametrization X as
$$X(u, v) = \big(x(u, v), y(u, v), z(u, v)\big).$$
Then the matrix
$$(dX)_{p_0} = \begin{pmatrix} x_u & x_v \\ y_u & y_v \\ z_u & z_v \end{pmatrix} (p_0)$$
has rank two and, so, has a non-null minor of order two. We will assume that this minor is formed just by the two first rows of the matrix. Let $\pi : \mathbb{R}^3 \to \mathbb{R}^2$ be the orthogonal projection onto the plane xy, which is given by $\pi(x, y, z) = (x, y)$. Then the composite map $\pi \circ X$ is a differentiable map of U to $\mathbb{R}^2$ and
$$d(\pi \circ X)_{p_0} = \begin{pmatrix} x_u & x_v \\ y_u & y_v \end{pmatrix} (p_0)$$
is regular. Applying the inverse function theorem, we find a set $V \subset U$ verifying the required assertion. $\qquad\square$

Remark 2.22. Observe that $Y = X \circ (\pi \circ X)^{-1} : W \to S$ is a parametrization of S which covers p and that $\pi \circ Y$ is the identity map on W. That is, one may always cover a given point on a surface by a parametrization which is the inverse map of an orthogonal projection onto a coordinate plane. From that, we deduce immediately the following result.

Proposition 2.23. *Let S be a surface and take a point $p \in S$. We can find a parametrization of S near p whose image is the graph of a differentiable function defined on some of the three coordinate planes of $\mathbb{R}^3$.*

Exercise 2.24. ↑ Demonstrate that the one-sheeted cone defined by $C = \{(x, y, z) \in \mathbb{R}^3 \mid z^2 = x^2 + y^2, \ z \geq 0\}$ is not a surface.

Exercise 2.25. ↑ Let S be a surface and p one of its points. Show that there are an open set $O \subset \mathbb{R}^3$ such that $p \in O$ and a differentiable map $f : O \to \mathbb{R}$ having 0 as a regular value in such a way that $S \cap O = f^{-1}(\{0\})$.

We will proceed by thinking about the following situation. Let S be a surface and $X_i : U_i \to S$, $i = 1, 2$, two parametrizations of it such that $O = X_1(U_1) \cap X_2(U_2)$ is a non-empty set. Then the map $h = X_2^{-1} \circ X_1 : X_1^{-1}(O) \to X_2^{-1}(O)$ is a homeomorphism taking coordinates with respect to X_1 into coordinates with respect to X_2. We will say that h is a *change of parameters* or a *change of coordinates*.

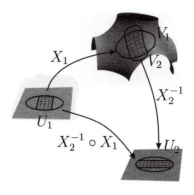

Figure 2.6. *Change of parameters*

Theorem 2.26. *Changes of parameters are diffeomorphisms.*

Proof. Since the change of parameters h is a bijective map and its inverse map is another change of parameters, to show the theorem, it is enough to see that h is differentiable at each point of its domain $X_1^{-1}(O)$. Let $q_i \in X_i^{-1}(O)$ with $X_i(q_i) = p \in S$, $i = 1, 2$. Applying Lemma 2.21, we find an open set $V \subset X_2^{-1}(O)$ and a projection π, which we can imagine to be just onto the plane xy, such that $\pi \circ X_2 : V \to (\pi \circ X_2)(V)$ is a diffeomorphism. Then $h^{-1}(V)$ is an open neighbourhood of q_1 and, in this neighbourhood, $h = (\pi \circ X_2)^{-1} \circ \pi \circ X_1$. Thus, h is differentiable on $h^{-1}(V)$, as can be expressed as a composition of the following differentiable maps: $X_1 : h^{-1}(V) \subset \mathbb{R}^2 \to \mathbb{R}^3$, $\pi : \mathbb{R}^3 \to \mathbb{R}^2$, and $(\pi \circ X_2)^{-1} : (\pi \circ X_2)(V) \subset \mathbb{R}^2 \to V \subset \mathbb{R}^2$. $\qquad\square$

2.4. Differentiable functions

We said at the beginning of this chapter that one of our goals is to transfer calculus from Euclidean open sets into surfaces. We want to make clear what it means for a function defined on a surface to be *differentiable*. The idea is simply to use parametrizations to reduce the concept of differentiability on a surface to that of differentiability on $\mathbb{R}^2$.

Definition 2.27 (Differentiable map). Let S be a surface and $O \subset \mathbb{R}^n$ an open set.

A: A function $f : S \to \mathbb{R}^m$ is said to be differentiable if, for any parametrization $X : U \to S$ of S, the composition $f \circ X : U \to \mathbb{R}^m$ is differentiable.

B: A map $f : O \to S$ will be differentiable if $i \circ f : O \to \mathbb{R}^3$ is differentiable, where $i : S \to \mathbb{R}^3$ is the inclusion map.

C: If S_1 is another surface, we will say that a map $f : S \to S_1$ is differentiable when $i_1 \circ f : S \to \mathbb{R}^3$ is differentiable in the sense of **A**, where i_1 represents the inclusion map of S_1 in $\mathbb{R}^3$.

Exercise 2.28. Show that one can deduce from **A**, **B**, and **C** of Definition 2.27 that a differentiable map is necessarily continuous.

Exercise 2.29. Let $f, g : S \to \mathbb{R}^m$ be differentiable vector valued functions and let $\lambda \in \mathbb{R}$. Prove that the functions $f + g, f - g, \lambda f : S \to \mathbb{R}^m$, and $\langle f, g \rangle : S \to \mathbb{R}$ are also differentiable. If $m = 1$ and $g(p) \neq 0$ for all $p \in S$, check that $f/g : S \to \mathbb{R}$ is also differentiable.

In what follows we will introduce the differentiable functions defined or taking their values on surfaces which will more frequently appear in this book.

Example 2.30 (Restrictions). Let $O \subset \mathbb{R}^3$ be an open set and $F : O \to \mathbb{R}^m$ a differentiable map in the sense of calculus. If $S \subset O$ is a surface, we immediately see that the map $f : S \to \mathbb{R}^m$ defined by $f = F_{|S}$ is differentiable in the sense of Definition 2.27 **A**. In particular, the *inclusion* $i : S \to \mathbb{R}^3$, the *identity* map $I_{|S} : S \to S$, and the *constant maps* defined on S are differentiable. In the following examples we will introduce other important functions on surfaces which are also differentiable since they are restrictions of differentiable maps of calculus.

Example 2.31 (Height function). Let P be a plane of $\mathbb{R}^3$ passing through the point p_0 with normal vector $a \in \mathbb{R}^3 - \{0\}$. If S is any surface, the function $h : S \to \mathbb{R}$ defined by

$$h(p) = \langle p - p_0, a \rangle, \qquad \forall p \in S,$$

is differentiable because it is the restriction to S of a differentiable function defined on the whole space $\mathbb{R}^3$. When $|a| = 1$, h measures the (oriented) height from the points of S to the plane P.

Example 2.32 (Square of the distance). If p_0 is any point of $\mathbb{R}^3$ and S is a surface, the function $f : S \to \mathbb{R}$ given by

$$f(p) = |p - p_0|^2, \qquad \forall p \in S,$$

gives the square of the Euclidean distance from the points of S to p_0. For the same reason as in Example 2.31 above, f is differentiable.

Example 2.33 (Distance function). Take $p_0 \in \mathbb{R}^3 - S$, where S is a surface. Then the function $f : S \to \mathbb{R}$ given by

$$f(p) = |p - p_0|, \qquad \forall p \in S,$$

measuring the Euclidean distance from the points of the surface S to the point p_0, is differentiable since it is the restriction to S of the differentiable function $p \mapsto |p - p_0|$ defined on the open set $\mathbb{R}^3 - \{p_0\}$ including S.

Example 2.34 (Square of the distance to a straight line). Let R be a straight line of $\mathbb{R}^3$ through the point p_0 with the unit vector a as a direction vector. The function $f : S \to \mathbb{R}$ given by

$$f(p) = |p - p_0|^2 - \langle p - p_0, a \rangle^2 = |(p - p_0) \wedge a|^2$$

for each $p \in S$, measuring the square of the distance from the points of S to the line R, is differentiable.

Example 2.35 (Differentiable curve on a surface). If $\alpha : I \to \mathbb{R}^3$ is a differentiable curve in $\mathbb{R}^3$ whose trace belongs to a surface S, the map $\alpha : I \to S$ is differentiable according to Definition 2.27, part **B**. We will say that $\alpha : I \to S$ is a *(differentiable) curve on the surface S*.

The following important result shows that, to use Definition 2.27, part **A**, of a differentiable function of a surface into a Euclidean space, we do not need to think about *all* the parametrizations of S.

Lemma 2.36. *Let S be a surface and $f : S \to \mathbb{R}^m$ a function defined on S. If, for each $p \in S$, there exists a parametrization $X_p : U_p \to S$ such that $p \in X_p(U_p)$ and such that $f \circ X_p : U_p \to \mathbb{R}^m$ is differentiable in the sense of calculus, then f is differentiable in the sense of Definition 2.27.*

Proof. Take any parametrization $X : U \to S$ of S. We have to prove that $f \circ X : U \to \mathbb{R}^m$ is differentiable. Let us see that this composition is differentiable at any point $q \in U$. By hypothesis, for the point $p = X(q) \in S$, there is a parametrization $X_p : U_p \to S$ such that $p \in X_p(U_p)$ and $f \circ X_p : U_p \to \mathbb{R}^m$ is differentiable. Let $O = X(U) \cap X_p(U_p)$, which is a non-empty set, since it contains p. Then $X^{-1}(O)$ is an open neighbourhood of q and, moreover, we know that the change of parameters $h : X^{-1}(O) \to X_p^{-1}(O)$ is a diffeomorphism. Since we have $f \circ X = (f \circ X_p) \circ h$, on $X^{-1}(O)$, we are finished. $\qquad\square$

Exercise 2.37. Let $X : U \to S$ be a parametrization of a surface S. Prove that the maps $X : U \to S$ and $X^{-1} : X(U) \to U \subset \mathbb{R}^2$ are differentiable, according to Definition 2.27.

A property that we should expect to obtain, if our definitions of differentiability are adequate, is that compositions of differentiable maps remain differentiable. In fact, we have

Theorem 2.38. *Suppose that $f : S_1 \to S_2$ and $g : S_2 \to S_3$ are differentiable maps, where S_1, S_2, S_3 are either surfaces or open sets of Euclidean spaces. Then the composite map $g \circ f : S_1 \to S_3$ is also differentiable.*

Proof. If S_2 is a Euclidean open set, the result is a direct consequence of Definition 2.27. If, instead, S_2 is a surface, it suffices to show it in the case where $S_1 = O$ is an open subset of $\mathbb{R}^n$ and $S_3 = \mathbb{R}^m$ is a Euclidean space. The remaining cases can be derived from this one and from our definitions. Let $q \in O$ and let $X : U \to S_2$ be a parametrization covering $p = X(q)$ such that $\pi \circ X = I_{|U}$, where $\pi : \mathbb{R}^3 \to \mathbb{R}^2$ is the orthogonal projection onto some coordinate plane; see Remark 2.22. Hence $f^{-1}\big(X(U)\big)$ is an open neighbourhood of q in which

$$g \circ f = (g \circ X) \circ \pi \circ f,$$

and this is a composition of differentiable maps between Euclidean open sets. Then, $g \circ f$ is differentiable at each point $q \in O$ and, so, we obtain the required result. $\qquad\square$

Example 2.39. Let $\alpha : I \to S_1$ be a differentiable curve on a surface S_1—see Example 2.35—and $f : S_1 \to S_2$ a differentiable map of S_1 into a surface or Euclidean open set S_2. Then $f \circ \alpha : I \to S_2$ is a differentiable curve on S_2.

Example 2.40. If $f : S \to \mathbb{R}^m$ is a differentiable map defined on a surface S and $S_1 \subset S$ is open, then $f_{|S_1} : S_1 \to \mathbb{R}^m$ is differentiable.

Example 2.41. Suppose that a surface S is a union $S = \bigcup_{i \in I} S_i$, where each S_i is open. If $f : S \to \mathbb{R}^m$ is a map such that each $f_{|S_i} : S_i \to \mathbb{R}^m$ is differentiable, then f is differentiable.

Example 2.42. Let $U \subset \mathbb{R}^2$ be an open set of the plane. By identifying $\mathbb{R}^2$ with a plane of $\mathbb{R}^3$, we can think of $\mathbb{R}^2$ as a surface and, so, U is one as well. Take a function $f : U \to \mathbb{R}^m$. Then, f is differentiable in the sense of Definition 2.27 as well as in the sense of calculus.

Once we dispose of a notion of a differentiable map between surfaces, we may imagine a concept of *equality* for these figures. In fact, the notion of a continuous map between topological spaces determines that of a homeomorphism between them. In a similar way, a *diffeomorphism* between two surfaces S_1 and S_2 will be, by definition, a differentiable bijective map $\phi : S_1 \to S_2$ whose inverse map $\phi^{-1} : S_2 \to S_1$ is also differentiable. For instance, the identity map $I_{|S} : S \to S$ on a given surface is a diffeomorphism, and, if $\phi : S_1 \to S_2$ is a diffeomorphism between surfaces, its inverse map $\phi^{-1} : S_2 \to S_1$ is a diffeomorphism as well. Two surfaces will be said to be *diffeomorphic* when there exists a diffeomorphism between them. In this case, we immediately see that they are homeomorphic—in fact, an involved result of differential topology asserts that two homeomorphic surfaces must necessarily be diffeomorphic.

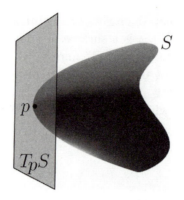

Figure 2.7. *Tangent plane*

Exercise 2.43. If $X : U \to S$ is a parametrization of a surface, show that $X : U \to X(U)$ is a diffeomorphism.

Exercise 2.44. Let $f : S_1 \to S_2$ and $g : S_2 \to S_3$ be two diffeomorphisms between surfaces. Prove that $g \circ f : S_1 \to S_3$ is another diffeomorphism. Then, the set of all the diffeomorphisms of a surface S is a group for the composition.

Exercise 2.45. Let $O_1, O_2 \subset \mathbb{R}^3$ be Euclidean open sets and $\phi : O_1 \to O_2$ a diffeomorphism. If $S \subset O_1$ is a surface, we know that $\phi(S)$ is another surface; see Example 2.12. Show that $\phi : S \to \phi(S)$ is a diffeomorphism.

Exercise 2.46. ↑ Prove that the following pairs of surfaces are diffeomorphic: a plane and an elliptic paraboloid; a one-sheeted hyperboloid and an elliptic cylinder.

2.5. The tangent plane

We want to define, at any point of a given surface, the best linear object approaching the surface near that point. Remember that this technique played a fundamental role in studying plane and space curves. We will construct that object from the notion of a tangent vector to a curve.

Definition 2.47 (Tangent plane). Let S be a surface and $p \in S$ a point. We will say that the vector $v \in \mathbb{R}^3$ is tangent to S at p if we can find a curve $\alpha : (-\varepsilon, \varepsilon) \to S$, $\varepsilon > 0$, such that $\alpha(0) = p$ and $\alpha'(0) = v$. The set consisting of all the vectors tangent to S at the point p will be represented by $T_p S$.

Lemma 2.48. *Let S be a surface, $p \in S$, and $X : U \to S$ a parametrization of S with $p \in X(U)$. Then*

$$T_p S = (dX)_{X^{-1}(p)}(\mathbb{R}^2).$$

Proof. Take $w \in \mathbb{R}^2$. Consider the line segment $\beta : (-\varepsilon, \varepsilon) \to U$, for a small ε, given by $\beta(t) = q + tw$, where $q = X^{-1}(p)$. The curve β verifies $\beta(0) = q$ and $\beta'(0) = w$. So, if we put $\alpha : (-\varepsilon, \varepsilon) \to S$ for the composition $\alpha = X \circ \beta$, we have that $\alpha(0) = p$ and $\alpha'(0) = (X \circ \beta)'(0) = (dX)_q(w)$. Hence, $(dX)_q(\mathbb{R}^2) \subset T_pS$, since α takes its values on S.

Conversely, now take a vector v tangent to S at p. By definition, there is a curve $\alpha : (-\varepsilon, \varepsilon) \to S$ such that $\alpha(0) = p$ and $\alpha'(0) = v$. Taking ε small enough, we may assume, by the continuity of α, that its trace is contained in $X(U)$. We define a curve on U by $\beta = X^{-1} \circ \alpha$. We have that $\beta(0) = q$ and $\alpha = X \circ \beta$. Thus, $v = \alpha'(0) = (X \circ \beta)'(0) = (dX)_q(\beta'(0))$, that is, any vector tangent to S at p lies in the image of $(dX)_q$. $\square$

Remark 2.49. As a first consequence of Lemma 2.48, the set T_pS formed by all the vectors tangent to a surface at a given point is a linear plane of $\mathbb{R}^3$ which will be called the *tangent plane of the surface S at the point p*.

Remark 2.50. If $X : U \to S$ is a parametrization of S which covers p and $q = X^{-1}(p)$, then, from Lemma 2.48, we have that $(dX)_q(\mathbb{R}^2)$ does not depend on X and that $\{X_u(q), X_v(q)\}$ is a basis of T_pS.

Example 2.51 (Inverse images). If $O \subset \mathbb{R}^3$ is open, $f : O \to \mathbb{R}$ is a differentiable function, and a is a regular value of f belonging to its image, we know—see Example 2.13—that $S = f^{-1}(\{a\})$ is a surface. Let us see that, for all $p \in S$, we have

$$T_pS = \ker\big((df)_p : \mathbb{R}^3 \longrightarrow \mathbb{R}\big).$$

In fact, if $v \in T_pS$, then there is a curve $\alpha : (-\varepsilon, \varepsilon) \to S$ such that $\alpha(0) = p$ and $\alpha'(0) = v$. Hence, $(f \circ \alpha)(t) = a$ for all t and, by differentiating at $t = 0$, $(df)_p(v) = (f \circ \alpha)'(0) = 0$. So, v is in the kernel of $(df)_p$. Since T_pS and $\ker(df)_p$ are linear subspaces of dimension two and one of them includes the other one, they have to coincide, as we wanted.

Example 2.52 (Planes). Let $P = \{p \in \mathbb{R}^3 | \ \langle p - p_0, a \rangle = 0\}$, with $p_0, a \in \mathbb{R}^3$ and $|a| = 1$, the plane which passes through p_0 and is normal to a. If we define $f : \mathbb{R}^3 \to \mathbb{R}$ by $f(p) = \langle p - p_0, a \rangle$, then one has that, for each $p \in P$, $T_pP = \ker(df)_p = \{v \in \mathbb{R}^3 | \ \langle v, a \rangle = 0\}$, that is, the tangent plane of P at any point coincides with its direction plane.

Example 2.53 (Spheres). Let $\mathbb{S}^2(r) = \{p \in \mathbb{R}^3 | \ |p - p_0|^2 = r^2\}$ be the sphere with centre p_0 and radius $r > 0$. Then $\mathbb{S}^2(r) = f^{-1}(\{r^2\})$ for the function $f : \mathbb{R}^3 \to \mathbb{R}$ given by $f(p) = |p - p_0|^2$ and, so, $T_p\mathbb{S}^2(r) = \ker(df)_p = \{v \in \mathbb{R}^3 | \ \langle p - p_0, v \rangle = 0\}$. Thus, $T_p\mathbb{S}^2(r)$ is the linear plane orthogonal to the position vector $p - p_0$ relative to p_0.

Example 2.54 (Open subsets). If S_1 is an open subset of a surface S and $p \in S_1$, then $T_pS_1 = T_pS$.

Given a point p of a surface S, we will call the straight line passing through p which is perpendicular to T_pS the *normal line of S at p*. By this definition, all the normal lines of a plane are parallel and all the normal lines of a sphere meet at its centre.

Exercise 2.55. Let $S = \{p \in \mathbb{R}^3 | \ |p|^2 - \langle p, a\rangle^2 = r^2\}$, with $|a| = 1$ and $r > 0$, be a right cylinder of radius r whose axis is the line passing through the origin with direction a. Prove that $T_pS = \{v \in \mathbb{R}^3 | \ \langle p, v\rangle - \langle p, a\rangle \langle a, v\rangle = 0\}$. Conclude that all the normal lines of S cut the axis perpendicularly.

2.6. Differential of a differentiable map

The notion of differential of a differentiable map defined on an open subset of $\mathbb{R}^n$ extends, in a natural way, to differentiable maps defined on surfaces. Furthermore, the classical properties of the differential of calculus will also be true in this new situation.

Definition 2.56. Let S be a surface and $f : S \to \mathbb{R}^m$ a differentiable vector valued function. For a point $p \in S$, we define the *differential of f at p*, which we will denote by $(df)_p$, in the following manner: Given $v \in T_pS$, we choose a curve $\alpha : (-\varepsilon, \varepsilon) \to S$ such that $\alpha(0) = p$ and $\alpha'(0) = v$, and then we put

$$(df)_p(v) = \frac{d}{dt}\Big|_{t=0} (f \circ \alpha)(t) = (f \circ \alpha)'(0).$$

Lemma 2.57. $(df)_p : T_pS \to \mathbb{R}^m$ *is a well-defined map, that is, $(df)_p(v)$ does not depend on the chosen curve and, moreover, it is a linear map.*

Proof. Let $X : U \to S$ be a parametrization with $p \in X(U)$. We know that $T_pS = (dX)_q(\mathbb{R}^2)$ where $q = X^{-1}(p)$ and so $(dX)_q : \mathbb{R}^2 \to T_pS$ is a linear isomorphism. Taking ε small enough, we may assume that $\alpha(-\varepsilon, \varepsilon) \subset X(U)$. The curve $X^{-1} \circ \alpha$ satisfies $(X^{-1} \circ \alpha)(0) = q$. Since $X \circ (X^{-1} \circ \alpha) = \alpha$, differentiating at $t = 0$, we have $(dX)_q\big[(X^{-1} \circ \alpha)'(0)\big] = v$, that is, $(X^{-1} \circ \alpha)'(0) = (dX)_q^{-1}(v)$. Now, using the chain rule of calculus, we get

$$(f \circ \alpha)'(0) = \frac{d}{dt}\Big|_{t=0} (f \circ X) \circ (X^{-1} \circ \alpha)$$

$$= d(f \circ X)_q\big((X^{-1} \circ \alpha)'(0)\big) = d(f \circ X)_q \circ (dX)_q^{-1}(v),$$

which proves the equality

$$(df)_p = d(f \circ X)_q \circ (dX)_q^{-1}$$

from which our assertion follows. □

Example 2.58. Let S be a surface contained in an open set $O \subset \mathbb{R}^3$, and let $F : O \to \mathbb{R}^m$ be a differentiable map. Denote by $f : S \to \mathbb{R}^m$ the restriction $f = F_{|S}$. Given $p \in S$ and $v \in T_pS$, for each curve $\alpha : (-\varepsilon, \varepsilon) \to S$ such that $\alpha(0) = p$ and $\alpha'(0) = v$, we have

$$(df)_p(v) = (f \circ \alpha)'(0) = (F \circ \alpha)'(0) = (dF)_p(v).$$

Hence, $(df)_p$ is the restriction to T_pS of $(dF)_p : \mathbb{R}^3 \to \mathbb{R}^m$.

Example 2.59. Let S be a surface, $p \in S$, $v \in T_pS$, and let $\alpha : (-\varepsilon, \varepsilon) \to S$ be a curve such that $\alpha(0) = p$ and $\alpha'(0) = v$.

A: (Height function) If $p_0 \in \mathbb{R}^3$ and $a \in \mathbb{R}^3 - \{0\}$, the differential of the height function $h : S \to \mathbb{R}$ given by $h(p) = \langle p - p_0, a \rangle$ is

$$(dh)_p(v) = \left. \frac{d}{dt} \right|_{t=0} \langle \alpha(t) - p_0, a \rangle = \langle \alpha'(0), a \rangle = \langle v, a \rangle .$$

B: (Square of the distance function) If $p_0 \in \mathbb{R}^3$, the function $f : S \to \mathbb{R}$ which is the square of the distance to the point p_0 has differential

$$(df)_p(v) = \left. \frac{d}{dt} \right|_{t=0} |\alpha(t) - p_0|^2 = 2 \langle \alpha'(0), \alpha(0) - p_0 \rangle = 2 \langle v, p - p_0 \rangle .$$

C: (Distance function) If $p_0 \in \mathbb{R}^3 - S$, the differential of the distance function $f : S \to \mathbb{R}$ to the point p_0 is given by

$$(df)_p(v) = \left. \frac{d}{dt} \right|_{t=0} |\alpha(t) - p_0| = \frac{\langle v, p - p_0 \rangle}{|p - p_0|}.$$

D: (Square of the distance to a straight line) For each $p_0, a \in \mathbb{R}^3$ with $|a| = 1$, the differential of the function $f : S \to \mathbb{R}$ defined by $f(p) = |p - p_0|^2 - \langle p - p_0, a \rangle^2$ is given by

$$(df)_p(v) = 2 \left[\langle v, p - p_0 \rangle - \langle p - p_0, a \rangle \langle v, a \rangle \right].$$

Example 2.60. We compute here the differential for inclusions, constant maps, and restrictions.

A: (Inclusions) If S is a surface and $i : S \to \mathbb{R}^3$ is the inclusion map, then

$$(di)_p(v) = v \qquad \forall v \in T_pS,$$

that is, $(di)_p$ is the inclusion of T_pS into $\mathbb{R}^3$.

B: (Constants) The differential of a constant function is, at any point, the identically zero linear map.

C: (Restrictions) If $S_1 \subset S$ is an open set in a surface S and $f : S \to \mathbb{R}^m$ is a differentiable map, then $d(f_{|S_1})_p = (df)_p$ for any $p \in S_1$.

Until now, we have extended differentials of differentiable maps between Euclidean open sets to differentiable vector valued functions defined on surfaces. We continue by considering the case of differentiable maps $f : O \to S$, where $O \subset \mathbb{R}^n$ is an open set and S is a surface. If $p \in O$ and $v \in \mathbb{R}^n$, then the curve $\alpha : (-\varepsilon, \varepsilon) \to O$ given by $\alpha(t) = p + tv$ with ε small enough satisfies $\alpha(0) = p$ and $\alpha'(0) = v$. Hence, $f \circ \alpha$ is a curve on S with $(f \circ \alpha)(0) = f(p)$ and $(f \circ \alpha)'(0) = (df)_p(v)$. Otherwise, the image of $(df)_p : \mathbb{R}^n \to \mathbb{R}^3$ lies in the linear plane $T_{f(p)}S$ and, from now on, we will deal with the restriction $(df)_p : \mathbb{R}^n \to T_{f(p)}S$. Analogously, if $f : S_1 \to S_2$ is a differentiable map between surfaces, the image of the differential of f at p, viewed as a map from S_1 to $\mathbb{R}^3$, is contained in the plane $T_{f(p)}S_2$. In such a situation, when we talk about the differential of f at $p \in S_1$, we will refer to the restriction $(df)_p : T_pS_1 \to T_{f(p)}S_2$.

By Theorem 2.38, we know that the composition of differentiable maps, in the sense of Definition 2.27, is also differentiable. The following result generalizes a classical property of calculus for the differential of a composite map.

Theorem 2.61 (Chain rule). *Let $f : S_1 \to S_2$ and $g : S_2 \to S_3$ be two differentiable maps, where S_1, S_2, S_3 stand for either surfaces or Euclidean open sets. Given $p \in S_1$, we have*

$$d(g \circ f)_p = (dg)_{f(p)} \circ (df)_p.$$

Proof. First, let us point out that the maps $(df)_p : T_pS_1 \to T_{f(p)}S_2$, $(dg)_{f(p)} : T_{f(p)}S_2 \to T_{(g \circ f)(p)}S_3$, and $d(g \circ f)_p : T_pS_1 \to T_{(g \circ f)(p)}S_3$, where $T_qS_i = \mathbb{R}^n$, when S_i is an open subset of $\mathbb{R}^n$, have domains and ranges which allow us to make all the required compositions.

If $v \in T_pS_1$, pick a curve $\alpha : (-\varepsilon, \varepsilon) \to S_1$ verifying $\alpha(0) = p$ and $\alpha'(0) = v$. Then $f \circ \alpha$ is a curve on S_2 with $(f \circ \alpha)(0) = f(p)$ and $(f \circ \alpha)'(0) = (df)_p(v)$. Thus,

$$d(g \circ f)_p(v) = [(g \circ f) \circ \alpha]'(0) = [g \circ (f \circ \alpha)]'(0) = (dg)_{f(p)}\big((df)_p(v)\big).$$

$\square$

Exercise 2.62. Let $\phi : S_1 \to S_2$ a diffeomorphism between surfaces and $p \in S_1$. Demonstrate that $(d\phi)_p : T_pS_1 \to T_{f(p)}S_2$ is a linear isomorphism and that $(d\phi)_p^{-1} = (d\phi^{-1})_{\phi(p)}$.

Given a differentiable function $f : S \to \mathbb{R}$ defined on a surface and a point $p \in S$, we will say that p is a *critical point* of f when $(df)_p = 0$. Proposition 2.63 below gathers some of the properties of differentiable functions relative to their critical points.

Proposition 2.63. *Let S be a surface. The following two assertions are true:*

> **A:** *If $f : S \to \mathbb{R}^m$ is differentiable, S is connected, and $(df)_p = 0$ for all $p \in S$, then f is constant.*

> **B:** *If $f : S \to \mathbb{R}$ is differentiable and at $p \in S$ there is a local extreme of f, then p is a critical point of f.*

Proof. Part **A**. Take $a \in f(S)$. Then the set $A = \{p \in S |\ f(p) = a\}$ is a non-empty closed subset of S. Let us check that it is also open. Choose $p \in A$ and let $X : U \to S$ be a parametrization such that $p \in X(U)$ for a connected U. For each $q \in U$ one has, from Theorem 2.38, $d(f \circ X)_q = (df)_{X(q)} \circ (dX)_q = 0$. Hence, from elementary calculus, $f \circ X$ is constant on U and, so, $f = (f \circ X) \circ X^{-1}$ is constant on $X(U)$. Since $f(p) = a$ and $p \in X(U)$, we have $X(U) \subset A$, that is, A is open. By the connection of S we deduce that $A = S$, i.e., f is constant.

Part **B**. Let p be a point of S where f presents a local extreme. If $v \in T_p S$ and $\alpha : (-\varepsilon, \varepsilon) \to S$ is a curve such that $\alpha(0) = p$ and $\alpha'(0) = v$, then the differentiable map $t \mapsto (f \circ \alpha)(t)$ has a local extreme at $t = 0$. Consequently, from elementary calculus, $(df)_p(v) = (f \circ \alpha)'(0) = 0$. Thus, p is a critical point of f. $\qquad\square$

Example 2.64 (Implicit function, transversality, and intersection of surfaces). The notion of the differential of a differentiable function defined on a surface and the chain rule that we have proved allow us to obtain an implicit function theorem in the context of the theory of surfaces.

Theorem 2.65. *Let S be a surface, $f : S \to \mathbb{R}$ a differentiable function, $p \in S$, and $a \in \mathbb{R}$. Suppose that $f(p) = a$ and $(df)_p \neq 0$, that is, p is not a critical point of f. Then there exist an open neighbourhood V of p in S, a real number $\varepsilon > 0$, and an injective regular curve $\alpha : (-\varepsilon, \varepsilon) \to \mathbb{R}^3$, which is a homeomorphism onto its image, such that*

$$\alpha(0) = p \qquad and \qquad f^{-1}(\{a\}) \cap V = \alpha(-\varepsilon, \varepsilon).$$

Thus, if a belongs to the image of f, the set $f^{-1}(\{a\})$ is a simple curve, in the sense of Example 2.18.

Proof. Let U be an open subset of $\mathbb{R}^2$ containing the origin, and let $X : U \to S$ be a parametrization of S with $X(0, 0) = p$. We define $g : U \to \mathbb{R}$ by $g = f \circ X$ and so g is differentiable with $g(0, 0) = f(X(0, 0)) = f(p) = a$ and, by the chain rule, $(dg)_{(0,0)} = (df)_p \circ (dX)_{(0,0)}$. Since $(dX)_{(0,0)}$ is injective and $(df)_p \neq 0$, we have that $(dg)_{(0,0)} \neq 0$, that is, $(g_u, g_v)(0, 0) \neq (0, 0)$. Assume that $g_v(0, 0) \neq 0$. Applying the implicit function theorem of calculus (see

Theorem 2.14), we have the existence of two real numbers $\varepsilon, \delta > 0$ and a differentiable function $h : (-\varepsilon, \varepsilon) \to (-\delta, \delta)$ such that

$$(-\varepsilon, \varepsilon) \times (-\delta, \delta) \subset U, \qquad h(0) = 0,$$

and moreover
(2.1)
$$\{(u, v) \in (-\varepsilon, \varepsilon) \times (-\delta, \delta) \mid g(u, v) = a\} = \{(u, h(u)) \in \mathbb{R}^2 \mid u \in (-\varepsilon, \varepsilon)\}.$$

Bearing this in mind, consider the neighbourhood $V = X((-\varepsilon, \varepsilon) \times (-\delta, \delta))$ of $(0, 0)$ and the curve $\alpha : (-\varepsilon, \varepsilon) \to \mathbb{R}^3$ given by $\alpha(t) = X(t, h(t))$, which is a homeomorphism onto its image, since it is the composition of a graph and X. Then $\alpha(0) = X(0, h(0)) = p$, $\alpha'(t) = (dX)_{(t,h(t))}(1, h'(t))$, which is non-null because $(dX)_{(t,h(t))}$ is injective. Furthermore, if we apply the map X to equality (2.1), we obtain

$$\{p \in V \mid f(p) = a\} = \{\alpha(t) \in \mathbb{R}^3 \mid t \in (-\varepsilon, \varepsilon)\},$$

that is, $f^{-1}(\{a\}) \cap V$ is the trace α.   □

An immediate geometrical application of this implicit function theorem is studying the intersection of surfaces near a common point. If S_1 and S_2 are two surfaces and $p \in S_1 \cap S_2$ is an intersection point, we will say that S_1 *and S_2 are tangent at p* when the corresponding tangent planes T_pS_1 and T_pS_2 coincide. Otherwise, that is, when $T_pS_1 \neq T_pS_2$, we will say that the two surfaces *intersect transversely* or that they are *transverse at p*. In this last situation, we may assert that, locally, the intersection is the trace of a curve.

Proposition 2.66. *If the surfaces S_1 and S_2 intersect transversely at a point p, then there are an open neighbourhood V of p in $\mathbb{R}^3$, an open interval $I \subset \mathbb{R}$, and a regular curve $\alpha : I \to \mathbb{R}^3$, which is a homeomorphism onto its image, such that $\alpha(I) = V \cap S_1 \cap S_2$.*

Proof. From Exercise 2.25, we know that there exist an open neighbourhood O of p in $\mathbb{R}^3$ and a differentiable function $g : O \to \mathbb{R}$ such that 0 is a regular value of g and such that $S_2 \cap O = g^{-1}(\{0\})$. We define $f : S_1 \cap O \to \mathbb{R}$ by $f = g_{|S_1 \cap O}$, which is differentiable on the surface $S_1 \cap O$ containing the point p. Moreover $f(p) = g(p) = 0$ and $(df)_p = (dg)_{p|T_pS_1}$. If p were a critical point of f, then we would have $T_pS_1 \subset \ker(dg)_p = T_pS_2$; see Example 2.51. But this is impossible because S_1 and S_2 meet transversely. Now it is enough to apply Theorem 2.65 above.   □

In order to arrive at a global knowledge of the intersection of surfaces, we will need a global version of the definition of transversality. We will say that *two surfaces S_1 and S_2 intersect transversely* when they are transverse at each common point. As a consequence of this definition and Proposition 2.66 we have the following assertion.

Theorem 2.67. *The intersection of two surfaces meeting transversely is either empty or a simple curve—with possibly more than one connected component.*

Example 2.68 (Diagonalization of a symmetric matrix)**.** Consider a symmetric matrix A of order three. For each $v, w \in \mathbb{R}^3$, it is clear that $\langle Av, w \rangle = \langle v, Aw \rangle$. We are going to show that the matrix A admits an orthonormal basis consisting of eigenvectors. For this, we define a differentiable function $f : \mathbb{S}^2 \to \mathbb{R}$ by $f(p) = \langle Ap, p \rangle$, for all $p \in \mathbb{S}^2$. We will show first that the following assertion is true.

> *A point $p \in \mathbb{S}^2$ is a critical point of f if and only $Ap = f(p)p$.*

In fact, the differential of f at the point $p \in \mathbb{S}^2$ is given by

$$(df)_p(v) = 2 \langle Ap, v \rangle \quad \forall v \in T_p\mathbb{S}^2 = \{v \in \mathbb{R}^3 | \ \langle p, v \rangle = 0\}.$$

Hence, $(df)_p = 0$ if and only if $Ap = \lambda p$ with $\lambda \in \mathbb{R}$. Taking the scalar product with p, we have $\lambda = f(p)$.

From this, if f is constant, then A is a scalar multiple of the identity matrix and the result that we seek is trivial. If f is not constant, there exist $p_1, p_2 \in \mathbb{S}^2$ in which f attains its maximum and its minimum, respectively. Then, $Ap_1 = f(p_1)p_1$, $Ap_2 = f(p_2)p_2$, and $f(p_1) \neq f(p_2)$. Hence,

$$\big(f(p_1) - f(p_2)\big) \langle p_1, p_2 \rangle = \langle f(p_1)p_1, p_2 \rangle - \langle p_1, f(p_2)p_2 \rangle$$
$$= \langle Ap_1, p_2 \rangle - \langle p_1, Ap_2 \rangle = 0.$$

So, p_1 and p_2 are orthogonal. Let $p_3 \in \mathbb{S}^2$ such that $\{p_1, p_2, p_3\}$ is an orthonormal basis of $\mathbb{R}^3$. It suffices to show that p_3 is an eigenvector of A. In fact, for $i = 1, 2$,

$$\langle Ap_3, p_i \rangle = \langle p_3, Ap_i \rangle = f(p_i) \langle p_3, p_i \rangle = 0.$$

Thus, Ap_3 is proportional to p_3.

Example 2.69 (Normal lines)**.** Let S be a surface, $p_0 \in \mathbb{R}^3$, and $f : S \to \mathbb{R}$ the function square of the distance to the point p_0. Taking Example 2.59 into account, we can deduce the following.

> *A point $p \in S$ is a critical point of f if and only if the normal line of S at p passes through the point p_0.*

If S is compact, the function f has at least a critical point. Hence, the following is true.

> *From any point of $\mathbb{R}^3$ we may draw a normal line to a given compact surface.*

If S is connected and all its points are critical for f, then f must be constant and, so, S is included in a sphere with centre p_0. Accordingly, we have the following.

> If all the normal lines of a connected surface meet at a
> given point, the surface is contained in a sphere centred at
> this point.

By arguing in a similar way, that is, looking carefully for a convenient function defined on the given surface and applying Proposition 2.63, we may solve the following exercises.

Exercise 2.70. ↑ Let S be a surface and $a \in \mathbb{R}^3, |a| = 1$. Show the following.

- If S is compact, there exists a point of S whose normal line is parallel to the vector a.

- If S is connected and all its normal lines are parallel to the vector a, then S is contained in some plane orthogonal to a.

Exercise 2.71. ↑ Let S be a surface and R a straight line of $\mathbb{R}^3$. Prove the following.

- If S is compact, there is a point of S whose normal line intersects the line R perpendicularly.

- If S is connected and all its normal lines intersect R orthogonally, then S is a subset of a right circular cylinder with axis R.

Exercise 2.72. ↑ Let S be a compact surface. Show that there exists a straight line cutting S perpendicularly at, at least, two points.

Exercise 2.73. ↑ Let S be a surface which is closed as a subset of $\mathbb{R}^3$. If $a \in \mathbb{R}^3$, prove that the distance function to the point a reaches its infimum at some point of S, that is, f has a minimum.

Exercise 2.74. ↑ Assume that S is entirely on one side of some plane P. Show that S and P are tangent at each point of $S \cap P$.

One of the more relevant results of calculus is the inverse function theorem relative to differentiable maps between Euclidean open sets. Let us see now that it is also true for differentiable maps defined on surfaces.

Theorem 2.75 (Inverse function theorem). *Let $f : S_1 \to S_2$ be a differentiable map between surfaces and let $p \in S_1$. If $(df)_p : T_pS_1 \to T_{f(p)}S_2$ is a linear isomorphism, then there exist an open neighbourhood V_1 of p in S_1 and an open neighbourhood V_2 of $f(p)$ in S_2 such that $f(V_1) = V_2$ and $f_{|V_1} : V_1 \to V_2$ is a diffeomorphism.*

Proof. Let $X_i : U_i \to S_i$, with $i = 1, 2$, be two parametrizations such that $p \in X_1(U_1)$, $f(p) \in X_2(U_2)$, and $f(X_1(U_1)) \subset X_2(U_2)$. Let $q_i \in U_i$, $i = 1, 2$, be points such that $X_1(q_1) = p$ and $X_2(q_2) = f(p)$. The map $X_2^{-1} \circ f \circ X_1 : U_1 \to U_2$ is differentiable and $d(X_2^{-1} \circ f \circ X_1)_{q_1} = (dX_2)_{q_2}^{-1} \circ (df)_p \circ (dX_1)_{q_1}$ is a linear isomorphism because it is a composition of isomorphisms. Then, we are able to apply the inverse function theorem of calculus. From it, there are open neighbourhoods $W_i \subset U_i$ with $i = 1, 2$ of q_i such that $(X_2^{-1} \circ f \circ X_1)(W_1) = W_2$ and such that $X_2^{-1} \circ f \circ X_1 : W_1 \to W_2$ is a diffeomorphism. Consider the open subsets $V_i = X_i(W_i)$ of S_i. V_1 is an open neighbourhood of p in S_1 and V_2 is an open neighbourhood of $f(p)$ in S_2. Furthermore, $f(V_1) = V_2$ and $f_{|V_1} = X_2 \circ (X_2^{-1} \circ f \circ X_1) \circ X_1^{-1} : V_1 \to V_2$ is a diffeomorphism since it is a composition of diffeomphisms. $\square$

As an immediate illustration of the usefulness of this theorem, we are going to obtain an improvement of Proposition 2.23. We proved there that any surface is locally a graph over some of the coordinate planes. Our goal now is to show that, near any of its points, the surface can be viewed as a graph over the tangent plane at that point. This property will be very fruitful in what follows.

Corollary 2.76. *Let p_0 be any point of a surface S, and let P_0 be the (affine) tangent plane at this point, that is, the plane passing through p_0 with direction plane $T_{p_0} S$. There are open neighbourhoods U and V of the point p_0 in P_0 and S, respectively, and a differentiable funtion $h : U \to \mathbb{R}$ such that V is the graph of h.*

Proof. Take a unit vector $a \in \mathbb{R}^3$ perpendicular to the tangent plane $T_{p_0} S$. Represent by $f : S \to \mathbb{R}^3$ the orthogonal projection over P_0, given by

$$f(p) = p - \langle p - p_0, a \rangle a, \qquad \forall p \in S.$$

Then $f(p_0) = p_0$ and $f(p) - p_0 \perp a$ for $p \in S$, that is, $f(S) \subset P_0$ and so $f : S \to P_0$ is a differentiable map between the surfaces S and P_0. A straightforward computation from Definition 2.56 gives us

$$(df)_p(v) = v - \langle v, a \rangle a, \qquad \forall v \in T_p S.$$

Hence $(df)_{p_0} : T_{p_0} S \to T_{p_0} P_0 = T_{p_0} S$ is the identity map. By the previous inverse function theorem, we get an open neighbourhood V of p_0 in S and another one, U, of $f(p_0) = p_0$ in P_0 so that $f(V) = U$ and $f_{|V} : V \to U$ is a diffeomorphism. Defining $h : U \to \mathbb{R}$ by

$$h(q) = \langle f^{-1}(q) - p_0, a \rangle, \qquad \forall q \in U,$$

we can finish the proof. $\square$

Let $f : S_1 \to S_2$ a differentiable map between surfaces. We will say that f is a *local diffeomorphism* if, for each $p \in S_1$, there is an open neighbourhood V_1 of p in S_1 and an open neighbourhood V_2 of $f(p)$ in S_2 such that $f(V_1) = V_2$ and $f : V_1 \to V_2$ is a diffeomorphism. Notice that a bijective local diffeomorphism is a diffeomorphism.

Proposition 2.77. *Let S_1 and S_2 be two surfaces and $f : S_1 \to S_2$ a differentiable map between them. Then the following three statements are true:*

A: *f is a local diffeomorphism if and only if $(df)_p$ is an isomorphism for each $p \in S_1$.*

B: *If f is a local diffeomorphism, then f is an open map.*

C: *If S_1 is a subset of S_2, then S_1 is open in S_2.*

Proof. Part **A**. If f is a local diffeomorphism, we know, from Exercise 2.62, that, for each $p \in S_1$, $(df)_p$ is an isomorphism. Conversely, if $(df)_p$ is regular for all $p \in S_1$, the inverse function Theorem 2.75, applied at each p, implies that f is a local diffeomorphism.

Part **B**. Let $V \subset S_1$ be an open set and $p \in V$. By definition, there are open sets $V_i \subset S_i$ with $p \in V_1$, $f(p) \in V_2$ such that $f : V_1 \to V_2$ is a homeomorphism. In particular, $f(V_1 \cap V)$ is an open neighbourhood of $f(p)$ in S_2 included in $f(V)$. Hence, $f(V)$ is an open subset of S_2.

Part **C**. The inclusion $i : S_1 \to S_2$ is a local diffeomorphism—see Example 2.60—and then, using part **B**, it is an open map. Thus, $S_1 = i(S_1)$ is an open subset of S_2. $\qquad\qquad\Box$

Example 2.78. As an immediate consequence of the above result, we may modify the assertions made in Example 2.69 and in Exercises 2.70 and 2.71.

- If all the normal lines of a connected surface S pass through a point p_0, then S is an open subset of a sphere centred at p_0. If, moreover, S is compact, then S is a sphere with centre p_0.

- If all the normal lines of a connected surface S are parallel, then S is an open subset of a plane.

- If all the normal lines of a connected surface S cut a straight line R perpendicularly, then S is an open subset of a right circular cylinder with axis R.

Example 2.79. Let S be a surface and $p_0 \notin S$. Consider the differentiable function $f : S \to \mathbb{S}^2$ defined by

$$f(p) = \frac{p - p_0}{|p - p_0|}, \qquad \forall p \in S.$$

For $v \in T_p S$, choose a curve $\alpha : (-\varepsilon, \varepsilon) \to S$ with $\alpha(0) = p$ and $\alpha'(0) = v$. Then

$$(2.2) \quad (df)_p(v) = \frac{d}{dt}\bigg|_{t=0} \frac{\alpha(t) - p_0}{|\alpha(t) - p_0|} = \frac{1}{|p - p_0|} v - \frac{\langle p - p_0, v \rangle}{|p - p_0|^3}(p - p_0).$$

If $p - p_0 \in T_p S$, we have

$$(df)_p(p - p_0) = \frac{1}{|p - p_0|}(p - p_0) - \frac{|p - p_0|^2}{|p - p_0|^3}(p - p_0) = 0,$$

that is, $p - p_0 \in \ker(df)_p$. On the other hand, if $v \in T_p S - \{0\}$ satisfies $(df)_p(v) = 0$, using (2.2),

$$p - p_0 = \frac{|p - p_0|^2}{\langle p - p_0, v \rangle} v,$$

and, so, $p - p_0 \in T_p S$.

Thus, we have just proved that $(df)_p$ has a non-trivial kernel if and only if $p - p_0 \in T_p S$. It results from part **A** of Proposition 2.77 that f *is a local diffeomorphism if and only if one cannot draw a line tangent to S from the point p_0.*

Exercises

(1) ↑ Prove that any surface has an empty interior when thought of as a subset of $\mathbb{R}^3$.

(2) ↑ Look for an example of a surface which is not a closed subset of $\mathbb{R}^3$.

(3) ↑ Let S be a surface satisfying $\phi(S) = S$ for any rotation ϕ of the space $\mathbb{R}^3$ whose axis R does not meet S. We will say that S is a *surface of revolution.* Prove that, for this type of surface, all the planes containing the rotation axis R intersect S transversely. If P is such a plane, show that $S \cap P$ is a curve which generates S. Give local parametrizations of S reflecting this fact.

(4) Let S be a compact surface, and let there be a differentiable function $f : S \to \mathbb{R}$ with at most three critical points. Show that S is connected.

(5) Take a compact surface S and a differentiable function $f : S \to \mathbb{R}$ defined on it. Estimate the number of connected components of S in terms of the number of critical points of f.

(6) ↑ Let S_1 and S_2 be two compact surfaces which do not intersect. Prove that there is a straight line that cuts at some point of S_1 and at another point of S_2 perpendicularly.

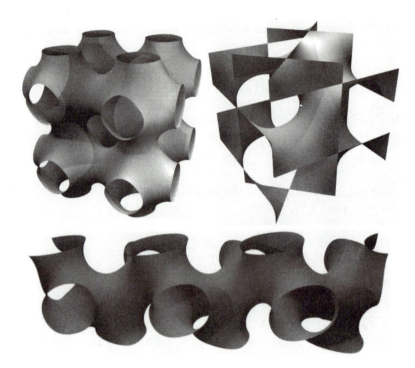

Figure 2.8. *Triply periodic surfaces*

(7) ↑ Put $S = \{(x, y, z) \in \mathbb{R}^3 |\; e^{x^2} + e^{y^2} + e^{z^2} = a\}$ with $a > 3$. Demonstrate that S is a surface. Find a diffeomorphism between S and the sphere $\mathbb{S}^2$.

(8) A surface is said to be *triply periodic* when it is invariant under three linearly independent translations of $\mathbb{R}^3$. This type of surface plays an important role in crystallography. Prove that the sets $\mathcal{P}$, $\mathcal{D}$, and $\mathcal{G}$, given, respectively, by the equations

$$\cos x + \cos y + \cos z = 0,$$
$$\cos(x - y - z) + \cos(x + y - z) - \cos(x + y + z) = 0,$$
$$\sin(x + y) + \sin(x - y) + \sin(y + z)$$
$$+ \sin(y - z) + \sin(x + z) + \sin(x - z) = 0$$

are samples of triply periodic surfaces. They are called, respectively, *primitive*, *diamond*, and *giroid*. Show also that, for these three examples, there are rigid motions of $\mathbb{R}^3$ interchanging the two regions determined by the surface.

(9) ↑ Let $\phi : S \to \mathbb{R}^3$ be a differentiable map defined on a surface which satisfies

- $(d\phi)_p : T_pS \longrightarrow \mathbb{R}^3$ is injective for all $p \in S$.
- $\phi : S \longrightarrow \phi(S)$ is a homeomorphism.

Prove that $\phi(S)$ is a surface and that $\phi : S \to \phi(S)$ is a diffeomorphism. If S is compact, show that the second requirement is equivalent to asking ϕ to be injective.

(10) $\uparrow$ Take a positive differentiable function $f : \mathbb{S}^2 \to \mathbb{R}$ on the unit sphere. Show that $S(f) = \{f(p)p \in \mathbb{R}^3 | \ p \in \mathbb{S}^2\}$ is a compact surface and that $\phi : \mathbb{S}^2 \to S(f)$ given by $\phi(p) = f(p)p$ for any $p \in \mathbb{S}^2$ is a diffeomorphism.

(11) $\uparrow$ Let S be a compact surface. We say that S is *star-shaped* with respect to the origin if $0 \notin S$ and one cannot draw from 0 a straight line tangent to S. Prove that S is star-shaped with respect to the origin if and only if it is one of the surfaces $S(f)$ of Exercise (10).

(12) $\uparrow$ Let S be a connected surface and suppose that each of its points has an open neighbourhood in S contained either in a plane or in a sphere. Prove that the entire surface is contained either in a plane or in a sphere.

(13) Let $f : S_1 \to S_2$ be a differentiable map between surfaces. If $p \in S_1$ and $\{e_1, e_2\}$ is an orthonormal basis of T_pS, we define the *absolute value of the Jacobian of f* at p as

$$|\text{Jac } f|(p) = |(df)_p(e_1) \wedge (df)_p(e_2)|.$$

a: Prove that this definition does not depend on the chosen orthonormal basis.

b: Prove that $|\text{Jac } f|(p) \neq 0$ if and only if $(df)_p$ is an isomorphism.

(14) Let S be a surface and $I \subset \mathbb{R}$ an open interval. A map $F : S \times I \to \mathbb{R}^m$ is said to be *differentiable* if, for any parametrization $X : U \to S$ of S, the composition $F \circ (X \times id_I) : U \times I \to \mathbb{R}^m$ is differentiable in the sense of calculus. Prove, in this situation, the analogue to Lemma 2.36.

(15) Let $F, G : S \times I \to \mathbb{R}^m$ be two differentiable maps and λ a real number. Prove that $F + G, F - G, \lambda F : S \times I \to \mathbb{R}^m$ are also differentiable and that $\langle F, G \rangle : S \times I \to \mathbb{R}$ is a differentiable function. When $m = 1$ and $G(p, t) \neq 0$ for all $(p, t) \in S \times I$, show that $F/G : S \times I \to \mathbb{R}$ is differentiable.

(16) If $F : S \times I \to \mathbb{R}^m$ is differentiable, we define, for each $t \in I$, $F_t : S \to \mathbb{R}^m$ by $F_t(p) = F(p, t)$ for all $p \in S$. Prove that F_t is differentiable for all $t \in I$. Analogously, if $p \in S$, we may define a curve $\alpha_p : I \to \mathbb{R}^m$ by $\alpha_p(t) = F(p, t)$ for each $t \in I$. Prove that α_p is a differentiable curve for all $p \in S$.

(17) $\uparrow$ Consider a differentiable map $F : S \times I \to \mathbb{R}^m$. For each $p \in S$ and each $t \in I$, we can define the *differential of F at (p, t)* as the map $(dF)_{(p,t)} : T_pS \times \mathbb{R} \to \mathbb{R}^m$ given by

$$(dF)_{(p,t)}(v, a) = \frac{d}{ds}\bigg|_{s=0} F\big(\alpha(s), t + sa\big), \quad \forall v \in T_pS, a \in \mathbb{R},$$

where $\alpha : (-\varepsilon, \varepsilon) \to S$ is any curve with $\alpha(0) = p$ and $\alpha'(0) = v$. Prove that $(dF)_{(p,t)}$ is a well-defined linear map.

(18) Let $O \subset \mathbb{R}^n$ be an open set, S a surface, and I an open interval of $\mathbb{R}$. A map $G : O \to S \times I$ is called *differentiable* when the composite maps $p_1 \circ G$ and $p_2 \circ G$ are differentiable, where $p_1 : S \times I \to S$ and $p_2 : S \times I \to I$ are the projections. Show that $G : O \to S \times I$ is differentiable if and only if $j \circ G : O \to \mathbb{R}^4$ is differentiable, where $j : S \times I \to \mathbb{R}^3 \times \mathbb{R} = \mathbb{R}^4$ is the inclusion map.

(19) ↑ (Inverse function theorem) Let $F : S \times I \to \mathbb{R}^3$ be a differentiable map and $(p_0, t_0) \in S \times I$ such that $(dF)_{(p_0,t_0)} : T_{p_0} S \times \mathbb{R} \to \mathbb{R}^3$ is a linear isomorphism. Prove that there exist an open neighbourhood V of p_0 in S, a real number $\varepsilon > 0$, and an open neighbourhood W of $F(p_0, t_0)$ in $\mathbb{R}^3$ such that $F\big(V \times (t_0 - \varepsilon, t_0 + \varepsilon)\big) \subset W$ and

$$F_{|V \times (t_0 - \varepsilon, t_0 + \varepsilon)} : V \times (t_0 - \varepsilon, t_0 + \varepsilon) \longrightarrow W$$

is a diffeomorphism.

(20) ↑ Two surfaces S_1 and S_2 cut each other transversely at a point $p \in S_1 \cap S_2$. Prove that, if $t \in \mathbb{R}$ is small enough, then $\phi_t(S_1) \cap S_2 \neq \emptyset$, where ϕ_t is the translation of $\mathbb{R}^3$ given by $\phi_t(p) = p + ta$, for each $p \in \mathbb{R}^3$ and where $a \in \mathbb{R}^3$, $|a| = 1$.

Hints for solving the exercises

Exercise 2.19: If the cone C were a surface, since $p = (0, 0, 0) \in C$, there would be a parametrization of C whose image would include the point p. By restricting its domain, we might find an open neighbourhood V of p in C mapped homeomorphically (by means of this parametrization) onto an open ball of $\mathbb{R}^2$. If we remove any point of this ball, the resulting punctured ball is connected as well. Thus, one obtains a contradiction by seeing that $V - \{p\}$ is not connected. Indeed, if P is the plane $z = 0$ and P^+ and P^- are the two open half-spaces into which it divides $\mathbb{R}^3$, since

$$C \cap P = \{(0, 0, 0)\},$$

we infer that $V \cap P^+$ and $V \cap P^-$ are two non-trivial disjoint open subsets of V.

Exercise 2.24: If the one-sheeted cone C were a surface, since $p = (0, 0, 0) \in C$, by Proposition 2.23, there would exist an open neighbourhood V of p in

C that is a graph of a differentiable function h on some of the three coordinates planes. Since the orthogonal projections onto the planes $x = 0$ and $y = 0$ are not one-to-one when restricted to C, V has to be a graph over the plane $z = 0$. But, since in fact

$$C = \{(x, y, z) \in \mathbb{R}^3 \mid z = +\sqrt{x^2 + y^2}\},$$

that is, C is the graph over $z = 0$ of the function $f(x, y) = +\sqrt{x^2 + y^2}$, the functions h and f must coincide on the domain of h, which is an open neighbourhood of the origin in $\mathbb{R}^2$. Since f is not differentiable at $(0, 0)$, we obtain a contradiction.

Exercise 2.25: According to Proposition 2.23 there are an open subset V of S with $p \in V$, an open set $U \subset \mathbb{R}^2$ and a differentiable function $h : U \to \mathbb{R}$ such that V is, up to a rearrangement of the coordinates x, y, z, the graph of h, that is,

$$V = \{(x, y, z) \in \mathbb{R}^3 \mid (x, y) \in U, \ z = h(x, y)\}.$$

Then, put $O = U \times \mathbb{R} \subset \mathbb{R}^3$, which is an open set, and $f : O \to \mathbb{R}$ the function defined by

$$f(x, y, z) = z - h(x, y).$$

As $f_z \equiv 1$, zero is a regular value of f and it is easily seen that $S \cap O = f^{-1}(\{0\})$.

Exercise 2.46: Any elliptic paraboloid is affinely equivalent to the paraboloid of revolution

$$\mathcal{P} = \{(x, y, z) \in \mathbb{R}^3 \mid z = x^2 + y^2\}.$$

Since each affine transformation of $\mathbb{R}^3$ is a diffeomorphism, according to Exercise 2.45, any elliptic paraboloid is diffeomorphic to $\mathcal{P}$. By analogous reasoning, any two planes of $\mathbb{R}^3$ are diffeomorphic. Thus, it suffices to see that $\mathcal{P}$ is diffeomorphic to the plane P of equation $z = 0$. We define $f : P \to \mathcal{P}$ by

$$f(x, y, z) = (x, y, x^2 + y^2).$$

This map is differentiable as a consequence of Example 2.30. Furthermore, the same argument shows that the map $g : \mathcal{P} \to P$ given by

$$g(x, y, z) = (x, y, 0)$$

is differentiable as well and is the inverse map of f. As regards the one-sheeted hyperboloid and the elliptic cylinder, arguing as before, one should show that the surface

$$\mathcal{H} = \{(x, y, z) \in \mathbb{R}^3 \mid x^2 + y^2 - z^2 = 1\}$$

is diffeomorphic to that given by

$$C = \{(x, y, z) \in \mathbb{R}^3 \mid x^2 + y^2 = 1\}.$$

Exercise 2.70: We consider the height function $h : S \to \mathbb{R}$ relative to the plane passing through the origin with normal direction a. In Example 2.59 we computed its differential and saw that $p \in S$ is critical for h if and only if a is perpendicular to the tangent plane $T_p S$. If S is compact, by continuity, h must have at least one extreme point p_0. This point will be critical for h, according to part **B** of Proposition 2.63. Then $a \perp T_{p_0} S$ and the normal line of S at p_0 is parallel to the vector a. On the other hand, if all the normal lines of S are parallel to a, then all the points of S are critical for h. By part **A** of Proposition 2.63, h will be constant, that is, $\langle p, a \rangle = c$ for all $p \in S$ and some $c \in \mathbb{R}$. Thus

$$S \subset P = \{p \in \mathbb{R}^3 \mid \langle p, a \rangle = c\},$$

and this P is clearly a plane perpendicular to the vector a.

Exercise 2.71: We work here in a way similar to Exercise 2.70, changing the height function to the square of the distance function to the given line R, whose differential was computed in Example 2.59.

Exercise 2.72: Consider the function $f : S \times S \to \mathbb{R}$ given by

$$f(p, q) = |p - q|^2, \qquad \forall p, q \in S.$$

Since S is compact, there is a pair $(a, b) \in S \times S$ where f reaches its maximum. It follows that $a \neq b$, for, otherwise, f would vanish identically and S would reduce to a point, giving a contradiction to the definition of a surface. Let R be the line through a and b. It is clear that R cuts S at the points a and b. Let us see that R is perpendicular to S at these points. By the same choice of a and b, the two differentiable functions $g, h : S \to \mathbb{R}$ given by

$$g(p) = |p - b|^2, \quad h(p) = |a - p|^2, \qquad \forall p \in S,$$

attain their maxima at a and b, respectively. Now one must use part **B** of Proposition 2.63 and the computation of the differential of the square of the distance function made in Example 2.59.

Exercise 2.73: The function $f : S \to \mathbb{R}$, which measures the distance to the point $a \in \mathbb{R}^3$, given by $f(p) = |p - a|$, for all $p \in S$, is a non-negative continuous function. Hence there exists

$$r = \inf_{p \in S} f(p) = \inf_{p \in S} |p - a| \geq 0.$$

By definition of infimum, we can find a sequence $\{p_n\}_{n \in \mathbb{N}}$ of points of S such that

$$\lim_{n \to \infty} f(p_n) = r.$$

Since it is convergent, the sequence $\{f(p_n)\}_{n \in \mathbb{N}}$ is bounded, that is, there is a number $C \in \mathbb{R}^+$ such that $|p_n - a| \leq C$ for all $n \in \mathbb{N}$. Then the sequence $\{p_n\}_{n \in \mathbb{N}}$ is inside of a sphere centred at a. Therefore, there is a partial sequence $\{p_{n_k}\}_{k \in \mathbb{N}}$ converging to a point $p \in \mathbb{R}^3$. Since the surface S is closed, we have $p \in S$, and, since f is continuous,

$$r = \lim_{k \to \infty} f(p_{n_k}) = f(p)$$

and, so, the infimum is attained at p.

Exercise 2.74: Let p_0 be a point of $S \cap P$ and assume that $a \in \mathbb{R}^3$ is a unit vector normal to P. Consider the height function to the plane P. By hypothesis, this function is either non-negative or non-positive and, so, it has an extreme at p_0. Using Proposition 2.63 and Example 2.59, one gets $a \perp T_{p_0}S$. Thus $T_{p_0}S = P$.

Exercise (1): Let us check that no points of S are interior. Indeed, if $p \in S$, from Exercise 2.25, there is an open subset O of $\mathbb{R}^3$ containing p and a function $f : O \to \mathbb{R}$ such that, up to a rearrangement of x, y, z, $f_z(p) \neq 0$ and $f^{-1}(\{0\}) = S \cap O$. Define $F : O \to \mathbb{R}^3$ as follows:

$$F(x, y, z) = (x, y, f(x, y, z)), \qquad (x, y, z) \in O.$$

After that, prove that, applying the inverse function theorem of calculus to the map F, there are two open subsets V and W of $\mathbb{R}^3$ such that $p \in V \subset O$ and $F(p) = (p_1, p_2, 0) \in W$, $F_{|V} : V \to W$ is a diffeomorphism, and such that

$$F(V \cap S) = W \cap P,$$

where P is the plane of equation $z = 0$. Then p is not an interior point of S because, otherwise, $F(p) = (p_1, p_2, 0)$ would be interior to the plane P. This is impossible since planes have an empty interior in Euclidean three-space. (See Lemma 4.1.)

Exercise (2): Let S be the right cylinder built over the trace of a logarithmic spiral as that of Exercise 1.13 contained in the plane $z = 0$. The map $X : \mathbb{R}^2 \to S \subset \mathbb{R}^3$ defined by

$$X(u, v) = (e^{-u} \cos u, e^{-u} \sin u, v), \qquad (u, v) \in \mathbb{R}^2,$$

is a bijective differentiable map and its differential is easily seen to be injective at each point. To see finally that X is a parametrization and that, therefore, S is a surface, it is enough to prove that, if R represents the z-axis, then $S \cap R = \emptyset$ and the restriction to S of the map $Y : \mathbb{R}^3 - R \to \mathbb{R}^2$ given by

$$Y(x, y, z) = (\log \frac{1}{\sqrt{x^2 + y^2}}, z)$$

is a continuous inverse map for X. Last, S is not a closed subset of $\mathbb{R}^3$ since it can easily be seen that $\overline{S} = S \cup R$. (Compare the proof that S is a surface with Exercise (9) below and see also Exercise (7) at the end of Chapter 3.)

Exercise (3): Let $p \in S \cap P$, where P is any plane containing the rotation axis R, which we will suppose to pass through the origin (and so $p \neq 0$). Let a be a unit vector in the direction of R, and let $b, c \in \mathbb{R}^3$ be two vectors such that a, b, c is a positively oriented orthonormal basis of three-space. Then the curve $\alpha : \mathbb{R} \to \mathbb{R}^3$ given by

$$\alpha(t) = \langle p, a \rangle a + (\langle p, b \rangle \cos t + \langle p, c \rangle \sin t) b + (-\langle p, b \rangle \sin t + \langle p, c \rangle \cos t) c$$

for all $t \in \mathbb{R}$ has a trace included in S since $\alpha(t) = \phi_t(p)$, where $\phi_t : \mathbb{R}^3 \to \mathbb{R}^3$ is the rotation of angle t around R. Moreover

$$\alpha(0) = p \qquad \text{and} \qquad \alpha'(0) = \langle p, c \rangle b - \langle p, b \rangle c = p \wedge a.$$

Thus $p \wedge a \in T_p S$. If S and P do not intersect transversely at p, then $T_p S = T_p P = P$ (see Example 2.52). So we would have that $p \wedge a \in P$. But also $a \in R \subset P$ and $p \in P$. This is impossible because P is a plane. Thus S cuts transversely each plane containing the straight line R. On the other hand, if P and Q are planes containing R, there is a rotation with axis R interchanging them. Then, if $P \cap S = \emptyset$, we would have $Q \cap S = \emptyset$. Since S cannot be empty, S intersects each plane P including R. From Theorem 2.67, each intersection $S \cap P$ is a simple curve that clearly generates by means of rotations around R the whole of the surface S. Finally, let P_0 be the plane passing through the origin perpendicular to the vector c and let $\alpha : I \to P_0$ be a regular curve defined on an open interval of $\mathbb{R}$ whose trace is contained in the simple curve $S \cap P_0$. We will have

$$\alpha(t) = x(t) a + y(t) b, \qquad \forall t \in I,$$

for some functions x and y. It can be shown that $X : I \times (\theta, \theta + 2\pi) \to \mathbb{R}^3$ given by

$$X(u, v) = x(u) a + y(u) \cos v\, b - y(u) \sin v\, c, \qquad (u, v) \in I \times (\theta, \theta + 2\pi),$$

is a parametrization of S, for all $\theta \in \mathbb{R}$.

Exercise (6): It can be solved like Exercise 2.72. The function $f : S_1 \times S_2 \to \mathbb{R}$ given by $f(p, q) = |p - q|^2$, for each $p \in S_1$ and $q \in S_2$, is continuous on a compact set and it is positive because $S_1 \cap S_2 = \emptyset$. Take a pair $(a, b) \in S_1 \times S_2$

where f reaches some extreme and see, as in Exercise 2.72, that the line joining a and b satisfies the requirement.

Exercise (7): Consider the differentiable function $f : \mathbb{R}^3 \to \mathbb{R}$ defined by

$$f(x, y, z) = e^{x^2} + e^{y^2} + e^{z^2}, \quad (x, y, z) \in \mathbb{R}^3,$$

whose gradient, given by

$$\nabla f_{(x,y,z)} = (2xe^{x^2}, 2ye^{y^2}, 2ze^{z^2}),$$

vanishes only at the origin. Since $f(0, 0, 0) = 3 < a$, this number a is a regular value of f. Therefore, by Proposition 2.15, either S is empty or it is a surface which does not contain the origin. For each $(x, y, z) \in \mathbb{R}^3 - \{(0, 0, 0)\}$, we set a function $h : \mathbb{R} \to \mathbb{R}$ given by

$$h(t) = f(tx, ty, tz) = e^{t^2 x^2} + e^{t^2 y^2} + e^{t^2 y^2}, \qquad t \in \mathbb{R}.$$

Since $h(0) = 3$ and $h(t)$ tends to infinity as $t \to +\infty$, there is a point of S in each half-line starting at the origin. Moreover, since

$$h'(t) = 2t(x^2 e^{t^2 x^2} + y^2 e^{t^2 y^2} + z^2 e^{t^2 z^2}) > 0 \quad \text{if } t > 0,$$

in such a half-line there will be exactly one point of S. Therefore, this shows that S is a surface and that the central projection $\phi : S \to \mathbb{S}^2$, defined in Example 2.79, is bijective. On the other hand, if $(x, y, z) \in S$, the position vector of this point of the surface does not belong to the tangent plane $T_{(x,y,z)}S$ for, otherwise, we would have, according to Example 2.51,

$$0 = (df)_{(x,y,z)}((x, y, z)) = x^2 e^{x^2} + y^2 e^{y^2} + z^2 e^{z^2},$$

which is not true. Taking Example 2.79 into account, we deduce that $(d\phi)_{(x,y,z)}$ is an isomorphism for each $(x, y, z) \in S$. By Proposition 2.77, ϕ is a local diffeomorphism that, since it is bijective, is a diffeomorphism.

Exercise (9): Prove that, if $X : U \to S$ is a parametrization of S, then $\phi \circ X : U \to \phi(S)$ is a parametrization. Such parametrizations cover $\phi(S)$, which will be, for this reason, a surface. It is also clear that $\phi : S \to \phi(S)$ is bijective. Furthermore, it is a diffeomorphism because, by Proposition 2.77, part **A**, and our first assumption, ϕ is a local diffeomorphism. Finally, if S is compact, since ϕ is continuous, it has to be a closed map. Since it is also a continuous bijective map from S onto $\phi(S)$, it must be a diffeomorphism.

Exercise (10): Under these assumptions, one should show that the map $\phi : \mathbb{S}^2 \to \mathbb{R}^3$ given by

$$\phi(p) = f(p)\, p, \qquad \forall p \in \mathbb{S}^2,$$

verifies the necessary conditions for Exercise (9) to be applied.

Exercise (11): It is clear that $0 \notin S(f)$. If, from $\phi(p) = f(p)\,p \in S(f)$, we could draw a tangent line passing through the origin, then we would have $p \in T_{\phi(p)}S(f) = (d\phi)_p(T_p\mathbb{S}^2)$, that is,

$$p = (d\phi)_p(v) = (df)_p(v)\,p + f(p)\,v$$

for some $v \in T_p\mathbb{S}^2$. Taking the scalar product by p and taking into account that $\langle p, v \rangle = 0$, we would deduce that $v = 0$, and this would give us a contradiction.

Exercise (12): For each $p \in S$ there exists an open subset V_p of S such that $p \in V_p$ and there exists either a sphere or a plane S_p such that $V_p \subset S_p$. If, for $p, q \in S$, the corresponding neighbourhoods meet, we have

$$\emptyset \neq V_p \cap V_q \subset S \cap S_p \cap S_q.$$

But if $S_p \neq S_q$, the intersection $S_p \cap S_q$ is empty or it has only a point or it is a circle or finally it is a straight line. Since $V_p \cap V_q$ is a non-empty open subset of S, using for instance a parametrization, we could get a subset of $S_p \cap S_q$ homeomorphic to an open ball of $\mathbb{R}^2$. This is not possible since the order of the connection of the points is not the same. Then

$$V_p \cap V_q \neq \emptyset \implies S_p = S_q.$$

Since $S \neq \emptyset$, for $a \in S$ the set

$$A = \{p \in S \mid S_p = S_a\}$$

is a non-empty subset of S. If $p \in A$, then $V_p \subset S_p = S_a$ and, so,

$$q \in V_p \Rightarrow q \in V_q \cap V_p \Rightarrow S_q = S_p = S_a,$$

and $V_p \subset A$. Hence A is an open subset of S. On the other hand, if $\{p_n\}_{n\in\mathbb{N}}$ is a sequence of points of A converging to a point $p \in S$, there exists $N \in \mathbb{N}$ such that

$$n \geq N \implies p_n \in V_p.$$

Then $p_N \in V_p \cap V_{p_N}$ and, as a consequence, $S_p = S_{p_N} = S_a$. Thus $p \in A$ and A is a closed subset of S. Since the surface S is connected, $S = A$ and so S is included either in a plane or in a sphere S_a.

Exercise (17): A proof similar to that of Lemma 2.48 can be given. Let $(p, t) \in S \times I$, and let $X : U \to S$ be a parametrization covering p. If $\varepsilon > 0$ is small enough, then $\alpha(-\varepsilon, \varepsilon) \subset X(U)$. Let $\beta : (-\varepsilon, \varepsilon) \to S \times I$ be the map given by $\beta(s) = (\alpha(s), t + sa)$, whose image is in $X(U) \times I$. We have

$$F \circ \beta = [F \circ (X \times id_I)] \circ (X^{-1} \times id_I) \circ \beta.$$

From that, according to the definition and applying the chain rule of calculus,

$$(dF)_{(p,t)} = d[F \circ (X \times id_I)]_{(q,t)}[d(X \times id_I)_{(q,t)}]^{-1},$$

where $q = X^{-1}(p)$. This equality proves the assertion.

Exercise (19): Let $X : U \to S$ be a parametrization of the surface S such that $X(q_0) = p_0$ for a certain $q_0 \in U$. The composition

$$G = F \circ (X \times id_I) : U \times I \subset \mathbb{R}^2 \times \mathbb{R} \longrightarrow \mathbb{R}^3$$

is differentiable in the sense of calculus and

$$(dG)_{(q_0,t_0)} = (dF)_{(p_0,t_0)} \circ d(X \times id_I)_{(q_0,t_0)}.$$

Thus the differential on the left side is an isomorphism. Using the inverse function of calculus, there are an open neighbourhood V' of q_0 contained in U, a number $\varepsilon > 0$, and an open neighbourhood W of $G(q_0, t_0)$ in $\mathbb{R}^3$ such that

$$G(V' \times (t_0 - \varepsilon, t_0 + \varepsilon)) = W \quad \text{and} \quad G : V' \times (t_0 - \varepsilon, t_0 + \varepsilon) \to W$$

is a diffeomorphism. Then $V = X(V')$, ε, and W are the required objects.

Exercise (20): Let X and Y be two parametrizations of S_1 and of S_2, respectively, both of them defined on the same open subset U of $\mathbb{R}^2$ containing the origin and such that $X(0,0) = Y(0,0) = p$. Define $F : U \times \mathbb{R} \to \mathbb{R}^3$ by

$$F(u, v, t) = X(u, v) - Y(u, v) - ta.$$

We may conclude by the implicit function Theorem 2.14, as in Exercise (17) at the end of Chapter 1.

Chapter 3

The Second Fundamental Form

3.1. Introduction and historical notes

Having defined the most convenient notion of a *surface* for our purposes and having developed the differential calculus corresponding to this class of geometrical objects, we will proceed now by starting the study of the *geometry of surfaces* that, first, will have a local nature. We intend to find, as we did in the case of curves with curvature and torsion, some functions or another type of objects which, in some sense, control their *shape*. The first mathematicians knowledgeable in this subject were likely EULER and MONGE. The latter developed intense activity as a teacher in the military school of Mézières and, after the Revolution—in fact, he was accidentally presiding over the Jacobin government on the day that Louis XVI was executed—as the organizer of the famous École Polytechnique. Indeed MONGE directed it during the entire empire of Napoleon. The members of the numerous schools of French geometers which arose from his teaching—which was already mentioned in Chapter 1— thought of a surface, in order to study its shape near a given point, as lots of curves passing through this point. In this way, in order to know how the surface bends in space, they restricted themselves to studying the geometrical behaviour of all these curves.

EULER already checked that the curvatures of all the plane curves resulting from intersections of a given surface with all the planes containing its normal line at a point are not organized in an anarchic way, but indeed only two of these sections are needed to know the curvatures of all of them. More precisely, it was one of the students of MONGE in the period of Mézières,

JEAN BAPTISTE MEUSNIER (1754–1793), who pointed out that all these curvatures are exactly—in modern language—the values that a quadratic form on the tangent plane takes. This is the *second fundamental form* of the surface at the given point, so named by way of contrast with the first fundamental form which is nothing more than the restriction of the Euclidean metric to this tangent plane.

As we pointed out at the beginning of Chapter 2, GAUSS, following some previous papers by EULER, was the first geometer who went beyond the results of the French school, when he pointed out that it is more convenient to think of surfaces as two-dimensional objects, that is, as pieces of the plane subjected to some deformations. In his book on surfaces cited earlier, he retrieves and makes full sense of the two most famous objects associated with the study of surfaces: *the Gauss map*, which EULER had already used, and *the total curvature or Gauss curvature*, introduced but not totally exploited by OLINDE RODRIGUES, a passionate follower of the SAINT-SIMON doctrines, who had been a student of MONGE in the École Polytechnique. In fact, the statement that this Gauss curvature does not depend on the way in which the surface bends in $\mathbb{R}^3$, but only on its first fundamental form—that is, it could be computed by two-dimensional geometers unable to fly over the surface—is the contents of the famous *Theorema Egregium* that one can find in the *Disquisitiones*. GAUSS defended firmly the importance of total curvature, once its intrinsic nature had been shown, although this importance was not admitted very much at that time. One of the mathematicians who argued against him by preferring another function which also measures the bending of the surface was SOPHIE GERMAIN (1776–1831), who had been led to define the so-called *mean curvature*—an extrinsic curvature depending strongly on the way in which the surface is *immersed* in $\mathbb{R}^3$—by the study of some elasticity problems. Even if we recognize the fundamental importance of the discovery of the intrinsic character of the Gauss curvature and its influence in the later birth of *Riemannian geometry*, from the point of view of the *geometry of submanifolds*, it is possible that SOPHIE GERMAIN were right on her part because, at least, problems concerning the mean curvature are much more interesting and difficult to solve than those relative to the Gauss curvature.

In this short review of the history of the geometry of surfaces, we must again mention G. DARBOUX (1842–1917), successor of CHASLES as professor of Géométrie Supérieure in the Sorbonne and of BERTRAND as secretary of the Académie des Sciences. Among his numerous accomplishments, we may list the *Leçons sur la théorie générale des surfaces*, a true compilation of all that was known about surfaces until that time and one of the best written books on geometry.

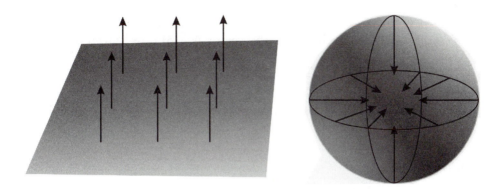

Figure 3.1. *Normal fields*

3.2. Normal fields. Orientation

In Chapter 2 we defined, at each point of a given surface of $\mathbb{R}^3$, the corresponding tangent plane. We know that the variation of the tangent vector—or of the normal vector—led us, in the case of plane curves, to the concept of curvature and to the consideration of their local geometry. We intend now to find out how the tangent plane of a surface varies when we move from a point to points close to it. This task is considerably easier if we focus on how the normal direction to the surface changes.

Definition 3.1 (Field, tangent field, and normal field)**.** Let S be a surface of Euclidean space $\mathbb{R}^3$. In general, a—differentiable—*vector field* on S will be, by definition, a differentiable map $V : S \to \mathbb{R}^3$. If the values taken by V at each point of the surface belong to the tangent plane of S at this point, that is, if $V(p) \in T_pS$ for all $p \in S$, we will say that V is a field *tangent* to the surface. If, instead, they are orthogonal to it at each point, that is, if $V(p) \perp T_pS$ for all $p \in S$, we will say that V is a field *normal* to the surface; see Figure 3.1. If moreover $|V(p)| = 1$ at each point p of S, V will be said to be a *unit normal field* defined on the surface S and we will represent it usually by N.

It is clear that, at any point $p \in S$, there are exactly two unit vectors of $\mathbb{R}^3$ perpendicular to the tangent plane T_pS. The following result informs us that one can determine continuously, even smoothly, a unit vector normal at each point, provided that we restrict ourselves to a small piece of the surface.

Lemma 3.2. *Let S be a surface and $X : U \to \mathbb{R}^3$ a parametrization of S. Then there is a unit normal field defined on the open set $V = X(U)$.*

Proof. At each $q \in U$ the vectors $X_u(q), X_v(q)$ form a basis of the tangent plane $T_{X(q)}S$. Therefore, if $q \in U$, the vector $N^X(q)$ of $\mathbb{R}^3$ given by

$$N^X(q) = \frac{X_u(q) \wedge X_v(q)}{|X_u(q) \wedge X_v(q)|} = \frac{X_u \wedge X_v}{|X_u \wedge X_v|}(q)$$

is a unit vector normal to the surface at the point $X(q)$. So, we may define a differentiable map

$$N^X : U \to \mathbb{R}^3$$

such that

$$N^X(q) \perp T_{X(q)}S \qquad \text{and} \qquad |N^X(q)|^2 = 1$$

for all $q \in U$. Finally, we denote by $N : V = X(U) \to \mathbb{R}^3$ the composite map $N = N^X \circ X^{-1}$. Then, N is the required unit normal field.   □

It seems intuitively clear that a unit field normal to a surface determines one *side* of it. We assert now that a connected surface has at most two sides.

Lemma 3.3. *If S is a connected surface and N_1 and N_2 are two unit normal fields on S, then either $N_1 = N_2$ or $N_1 = -N_2$.*

Proof. Indeed, if $p \in S$, we have that $N_1(p)$ and $N_2(p)$ are two unit vectors perpendicular to the same plane T_pS. Hence, either they coincide or they are opposite. Then $S = A \cup B$, where $A = \{p \in S \,|\, N_1(p) = N_2(p)\}$ and $B = \{p \in S \,|\, N_1(p) = -N_2(p)\}$, with A and B disjoint subsets of S which are closed because N_1 and N_2 are continuous. Since S is connected, we have that either $A = S$ and $B = \emptyset$ or $A = \emptyset$ and $B = S$.   □

Definition 3.4 (Orientable surfaces). If, on a given surface S, there is a unit normal field defined on the whole of the surface, we will say that S is *orientable*. Each unit normal field on an orientable surface S will be called an *orientation* of S. Then, Lemma 3.2 asserts that any surface is locally orientable and, when a surface is connected and orientable, Lemma 3.3 states that there are exactly two different orientations on it. The orientable surface S will be said to be *oriented* when a precise orientation has been chosen on it.

Example 3.5 (Planes are orientable). Planes are orientable as a direct consequence of Lemma 3.2. Also, one can explicitly construct a unit normal field on a given plane $P = \{p \in \mathbb{R}^3 \,|\, \langle p - p_0, a \rangle = 0\}$ which passes through the point p_0 with unit normal vector a. In fact, remember that, if $p \in P$, then

$$T_pP = \{v \in \mathbb{R}^3 \,|\, \langle v, a \rangle = 0\}.$$

Thus $N : P \to \mathbb{R}^3$ given by $N(p) = a$ for all $p \in P$ is a unit normal field defined on P.

Example 3.6 (Spheres are orientable). Let us consider the sphere $\mathbb{S}^2(r)$ with centre p_0 and radius r. We know that, if $p \in \mathbb{S}^2(r)$, the corresponding tangent plane is the orthogonal complement of the vector $p - p_0$. Hence, the map $N : \mathbb{S}^2(r) \to \mathbb{R}^3$ given by

$$N(p) = \frac{1}{r}(p - p_0), \qquad \forall p \in \mathbb{S}^2(r),$$

is a unit normal field defined on the sphere—the so-called *outer* normal field.

Example 3.7 (Graphs are orientable). This assertion follows Lemma 3.2. Lemma 3.2 allows us to construct an explicit unit normal field on a given graph. Let S be a surface which is the graph of the differentiable function $f : U \to \mathbb{R}$ defined on an open subset U of $\mathbb{R}^2$. We know that $X : U \to \mathbb{R}^3$ given by $X(u,v) = (u, v, f(u,v))$ is a parametrization which covers the whole of S. Therefore $N = N^X \circ X^{-1} : S \to \mathbb{R}^3$ with

$$N^X = \frac{X_u \wedge X_v}{|X_u \wedge X_v|} = \frac{1}{\sqrt{1 + |\nabla f|^2}}(-f_u, -f_v, 1)$$

is a unit normal field defined on S.

Example 3.8 (Inverse images are orientable). In fact, if O is an open subset of $\mathbb{R}^3$, $f : O \to \mathbb{R}$ is a differentiable function, $a \in \mathbb{R}$ is a regular value of f, and $S = f^{-1}(\{a\}) \neq \emptyset$ is the corresponding surface, we saw that, at each point $p \in S$,

$$T_p S = \ker(df)_p = \{v \in \mathbb{R}^3 \,|\, \langle (\nabla f)_p, v \rangle = 0\},$$

and, so, $\nabla f_{|S} = (f_x, f_y, f_z)$ is a normal field defined on all S. Since a is a regular value, ∇f does not vanish at the points of S. Then the map $N : S \to \mathbb{R}^3$ given by

$$N = \frac{1}{|\nabla f|}\nabla f_{|S} = \frac{1}{\sqrt{f_x^2 + f_y^2 + f_z^2}}(f_x, f_y, f_z)$$

is a unit normal field on S.

As a clear sample, take $O = \mathbb{R}^3$ and $f(x, y, z) = x^2 + y^2$. Then $S = f^{-1}(\{r^2\})$, $r > 0$, is a right circular cylinder of radius r whose axis is the z-axis. From the above $N(x, y, z) = (1/r)(x, y, 0)$ is a unit normal field on this cylinder.

Even though we proved that any surface is locally orientable and all the previous examples are orientable surfaces, now we want to show that *there exist non-orientable surfaces*. To do this, we will examine in detail a precise example: *the Moebius strip*; see Figure 3.2. Take a circle C centred at the origin with radius 2 inside the plane of equation $z = 0$ in $\mathbb{R}^3$ and a line segment L with length 2 whose centre is the point $(0, 2, 0) \in C$ and whose direction is given by the vector $(0, 0, 1)$. Let S be the set of the points of $\mathbb{R}^3$

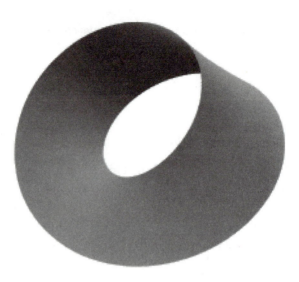

Figure 3.2. *Moebius strip*

swept out by the segment L while its center runs uniformly along C and it rotates around its centre in the plane determined by its position vector and the z-axis in such a way that, when the centre has gone back to the original position, L has completed only a half-revolution.

First, we propose to see that this subset S of $\mathbb{R}^3$, described in a cinematic way, is a surface. For this, note that, if we denote by $u \in \mathbb{R}$ the angle made by the position vector of the centre of the turning segment L with the original position, then the centre L has components given by

$$c(u) = (2 \sin u, 2 \cos u, 0).$$

When the centre of L arrives at the position $c(u)$, L has turned around its centre an angle of $u/2$. In this way, the direction vector $d(u)$ of the segment $L(u)$ results from applying to the original direction $(0,0,1)$ first a rotation of angle $u/2$ around the y-axis and, later, another rotation of angle u around the z-axis. Therefore

$$d(u) = \begin{pmatrix} \cos u & \sin u & 0 \\ -\sin u & \cos u & 0 \\ 0 & 0 & 1 \end{pmatrix} \begin{pmatrix} 1 & 0 & 0 \\ 0 & \cos \frac{u}{2} & -\sin \frac{u}{2} \\ 0 & \sin \frac{u}{2} & \cos \frac{u}{2} \end{pmatrix} \begin{pmatrix} 0 \\ 0 \\ 1 \end{pmatrix}$$

$$= \begin{pmatrix} -\sin u \sin \frac{u}{2} \\ -\cos u \sin \frac{u}{2} \\ \cos \frac{u}{2} \end{pmatrix}.$$

Consequently, the points of $\mathbb{R}^3$ where the moving segment L lies when its centre is at $c(u)$, after having turned an angle of u, are exactly

$$c(u) + vd(u) = (2\sin u, 2\cos u, 0) + v(-\sin u \sin \frac{u}{2}, -\cos u \sin \frac{u}{2}, \cos \frac{u}{2}),$$

where $v \in (-1, 1)$. From these considerations, we define a map $F : \mathbb{R}^2 \to \mathbb{R}^3$ by

$$(3.1) \qquad F(u, v) = \left((2 - v\sin \frac{u}{2})\sin u, (2 - v\sin \frac{u}{2})\cos u, v\cos \frac{u}{2}\right).$$

It is clear that F is differentiable and, according to what we have just seen, the subset S satisfies

$$S = F([0, 2\pi] \times (-1, 1)) = F(\mathbb{R} \times (-1, 1)).$$

The last equality is deduced from the fact that $F(u + 2\pi, -v) = F(u, v)$, as we may easily check by definition (3.1). Furthermore, F is clearly 4π-periodic in the variable u.

Now represent by R the closed rectangle $[0, 2\pi] \times [-1, 1]$ in $\mathbb{R}^2$ and let $K = F(R)$. Then $F : R \to K$ is an identification map because it is continuous, surjective, and closed, the latter because it is defined on a compact set. Let R_F be the equivalence relation induced on R by the map F. If $(u, v), (u', v') \in R$, we have $F(u, v) = F(u', v')$ if and only if

$$(3.2) \qquad \begin{aligned} (2 - v\sin \tfrac{u}{2})\sin u &= (2 - v'\sin \tfrac{u'}{2})\sin u', \\ (2 - v\sin \tfrac{u}{2})\cos u &= (2 - v'\sin \tfrac{u'}{2})\cos u', \\ v\cos \tfrac{u}{2} &= v'\cos \tfrac{u'}{2}. \end{aligned}$$

Squaring the two first equalities and adding them, we obtain that

$$(2 - v\sin \frac{u}{2})^2 = (2 - v'\sin \frac{u'}{2})^2.$$

Now then, since $|v| \leq 1$ and $0 \leq \sin \frac{u}{2} \leq 1$ because $0 \leq u/2 \leq \pi$, we have $|v\sin \frac{u}{2}| \leq 1$ and, hence, $2 - \sin \frac{u}{2} > 0$. An analogous result can be obtained for the pair (u', v'). Then

$$2 - v\sin \frac{u}{2} = 2 - v'\sin \frac{u'}{2}.$$

Therefore $(u, v)R_F(u', v')$ if and only if

$$(3.3) \qquad v\sin \frac{u}{2} = v'\sin \frac{u'}{2} \quad \text{and} \quad v\cos \frac{u}{2} = v'\cos \frac{u'}{2}.$$

Squaring the two equalities again and adding, we have $|v| = |v'|$. If this common absolute value vanishes, going back to (3.2), one has that $\sin u = \sin u'$ and that $\cos u = \cos u'$ and, so, u and u' differ in an integral number of complete turns. Recalling that $u, u' \in [0, 2\pi]$, we conclude that $u = u'$, or $u = 0$ and $u' = 2\pi$, or $u = 2\pi$ and $u' = 0$. Instead, if $|v| = |v'| \neq 0$, then either $v = v' \neq 0$ or $v' = -v \neq 0$. In the first case, using (3.3), we have that the angles $u/2$ and $u'/2$ have the same sines and the same cosines. Hence $u = u'$ because $u/2, u'/2 \in [0, \pi]$. In the second case, again by (3.3), we have that $u/2$ and $u'/2$ have opposite sines and cosines. Therefore, either $u = 0$ and $u' = 2\pi$ or $u = 2\pi$ and $u' = 0$. Thus, we may summarize all these calculations in the following expression:

$$(u, v) R_F (u', v') \iff \begin{cases} (u, v) = (u', v'), \\ u = 0, u' = 2\pi, v' = -v, \\ u = 2\pi, u' = 0, v' = -v. \end{cases}$$

Consequently, F induces a homeomorphism $\bar{F}$ from the quotient R/R_F to $K \subset \mathbb{R}^3$. But this quotient space R/R_F is usually called a compact Moebius strip. Hence K is topologically a compact Moebius strip. The restriction of $\bar{F}$ to $R' = [0, 2\pi] \times (-1, 1)$ gives us a homeomorphism between R'/R_F and $S = F([0, 2\pi] \times (-1, 1))$, which is an open subset of K. Thus S is topologically an *open* Moebius strip.

The same computations that we have just made above show that $S = F([a, a + 2\pi] \times (-1, 1))$ for each $a \in \mathbb{R}$, that F is an identification map between $[a, a + 2\pi] \times (-1, 1)$ and S, and that, on the open subset $U^a = (a, a + 2\pi) \times (-1, 1)$ of $\mathbb{R}^2$, it is injective. In conclusion $V^a = F(U^a)$ is an open subset of S and $X^a = F_{|U^a}$ is a homeomorphism between $U^a \subset \mathbb{R}^2$ and $V^a \subset S$. Therefore, the Moebius strip S will be a surface, according to our Definition 2.2, if we check that each X^a has an injective differential at each point of its domain, or, equivalently, if $(dF)_{(u,v)}$ is injective at each $(u, v) \in \mathbb{R} \times (-1, 1)$. Moreover, in this case, the map X^a will be a parametrization. In fact, it follows from the very definition (3.1) that

$$F_u = \left(\left(2 - v \sin \frac{u}{2}\right) \cos u - \frac{1}{2} v \sin u \cos \frac{u}{2}, \right.$$
$$\left. -\left(2 - v \sin \frac{u}{2}\right) \sin u - \frac{1}{2} v \cos u \cos \frac{u}{2}, -\frac{1}{2} v \sin \frac{u}{2} \right),$$

$$F_v = \left(-\sin u \sin \frac{u}{2}, -\cos u \sin \frac{u}{2}, \cos \frac{u}{2} \right).$$

It follows from this that

$$|F_u \wedge F_v|^2 = \left(2 - v \sin \frac{u}{2}\right)^2 + \frac{1}{4} v^2 > 0$$

for all $(u, v) \in \mathbb{R}^2$. So, S is a surface and each X^a, with $a \in \mathbb{R}$, is a parametrization of S.

Let us show now that our Moebius strip S is an example of a non-orientable surface. In order to prove that, it is convenient to consider and solve the following two exercises.

Exercise 3.9. ↑ Let $\mathcal{P}$ be a property which can be said to be of the unit vectors normal to a given surface S, in such a manner that, for each unit vector v normal to the surface at a point p, either v or its opposite $-v$ satisfies $\mathcal{P}$, and, moreover, the two cannot happen simultaneously. Suppose also that, for each v obeying $\mathcal{P}$, there exist an open neighbourhood V of p in S and a differentiable unit normal field N defined on V such that $N(q)$ satisfies $\mathcal{P}$ for each $q \in V$. Show that S is an orientable surface.

Exercise 3.10. ↑ Let $\alpha : [a, b] \to S$ be a continuous curve on a surface S, and let $n \in \mathbb{R}^3$ be a unit vector normal to S at the extreme $\alpha(a)$ of the curve. Prove that there is a unique continuous map $N : [a, b] \to \mathbb{R}^3$ such that $N(a) = n$ and

$$|N(t)|^2 = 1 \quad \text{and} \quad N(t) \perp T_{\alpha(t)}S, \qquad \forall t \in [a, b].$$

Demonstrate that, if α is a closed curve and S is orientable, then $N(b) = n$. Conversely, if, for each closed curve α, the corresponding *normal field* N *along* α with $N(a) = n$ satisfies $N(b) = n$, then S is an orientable surface.

Let us consider the curve $\alpha : [0, 2\pi] \to S$ on the Moebius strip given by $\alpha(t) = F(t, 0)$ which is clearly differentiable and which is closed because of the periodicity of F. Furthermore, the map $N : [0, 2\pi] \to \mathbb{R}^3$ given by $N(t) = (F_u \wedge F_v)(t, 0)$ is a differentiable normal field along α and it is obvious that $N(0)$ and $N(2\pi)$ are opposite each other, since $F(u+2\pi, -v) = F(u, v)$ and so

$$F_u(0, 0) = F_u(2\pi, 0) \quad \text{and} \quad F_v(0, 0) = -F_v(2\pi, 0).$$

Hence, taking Exercise 3.10 into account, S is a non-orientable surface.

Remark 3.11. The proof that we gave to show that the open Moebius strip is a non-orientable surface consists geometrically in proving that, if we travel along the central circle of the strip—the trace of the centre of the moving segment—starting on a side—remember that it always makes sense to talk about a *side* locally—when we complete a turn, we will find ourselves on the opposite side.

Remark 3.12. In Chapter 4, we will prove the important Theorem 4.21, asserting that any *compact* surface of $\mathbb{R}^3$ is orientable. In fact, this result is true if we weaken the hypothesis to suppose that the surface is only *closed* as a subset of Euclidean space.

Exercise 3.13. ↑ Suppose that a surface S is a union $S = S_1 \cup S_2$, where S_1 and S_2 are two orientable surfaces such that $S_1 \cap S_2$ is connected. Prove that S is also orientable.

Exercise 3.14. ↑ Let S, S_1, S_2 be three surfaces such that $S = S_1 \cup S_2$, S_1 and S_2 are connected, and $S_1 \cap S_2$ has exactly two connected components, namely A and B. If S_1 and S_2 can be oriented so that the orientations induced on A coincide and those induced on B are opposite, then S is a non-orientable surface. Show that this is the case for the Moebius strip.

Exercise 3.15. ↑ Let $f : S_1 \to S_2$ be a local diffeomorphism between two surfaces S_1 and S_2 from which S_2 is orientable. Let N_2 be a unit normal field on S_2. We define a map $N_1 : S_1 \to \mathbb{R}^3$ as follows: if $p \in S_1$, we put

$$N_1(p) = \frac{a \wedge b}{|a \wedge b|},$$

where a, b is a basis of $T_p S_1$ satisfying

$$\det((df)_p(a), (df)_p(b), N_2(f(p))) > 0.$$

Show that N_1 is a unit normal field on S_1 and that, consequently, S_1 is also orientable.

Exercise 3.16. Consider two diffeomorphic surfaces S_1 and S_2. Show that S_1 is orientable if and only if S_2 is orientable.

Exercise 3.17. If S is an oriented surface, N is the corresponding unit normal field, and $p \in S$ is a point of S, we will say that a basis $\{a, b\}$ of the tangent plane $T_p S$ is *positively oriented* when $\det(a, b, N(p)) > 0$. Otherwise, we will say that it is *negatively oriented*. If S_1 and S_2 are two oriented surfaces, we will say that a local diffeomorphism $f : S_1 \to S_2$ *preserves orientation* if its differential at each point of S_1 takes positively oriented bases on S_1 into positively oriented bases on S_2. We define a function $\operatorname{Jac} f : S_1 \to \mathbb{R}$ which will be named the *Jacobian* of f—compare with Exercise (13) at the end of Chapter 2—by the equation

$$(\operatorname{Jac} f)(p) = \det((df)_p(e_1), (df)_p(e_2), N_2(f(p))),$$

where $\{e_1, e_2\}$ is a positively oriented orthonormal basis of $T_p S_1$. Prove that, if S_1 and S_2 are connected, f preserves orientation if and only if its Jacobian is positive everywhere.

3.3. Gauss map and the second fundamental form

Henceforth, in this chapter, we will assume all the surfaces to be orientable. Anyway, according to Lemma 3.2, all the results that we will obtain can be applied to small enough pieces of an arbitrary surface. Then, let S be a surface and N a unit normal field on S. Since $|N(p)|^2 = 1$ for all $p \in S$, if $\mathbb{S}^2$ represents, as usually, the sphere centred at the origin of $\mathbb{R}^3$ with radius 1, we have $N(S) \subset \mathbb{S}^2$ and, consequently, *each unit normal field N on S can be thought of as a differentiable map $N : S \to \mathbb{S}^2$ of the surface into the unit sphere $\mathbb{S}^2$.* This map taking each point on the surface to a unit vector orthogonal to the surface at this point will be called a *Gauss map* on S.

By analogy with the case of curves, we expect that the variation of the Gauss map N on the surface S when we move in a neighbourhood of a given point will give us some information about the shape of the surface near this point. In this case, the variation will no longer be controlled by a unique function—like the curvature in the case of curves—but by the differential $(dN)_p : T_p S \to T_{N(p)} \mathbb{S}^2$. But—see Example 2.53—the tangent plane to the sphere $\mathbb{S}^2$ at the point $N(p)$ is the orthogonal complement of the vector $N(p)$, that is, $T_{N(p)} \mathbb{S}^2 = T_p S$. It follows that *the differential of a Gauss map at each point of a surface is an endomorphism of the tangent plane at this point.* Furthermore, an endomorphism of a two-dimensional vector space has only two associated invariants, namely, its determinant and its trace. We are led in a natural way to consider the following two functions defined on the surface:

$$(3.4) \qquad K(p) = \det(dN)_p \quad \text{and} \quad H(p) = -\frac{1}{2}\operatorname{trace}(dN)_p, \qquad p \in S,$$

which are, respectively, the *Gauss curvature* and the *mean curvature* of the surface at this point. It is clear that, since we are in a two-dimensional context, the Gauss curvature K does not change its sign when we reverse the orientation of the surface and, so, it can also be globally defined on non-orientable surfaces. Instead, the mean curvature H depends on the orientation chosen on the surface and changes its sign when we reverse it. Gradually, in this chapter and in the following ones, we will give different geometrical meanings to these two functions defined on surfaces of $\mathbb{R}^3$.

From the endomorphism $(dN)_p$ of $T_p S$, $p \in S$, and since we dispose on the vector plane $T_p S$ of the Euclidean scalar product, we may define a bilinear form σ_p:

$$\sigma_p : T_p S \times T_p S \longrightarrow \mathbb{R}, \qquad p \in S,$$
$$\sigma_p(v, w) = -\langle (dN)_p(v), w \rangle, \qquad v, w \in T_p S.$$

This is the so-called *second fundamental form of the surface S at the point p*, in terms of which we can rewrite (3.4) as

$$(3.5) \qquad K(p) = \det \sigma_p \quad \text{and} \quad H(p) = \frac{1}{2}\operatorname{trace} \sigma_p, \qquad p \in S.$$

There must exist by way of contrast a *first fundamental form* at each point of the surface even though we have not defined it explicitly. Indeed, it is nothing more than the restriction to each tangent plane of the Euclidean scalar product of the space $\mathbb{R}^3$.

Lemma 3.18. *The second fundamental form σ_p of a surface S at any of its points $p \in S$ is a symmetric bilinear form on the corresponding tangent plane $T_p S$. This can also be expressed by saying that the differential $(dN)_p$ of* *the Gauss map at each point is a self-adjoint endomorphism on the tangent plane at this point.*

Proof. Let U be an open subset of $\mathbb{R}^2$, and let $X : U \to S$ be a parametrization of S. Then, as we know, at each point $(u, v) \in U$, the vectors $X_u(u, v)$ and $X_v(u, v)$ lie in the tangent plane $T_{X(u,v)}S$ and the vector $(N \circ X)(u, v)$ is perpendicular to this plane. We obtain from this the following equalities of functions defined on $U \subset \mathbb{R}^2$:

$$\langle N \circ X, X_u \rangle = 0 \qquad \text{and} \qquad \langle N \circ X, X_v \rangle = 0.$$

Differentiating with respect to the v-variable in the first equality and with respect to u in the second one and subtracting the two resulting expressions, we get

$$\langle (N \circ X)_v, X_u \rangle = \langle (N \circ X)_u, X_v \rangle,$$

where we have taken into account that $X_{uv} = X_{vu}$ by the Schwarz theorem of elementary calculus. Then, by the chain rule Theorem 2.61,

$$(N \circ X)_u(u, v) \quad = \quad d(N \circ X)_{(u,v)}(1, 0)$$

$$= \quad (dN)_{X(u,v)}[(dX)_{(u,v)}(1, 0)] = (dN)_{X(u,v)}[X_u(u, v)]$$

for all $(u, v) \in U$ and, similarly,

$$(N \circ X)_v(u, v) = (dN)_{X(u,v)}[X_v(u, v)].$$

Thus, by the definition of the second fundamental form, one obtains that

$$\sigma_{X(q)}(X_u(q), X_v(q)) = \sigma_{X(q)}(X_v(q), X_u(q))$$

for all $q \in U$. Since the $\{X_u(q), X_v(q)\}$ form a basis of $T_{X(q)}S$, the bilinear form $\sigma_{X(q)}$ is symmetric. $\qquad\square$

One of the most important features of an endomorphism of a Euclidean vector space, arising from being self-adjoint, is the fact that it can be diagonalized. In this way, as a consequence of Lemma 3.18, we may establish the following definition. For each point p of the surface S, the endomorphism $-(dN)_p$ has two real eigenvalues that we will denote by $k_1(p)$ and $k_2(p)$ and that we will suppose to be ordered so that $k_1(p) \leq k_2(p)$. They will be called the *principal curvatures of S at the point p* and, since they are the roots of the characteristic polynomial of $-(dN)_p$, the functions k_1 and k_2 defined on the surface must satisfy

$$K = k_1 k_2, \quad H = \frac{1}{2}(k_1 + k_2), \quad \text{and} \quad k_i^2 - 2Hk_i + K = 0, \qquad i = 1, 2.$$

Since the discriminant of this second degree equation has to be non-negative so that two real roots exist, we have

$$(3.6) \qquad\qquad K(p) \leq H(p)^2, \qquad \forall p \in S,$$

and equality occurs at a given point if and only if the two principal curvatures coincide at this point. Associated to these two eigenvalues of the endomorphism $-(dN)_p$ we have the corresponding eigenspaces, which are two orthogonal lines when $k_1(p) \neq k_2(p)$. In this case, we will call the corresponding directions of these two tangent lines the *principal directions of S at the point p*. (See Remark 3.29 below.)

Example 3.19 (Planes). In Example 3.5, we saw that, if P is a plane of $\mathbb{R}^3$ with unit normal vector a, then a Gauss map for P is the constant map $N \equiv a$. Hence $(dN)_p = 0$ and so $\sigma_p = 0$ for each $p \in P$. In other words, *planes have null second fundamental form at each point*. As an immediate consequence of (3.4) and (3.5), we have that *the Gauss and mean curvatures of a plane are identically zero*. The same is true for the corresponding principal curvatures.

Example 3.20 (Spheres). If $\mathbb{S}^2(r)$ is a sphere with centre $p_0 \in \mathbb{R}^3$ and radius $r > 0$, we know from Example 3.6 that its outer Gauss map is given by

$$N(p) = \frac{1}{r}(p - p_0), \qquad \forall p \in \mathbb{S}^2(r).$$

Then, it follows from the definition of a differential that

$$(dN)_p(v) = \frac{1}{r}v, \qquad \forall v \in T_p\mathbb{S}^2(r),$$

and thus

$$\sigma_p(v, w) = -\frac{1}{r}\langle v, w \rangle, \qquad \forall v, w \in T_p\mathbb{S}^2(r).$$

Therefore, at each point of a sphere, the first and second fundamental forms are proportional. Moreover, using definitions (3.4) and (3.5) again, we obtain

$$K(p) = \frac{1}{r^2} \quad \text{and} \quad H(p) = -\frac{1}{r}, \qquad p \in \mathbb{S}^2(r).$$

That is, *the Gauss curvature of a sphere is constant and its value is the inverse of the square of the radius* and, on the other hand, *the mean curvature of a sphere relative to its outer normal is constant and its value is exactly the opposite of the inverse of the radius*. Furthermore, the principal curvatures coincide at each point and equal $-1/r$.

Example 3.21 (Cylinder). We also determined, in Example 3.8, an outer Gauss map N for the right circular cylinder S with radius $r > 0$ and with revolution axis the coordinate z-axis, which was described by the implicit equation $x^2 + y^2 = r^2$. In fact, we saw that $N(x, y, z) = (1/r)(x, y, 0)$ for all $(x, y, z) \in S$. Thus

$$(dN)_{(x,y,z)}(v_1, v_2, v_3) = \frac{1}{r}(v_1, v_2, 0),$$

and, consequently,

$$\sigma_{(x,y,z)}((v_1, v_2, v_3), (w_1, w_2, w_3)) = -\frac{1}{r}(v_1 w_1 + v_2 w_2),$$

for each $(x, y, z) \in S$ and each $(v_1, v_2, v_3), (w_1, w_2, w_3) \in T_{(x,y,z)}S$. It is not difficult to deduce from this that the Gauss and mean curvatures of the cylinder S are both constant and take, respectively, the values

$$K(x, y, z) = 0 \quad \text{and} \quad H(x, y, z) = -\frac{1}{2r}, \qquad (x, y, z) \in S.$$

The corresponding principal curvatures are $k_1 \equiv -(1/r)$ and $k_2 \equiv 0$, whose respective principal directions are those of the circle of radius r and of the line parallel to the axis passing through the given point.

Exercise 3.22. Consider as a surface S the hyperbolic paraboloid given by

$$S = \{(x, y, z) \in \mathbb{R}^3 \mid z = x^2 - y^2\}.$$

Study the second fundamental form of S at the point $(0, 0, 0)$ and show that it is not a semi-definite bilinear form. In other words, show that its Gauss curvature is negative at this point.

Exercise 3.23. ↑ Show that an orientable connected surface whose Gauss and mean curvatures are identically zero must be an open subset of a plane.

Exercise 3.24. ↑ Let S be the regular quadric—see Example 2.16—given by the implicit second degree equation

$$0 = (1, p^t) A \begin{pmatrix} 1 \\ p \end{pmatrix} = \langle Bp, p \rangle + 2\langle b, p \rangle + c, \qquad p \in \mathbb{R}^3,$$

where A is a symmetric matrix of order four, B is a non-null symmetric matrix of order three, $b \in \mathbb{R}^3$, and $c \in \mathbb{R}^3$. Show that

$$N(p) = \frac{Bp + b}{|Bp + b|}, \qquad \forall p \in S,$$

is a Gauss map defined on S. As a consequence, check that the corresponding second fundamental form is given by

$$\sigma_p(v, v) = -\frac{1}{|Bp + b|} \langle Bv, v \rangle, \qquad p \in S, v \in T_p S.$$

Therefore, an ellipsoid has positive Gauss curvature at each of its points.

Exercise 3.25 (Invariance under rigid motions). Let S be an orientable surface and $\phi : \mathbb{R}^3 \to \mathbb{R}^3$ the rigid motion given by $\phi(p) = Ap + b$ where $A \in O(3)$ and $b \in \mathbb{R}^3$. If N is a Gauss map for the surface S, prove that

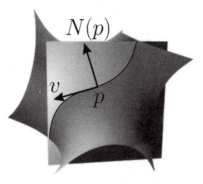

Figure 3.3. *Normal sections*

$N' = A \circ N \circ \phi^{-1}$ is a Gauss map for the image surface $S' = \phi(S)$. Conclude that

$$(dN')_{\phi(p)} = A \circ (dN)_p \circ A^{-1},$$

$$\sigma'_{\phi(p)}((d\phi)_p(v), (d\phi)_p(w)) = \sigma'_{\phi(p)}(Av, Aw) = \sigma_p(v, w)$$

for each $p \in S$, $v, w \in T_pS$, where σ and σ' stand, respectively, for the second fundamental forms of S and S'. Also find the relation between the Gauss and mean curvatures of S and S'.

3.4. Normal sections

We will work at a fixed point p of an orientable surface S on which we have chosen a Gauss map $N : S \rightarrow \mathbb{S}^2$. For each direction $v \in T_pS$ tangent to S at p, we consider the plane P_v passing through the point p and having as a direction plane the plane spanned in $\mathbb{R}^3$ by the vectors $N(p)$ and v. In this way, $\{P_v \,|\, v \in T_pS\}$ is the pencil of planes whose axis is the normal line of the surface S at the point p. Each plane in this pencil cuts the surface at the point p transversely. Indeed, $p \in S \cap P_v$ for all $v \in T_pS$ and, if the equality $T_pS = T_pP_v = \text{span}\,\{N(p), v\}$ holds, then we would have $N(p) \in T_pS$, which is impossible. Applying Theorem 2.67 to this particular situation, we see that, for each $v \in T_pS$, the intersection $S \cap P_v$, near the point p, is the trace of a regular curve. This intersection is usually known as the *normal section* of the surface S at the point p corresponding to the direction v; see Figure 3.3.

Proposition 3.26. *Let S be an orientable surface, N a Gauss map, and σ_p the second fundamental form at $p \in S$. Then, if $v \in T_pS$ is a unit vector tangent to S at p, the curvature of the section $S \cap P_v$ at the point p, considered as the image of a plane curve with velocity v and where $\{v, N(p)\}$ determines a positive orientation for P_v, is just $\sigma_p(v, v)$.*

Proof. As mentioned earlier, there exists a real number $\delta_v > 0$ and a curve $\alpha_v : (-\delta_v, \delta_v) \to \mathbb{R}^3$—which we may assume to be p.b.a.l.—such that $\alpha_v(0) = p$ and $\alpha_v(-\delta_v, \delta_v) \subset S \cap P_v$. Hence, reversing the orientation of α if necessary, we will also suppose that $\alpha'_v(0) = v$. Since $\alpha'_v(t) \in T_{\alpha_v(t)}S$, we have

$$\langle \alpha'_v(t), (N \circ \alpha_v)(t) \rangle = 0, \qquad \forall t \in (-\delta_v, \delta_v).$$

Taking the derivative with respect to t at the time $t = 0$, we obtain

$$\langle \alpha''_v(0), N(p) \rangle = -\langle \alpha'_v(0), (N \circ \alpha_v)'(0) \rangle = -\langle v, (dN)_p(v) \rangle.$$

Now it is enough to remember the definitions of the curvature of a plane curve and of the second fundamental form. $\qquad\qquad\square$

Remark 3.27. This geometrical interpretation of the second fundamental form can be viewed as a reformulation in modern language of the so-called *Euler theorem*, generalized later by Meusnier, that is, the discovering of the fact that the curvatures of the infinitely many normal sections of a surface at a given point are not a random set of numbers, but they are organized as the values taken by a second degree polynomical function in two variables, that is, of a quadratic form. In fact, the most common formulation of this result is that $k_v = \langle v, e_1 \rangle^2 k_1(p) + \langle v, e_2 \rangle^2 k_2(p)$, where k_v is the curvature of the normal section at $p \in S$ corresponding to the direction v and where $k_1(p), k_2(p)$ are the principal curvatures at this point. This can be easily inferred from Proposition 3.26 above.

Remark 3.28. Proposition 3.26 interprets the values taken by the second fundamental form in terms of curvatures of normal sections. This interpretation, together with the comparison result of plane curves contained in Exercise (8) at the end of Chapter 1, allows us to give a geometrical meaning to the sign that the second fundamental form takes for a given direction. We have in fact the following.

- If $v \in T_pS$ is a tangent direction at a point p of an oriented surface S, with Gauss map N, and $\sigma_p(v, v) > 0$, the corresponding normal section $S \cap P_v$ lies locally on the side of the—affine—tangent plane towards which the normal vector $N(p)$ points and, furthermore, touches this plane only at p.

- If the normal section $S \cap P_v$ of a surface S at a point p corresponding to a tangent direction $v \in T_pS$ has a neighbourhood of the point p lying on the side of the tangent plane towards which the normal vector $N(p)$ points, then $\sigma_p(v, v) \geq 0$.

Of course, the analogous assertions for the case of a negative sign are valid too. It is a matter of elementary linear algebra that the sign of a quadratic form defined on a real vector plane is mainly controlled by its determinant.

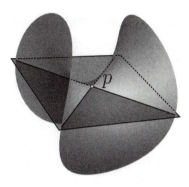

Figure 3.4. *Hyperbolic point*

In the particular case of the second fundamental form σ_p of S at the point p, this determinant is, according to (3.5), the Gauss curvature $K(p)$. When this curvature is positive, σ_p must be definite—positive or negative. This is why the points of a surface with positive Gauss curvature will be called *elliptic points*. We now give a consequence of what we pointed out earlier.

All the normal sections at an elliptic point of a surface lie on the same side of the tangent plane at this point.

Notice that this assertion is not enough to infer that some neighbourhood of the given point is entirely on a side of the tangent plane. However, this is also true, as will be proved in Proposition 3.38. On the contrary, when $K(p) < 0$ at a point $p \in S$, the second fundamental form σ_p is not semi-definite, that is, it is a Lorentz scalar product on T_pS. Then, the points of a surface with negative Gauss curvature will be called *hyperbolic points*. In this case, the two principal curvatures of S at p have different signs: $k_1(p) < 0$ and $k_2(p) > 0$ and, so, the normal sections corresponding to the respective principal directions lie on different sides of the tangent plane. Therefore we obtain the following statement, which will also be improved in the following section:

Each neighbourhood of a hyperbolic point of a surface intersects the two open half-spaces determined by the tangent plane at this point; see Figure 3.4.

The points $p \in S$ of the surface where the Gauss curvature vanishes will be called *parabolic* when $\sigma_p \neq 0$ and *planar* when $\sigma_p = 0$.

Remark 3.29. Bearing in mind that the eigenvalues of a bilinear symmetric form—or of a quadratic form—on a two-dimensional Euclidean vector space can be obtained as the maximum and the minimum values taken by the form on the circle of unit vectors, the principal curvatures and directions of the surface S at a point $p \in S$, defined in Section 3.3, can be characterized as follows:

> If $k_1(p) = k_2(p)$, all the normal sections of S at p have
> the same curvature. Otherwise, the principal directions of
> S at p correspond to the two normal sections of maximum
> and minimum curvature and, so, they are orthogonal.

For this reason, the points of a surface where the two principal curvatures coincide or, equivalently, according to (3.6), where the equality $K = H^2$ holds, are called *umbilical points*. This is clearly equivalent to the fact that the second fundamental form is proportional to the first fundamental form— the Euclidean scalar product—at the point, and that the differential of the Gauss map at this point is a similarity. In short, from an umbilical point, the surface is seen to bend in the same way along any direction. When all the points of a surface S share this property, we will say that S is a *totally umbilical* surface. Examples 3.19 and 3.20 show that planes and spheres and, hence, any open subset of them, are totally umbilical surfaces. The following result says that, in fact, we cannot expect to find more examples.

Theorem 3.30 (Classification of totally umbilical surfaces). *The only connected surfaces of $\mathbb{R}^3$ which are totally umbilical are the open subsets of planes and spheres.*

Proof. Let U be an open connected subset of $\mathbb{R}^2$ and $X : U \to S$ a parametrization of the surface S. We dispose, according to Lemma 3.2, of a local Gauss map N on the open set $X(U) \subset S$. In the proof of Lemma 3.18 we saw that

$$(dN)_{X(u,v)}[X_u(u,v)] = (N \circ X)_u(u,v),$$

$$(dN)_{X(u,v)}[X_v(u,v)] = (N \circ X)_v(u,v).$$

for each $(u, v) \in U$. Since S is totally umbilical, the differential $(dN)_{X(u,v)}$ is proportional to the identity map on the plane $T_{X(u,v)}S$. Then, there exists a function $f : U \to \mathbb{R}$ such that

$$(dN)_{X(u,v)} = f(u,v)\, I_{T_{X(u,v)}S}, \qquad \forall (u,v) \in U.$$

Combining the three last equations, we have

(3.7) $\qquad\qquad (N \circ X)_u = f\, X_u \quad \text{and} \quad (N \circ X)_v = f\, X_v.$

From the first of the equalities in (3.7), and after scalar multiplication by the vector valued function X_u defined on the open set U—which does not vanish anywhere—we may see that the function f is differentiable. Once we have checked this point, we will able to take partial derivatives of the two equalities relative to v and u respectively. We obtain

$$(N \circ X)_{uv} = f_v\, X_u + f\, X_{uv} \quad \text{and} \quad (N \circ X)_{vu} = f_u\, X_v + f\, X_{vu}.$$

Finally, applying the Schwarz theorem of elementary calculus, we get the following equality of functions defined on U:

$$f_u \, X_v = f_v \, X_u.$$

Since $X_u(u,v), X_v(u,v)$ are linearly independent at each point of U, it follows that $f_u = f_v \equiv 0$. Since we have taken U to be connected, it turns out that the function f is constant. If this constant is zero, we deduce from (3.7) that N is constant on $X(U)$ and so—see Exercise 2.70—$X(U)$ is contained in a plane of $\mathbb{R}^3$. Suppose then that f is a non-null constant function. Using (3.7) again, we can deduce that $X - (1/f)(N \circ X) \equiv c$ for some $c \in \mathbb{R}^3$. Therefore

$$|X - c|^2 = |\frac{1}{f}(N \circ X)|^2 = \frac{1}{f^2},$$

that is, $X(U)$ is contained in the sphere with centre c and radius $1/|f|$. In conclusion, we have proved that, under our hypotheses, the surface S is covered by open connected subsets each of which is contained either in a plane or in a sphere. Since the surface is connected, we conclude using Exercise (12) at the end of Chapter 2. $\square$

Corollary 3.31. *The only connected totally umbilical surfaces which are closed as subsets of $\mathbb{R}^3$ are planes and spheres.*

Exercise 3.32. ↑ Suppose that a surface S and a plane P are tangent along the trace of a regular curve. Show that all the points of this curve are either parabolic or planar points of S.

Exercise 3.33. ↑ If a surface contains a straight line, then the surface has non-positive Gauss curvature at all the points of this line.

3.5. Height function and the second fundamental form

In Section 3.4, we obtained a certain interpretation of the second fundamental form by considering the surface as consisting locally of a pencil of regular curves passing through a given point. Now, we will change our point of view and think of the surface as a two-dimensional object. In this way, another useful geometrical interpretation of this quadratic form associated to each point of the surface will be seen.

Definition 3.34 (Hessian at a critical point). Suppose that S is a surface and that $f : S \to \mathbb{R}$ is a differentiable function on S. If $p \in S$ is a critical point of f—as introduced in Section 2.6—we may define the *Hessian* of f at the point p as a map of T_pS into $\mathbb{R}$, which we will denote by $(d^2 f)_p$, defined as:

$$(d^2 f)_p(v) = \frac{d^2}{dt^2}\bigg|_{t=0} (f \circ \alpha)(t), \qquad v \in T_pS,$$

where $\alpha : (-\varepsilon, \varepsilon) \to S$ is a curve on the surface with $\alpha(0) = p$ and $\alpha'(0) = v$.

Proposition 3.35. *The map* $(d^2 f)_p : T_p S \to \mathbb{R}$ *is well-defined by the previous expression at any critical point* $p \in S$ *of the function* f, *that is, this expression does not depend on the chosen curve. Furthermore, the following properties are true:*

A: $(d^2 f)_p$ *is a quadratic form on the tangent plane* $T_p S$.

B: *If the function* f *has a local maximum—respectively, minimum— at the point* p, *then the map* $(d^2 f)_p$ *is semi-definite negative— respectively, positive.*

C: *If* $(d^2 f)_p$ *is a quadratic form definite negative—respectively, positive—then* f *has an isolated local maximum—respectively, minimum—at* p.

Proof. Let us choose a parametrization $X : U \to S$ with $p = X(q)$ and $q = (a, b) \in U$. Taking $\varepsilon > 0$ small enough, we may assume that $\alpha(-\varepsilon, \varepsilon) \subset X(U)$. If we represent by β the composite curve $X^{-1} \circ \alpha$ and write explicitly its components $\beta(t) = (u(t), v(t))$, we have

$$(f \circ \alpha)(t) = (f \circ X) \circ (X^{-1} \circ \alpha)(t) = (f \circ X)(u(t), v(t)).$$

Computing the second derivative with respect to t at $t = 0$ and applying the chain rule of calculus, one has

$$(f \circ \alpha)''(0) = (f \circ X)_{uu}(a, b)u'(0)^2 + 2(f \circ X)_{uv}(a, b)u'(0)v'(0)$$

$$+(f \circ X)_{vv}(a, b)v'(0)^2 + (f \circ X)_u(a, b)u''(0) + (f \circ X)_v(a, b)v''(0).$$

On the other hand,

$$(f \circ X)_u(a, b) = d(f \circ X)_{(a,b)}(1, 0) = (df)_p[(dX)_{(a,b)}(1, 0)] = 0$$

because p is a critical point of the function f. Similarly, $(f \circ X)_v(a, b) = 0$. Hence,

$$(f \circ \alpha)''(0) =$$

$$(f \circ X)_{uu}(a, b)u'(0)^2 + 2(f \circ X)_{uv}(a, b)u'(0)v'(0) + (f \circ X)_{vv}(a, b)v'(0)^2.$$

Now, it suffices to realize that $(u'(0), v'(0)) = \beta'(0) = (dX^{-1})_p(v)$ to conclude that the right-hand side of this equality depends only on v and does not depend on the curve α and that it is a quadratic form in v. This proves assertion **A** of the proposition.

With respect to **B**, suppose that f has a local maximum at p, that $v \in T_p S$, and that $\alpha : (-\varepsilon, \varepsilon) \to S$ is a curve with $\alpha(0) = p$ and $\alpha'(0) = v$. Thus, the function $f \circ \alpha$ has a local maximum at 0 and, so, $(d^2 f)_p(v) = (f \circ \alpha)''(0) \leq 0$, that is, $(d^2 f)_p$ is semi-definite negative.

Concerning assertion **C**, the calculations made to show **A** may convince us that, if $(d^2f)_p$ is definite negative at the critical point p, then $q = (a, b)$ is a critical point of $f \circ X : U \subset \mathbb{R}^2 \to \mathbb{R}$ and its Hessian matrix at this point is definite negative. Using a well-known result of elementary calculus, one can finish the proof. □

Example 3.36 (Height function). Let $h : S \to \mathbb{R}$ be the height function relative to the plane which passes through $p_0 \in \mathbb{R}^3$ with unit normal vector a. Let $p \in S$ be a critical point of h. We proved in Example 2.59 that this is equivalent to $a \perp T_pS$ and, so, we will suppose that $N(p) = a$ for a suitable local Gauss map N. If $v \in T_pS$ and α is a curve on S with velocity v, we have

$$(d^2h)_p(v) = \frac{d^2}{dt^2}\bigg|_{t=0} \langle \alpha(t) - p_0, a \rangle = \langle \alpha''(0), a \rangle = \langle \alpha''(0), N(p) \rangle.$$

By Proposition 3.26 we infer that

$$(d^2h)_p(v) = \sigma_p(v, v).$$

Note that a point of a surface is always critical for the height function relative to the tangent plane at this point. This is why we may assert the following result.

Proposition 3.37. *The second fundamental form σ_{p_0} of the surface S at the point p_0 with respect to the choice of normal field $N(p_0)$ at this point is nothing more than the Hessian $(d^2h)_{p_0}$ of the height function $h(p) = \langle p - a, N(p_0) \rangle$, $p \in S$, $a \in \mathbb{R}^3$, relative to any plane parallel to the tangent plane at this point.*

Now, if we combine Propositions 3.35 and 3.37, we get the following information about elliptic and hyperbolic points of a surface, which improves what we obtained in Remark 3.28.

Proposition 3.38. *Let S be a surface, K its Gauss curvature function, and $p \in S$. Then the following hold.*

    A: *If $K(p) > 0$, that is, if p is an elliptic point, there exists a neighbourhood of p in S lying on one of the sides of the affine tangent plane of S at p. Moreover, p is the only contact point, in this neighbourhood, between the surface and the plane.*

    B: *If $K(p) < 0$, that is, if p is a hyperbolic point, in each neighbourhood of p in S, we find points of the surface on both open half-spaces determined by the affine tangent plane of S at p.*

Example 3.39 (Square of the distance). We consider the function f measuring the square of the distance from each point of S to the point $p_0 \in \mathbb{R}^3$, that is, $f(p) = |p - p_0|^2$, $p \in S$. We already saw, in Example 2.69, that p is a critical point of f if and only if p_0 belongs to the normal line of S at p. Suppose that this is the case, that is,

$$p_0 = p + \lambda N(p) \quad \text{for some } \lambda \in \mathbb{R},$$

where N is a Gauss map for S. Take now $v \in T_p S$, and let α be a curve in S with velocity v. We have

$$(d^2 f)_p(v) = \frac{d^2}{dt^2}\bigg|_{t=0} |\alpha(t) - p_0|^2$$

$$= 2|v|^2 + 2\langle \alpha''(0), p - p_0 \rangle = 2|v|^2 - 2\lambda\langle \alpha''(0), N(p)\rangle.$$

Using again Proposition 3.26, we have

$$(d^2 f)_p(v) = 2[|v|^2 - \lambda\sigma_p(v, v)], \qquad p_0 = p + \lambda N(p).$$

We can infer from this some interesting results similar to what we obtained in the previous example for the height function. Two of them are proposed to the reader in the following exercises and the remaining ones are in the exercises at the end of this chapter.

Exercise 3.40. ↑ Show that there are no compact surfaces that have negative Gauss curvature everywhere.

Exercise 3.41. ↑ Let S be a surface and $p \in S$. Prove that p is an elliptic point if and only if there exists a point $p_0 \in \mathbb{R}^3$ such that p is a local maximum of the square of the distance to the point p_0.

Exercise 3.42. ↑ Show that there are no compact surfaces that have non-positive Gauss curvature everywhere.

3.6. Continuity of the curvatures

We defined, on the tangent plane of the surface at any of its points, a quadratic form, namely the second fundamental form, enclosing, as we have begun to see, non-trivial information about the geometry of the surface near the point. In the case of curves, this role was played by the curvature and the torsion functions. They both appeared naturally as differentiable functions. However, until now, we have not said anything about the way in which the second fundamental form—or the curvatures arising from it—depend on the points of the surface. We proceed in this section with the study of this dependence and some of its consequences.

Proposition 3.43. *The Gauss and mean curvatures of any surface are differentiable functions. On the other hand, the principal curvatures are continuous functions and they are differentiable on the open subset consisting of the non-umbilical points.*

Proof. Call S the given surface, and let $X : U \to S$ be a parametrization of S. In Lemma 3.2, we constructed a Gauss map $N = N^X \circ X^{-1}$ on the open subset $X(U)$, where $N^X = X_u \wedge X_v / |X_u \wedge X_v|$. Let us work at an arbitrary point of U without representing it explicitly. Thus, since the vectors $\{X_u, X_v\}$ form a basis of the plane $T_X S$, the first fundamental form (i.e., the scalar product), the second fundamental form, and the differential of the Gauss map can be represented with respect to this basis by means of symmetric matrices of order two. We will denote them by M, Σ, and A respectively. Writing the corresponding entries, we have

$$(3.8) \quad M = \begin{pmatrix} E & F \\ F & G \end{pmatrix}, \qquad E = |X_u|^2, \quad F = \langle X_u, X_v \rangle, \quad G = |X_v|^2.$$

The functions $E, F, G : U \subset \mathbb{R}^2 \to \mathbb{R}$ are obviously differentiable. As for the second fundamental form,

$$(3.9)$$
$$\Sigma = \begin{pmatrix} e & f \\ f & g \end{pmatrix}, \quad e = \sigma_X(X_u, X_u), \ f = \sigma_X(X_u, X_v), \ g = \sigma_X(X_v, X_v).$$

Let us see that these three new functions $e, f, g : U \subset \mathbb{R}^2 \to \mathbb{R}$ are also differentiable. This is a consequence of the definition of the second fundamental form and of the proof of Lemma 3.18. In fact,

$$e = \sigma_X(X_u, X_u) = -\langle (dN)_X X_u, X_u \rangle = -\langle (N \circ X)_u, X_u \rangle = \langle N \circ X, X_{uu} \rangle,$$

because $\langle N \circ X, X_u \rangle \equiv 0$. Therefore, if we make the same calculation for f and g, we obtain

$$
\begin{aligned}
(3.10) \quad
e &= \langle N^X, X_{uu} \rangle = \frac{1}{|X_u \wedge X_v|} \langle X_u \wedge X_v, X_{uu} \rangle \\
&= \frac{1}{|X_u \wedge X_v|} \det(X_u, X_v, X_{uu}), \\
f &= \langle N^X, X_{uv} \rangle = \frac{1}{|X_u \wedge X_v|} \langle X_u \wedge X_v, X_{uv} \rangle \\
&= \frac{1}{|X_u \wedge X_v|} \det(X_u, X_v, X_{uv}), \\
g &= \langle N^X, X_{vv} \rangle = \frac{1}{|X_u \wedge X_v|} \langle X_u \wedge X_v, X_{vv} \rangle \\
&= \frac{1}{|X_u \wedge X_v|} \det(X_u, X_v, X_{vv}),
\end{aligned}
$$

where the expressions on the right-hand sides are clearly differentiable functions on U. Finally, the matrix A representing the endomorphism $(dN)_X$ in the basis $\{X_u, X_v\}$ that we are working with is

$$(3.11) \qquad A = -M^{-1}\Sigma = -\begin{pmatrix} E & F \\ F & G \end{pmatrix}^{-1} \begin{pmatrix} e & f \\ f & g \end{pmatrix}.$$

Then, since $\text{trace}\,(dN)_X = -\text{trace}\,A$ and $\det(dN)_X = \det A$, we have by definition (3.4)

$$(3.12) \qquad K \circ X = \frac{eg - f^2}{EG - F^2} \quad \text{and} \quad H \circ X = \frac{1}{2}\frac{eG + gE - 2fF}{EG - F^2},$$

where the six functions E, F, G and e, f, g are given by (3.8), (3.9), and (3.10), and so they are differentiable, and where $EG - F^2 = |X_u \wedge X_v|^2 > 0$. Since this occurs for any parametrization X, the functions K and H are differentiable on S. Concerning the principal curvatures $k_1, k_2 : S \to \mathbb{R}$, we know by the definition preceding inequality (3.6) that they can be expressed in terms of K and H as:

$$k_i = H \pm \sqrt{H^2 - K}, \qquad i = 1, 2.$$

We may finish from this equality. $\qquad\qquad\qquad\qquad\qquad\qquad\qquad\qquad$ $\square$

Remark 3.44. This proof supplies us with the two equalities in (3.12), which allow us to compute the Gauss and mean curvatures and the principal curvatures of a surface, on the open subset covered by a parametrization, in terms of easy functions arising from the parametrization itself and from its first and second order partial derivatives. Note that, on a large number of surfaces, the image of only one parametrization is dense on the surface. Therefore, although the whole of the surface is not entirely covered by the parametrization, since all these curvatures are continuous, in such a case, we will be able to compute them using only one parametrization.

Example 3.45 (Paraboloid). Let S be the elliptic paraboloid given by

$$S = \{(x, y, z) \in \mathbb{R}^3 \mid 2z = x^2 + y^2\}.$$

This surface is the graph of the function $f : \mathbb{R}^2 \to \mathbb{R}$ given by $f(x, y) = (1/2)(x^2 + y^2)$. Then the whole of S is covered by the following parametrization (see Examples 2.8 and 3.7):

$$X(u, v) = (u, v, \tfrac{1}{2}(u^2 + v^2)), \qquad (u, v) \in \mathbb{R}^2.$$

The first and second partial derivatives of X are

$$X_u(u, v) = (1, 0, u) \quad \text{and} \quad X_v(u, v) = (0, 1, v),$$

$$X_{uu}(u, v) = (0, 0, 1), \quad X_{uv}(u, v) = (0, 0, 0), \quad \text{and} \quad X_{vv}(u, v) = (0, 0, 1).$$

Thus, the coefficients of the first and second fundamental forms, given in (3.8) and (3.10), are

$$E(u, v) = 1 + u^2, \quad F(u, v) = uv, \quad G(u, v) = 1 + v^2,$$

$$e(u, v) = \frac{1}{\sqrt{1 + u^2 + v^2}}, \quad f(u, v) = 0, \quad g(u, v) = \frac{1}{\sqrt{1 + u^2 + v^2}}.$$

It follows, using (3.12), that

$$K(X(u,v)) = \frac{1}{(1+u^2+v^2)^2} \quad \text{and} \quad H(X(u,v)) = \frac{1}{2}\frac{2+u^2+v^2}{(1+u^2+v^2)^{\frac{3}{2}}}.$$

Finally, if $(x,y,z) \in S$, one has

$$K(x,y,z) = \frac{1}{(1+2z^2)^2} \quad \text{and} \quad H(x,y,z) = \frac{1+z^2}{(1+2z^2)^{\frac{3}{2}}}.$$

Example 3.46 (Helicoid). The *helicoid* is defined as the surface $S = X(\mathbb{R}^2)$, where $X : \mathbb{R}^2 \to \mathbb{R}^3$ is the map

$$X(u,v) = (v\cos u, v\sin u, au), \qquad a \in \mathbb{R}^*, \quad (u,v) \in \mathbb{R}^2.$$

It can be easily shown that S is indeed a surface and that X is a parametrization which covers it entirely. Its first and second partial derivatives are

$$X_u(u,v) = (-v\sin u, v\cos u, a) \quad \text{and} \quad X_v(u,v) = (\cos u, \sin u, 0)$$

$$X_{uv}(u,v) = (-\sin u, \cos u, 0),$$
$$X_{uu}(u,v) = (-v\cos u, -v\sin u, 0), \quad \text{and} \quad X_{vv} = (0,0,0).$$

Repeating the process followed in above Example 3.45, we see that

$$K(X(u,v)) = -\left(\frac{a}{a^2+v^2}\right)^2 \quad \text{and} \quad H(X(u,v)) = 0.$$

The fact that the principal curvatures of a surface depend continuously on its points will lead us to prove two of the first results of global geometry of surfaces in this book. They are known as the Liebmann theorem (1899) and the Jellett theorem (1853), and both are characterizations of the sphere among all the compact surfaces in terms of the behaviour of their curvatures. We will obtain them as consequences of a theorem due to Hilbert (1945), for which we will provide a simpler proof than the usual ones.

Theorem 3.47 (Hilbert's theorem). *Let S be an oriented surface and $k_1 \leq k_2 : S \to \mathbb{R}$ the corresponding principal curvature functions. If, at a point $p \in S$, the following three conditions hold simultaneously:*

> **A:** *the Gauss curvature at p is positive,*

> **B:** k_1 *has a local minimum at p,*

> **C:** k_2 *has a local maximum at p,*

then p is an umbilical point.

Proof. We place the origin of $\mathbb{R}^3$ at the point $p \in S$ and choose the coordinate axes so that

$$e_1 = (1,0,0), \quad e_2 = (0,1,0), \quad \text{and} \quad N(p) = (0,0,1),$$

where N is the Gauss map of S and $e_1, e_2 \in T_p S$ are principal directions at the point p. In this affine reference the affine tangent plane of S at $p = (0, 0, 0)$ is the coordinate plane $z = 0$. By Proposition 2.23, we can parametrize S near p as a graph over this plane $z = 0$. To be exact, there is a parametrization X of S such that $X(u, v) = (u, v, f(u, v))$, where $f : U \to \mathbb{R}$ is a differentiable function on the open subset U of $\mathbb{R}^2$ with $(0, 0) \in U$ and $X(0, 0) = p$. By our choice of reference we have

$$f(0, 0) = 0, \quad f_u(0, 0) = 0, \quad \text{and} \quad f_v(0, 0) = 0.$$

This can be written in terms of X as

$$X(0, 0) = (0, 0, 0), \quad X_u(0, 0) = (1, 0, 0) = e_1, \quad \text{and} \quad X_v(0, 0) = (0, 1, 0) = e_2.$$

On the other hand, using the expressions in (3.10), providing the coefficients of the matrix of the second fundamental form σ_X of S at any point X covered by the parametrization, we have

$$e = \frac{f_{uu}}{\sqrt{1 + f_u^2 + f_v^2}}, \quad f = \frac{f_{uv}}{\sqrt{1 + f_u^2 + f_v^2}}, \quad \text{and} \quad g = \frac{f_{vv}}{\sqrt{1 + f_u^2 + f_v^2}}.$$

Since the vectors X_u, X_v at the point $(0, 0) \in U$ are principal directions, one has

$$f_{uu}(0, 0) = k_1(p), \quad f_{uv}(0, 0) = 0, \quad \text{and} \quad f_{vv}(0, 0) = k_2(p).$$

Let α and β be the curves on the open subset of S covered by X given by

$$\alpha(u) = X(u, 0) \quad \text{and} \quad \beta(v) = X(0, v),$$

and let E_1 and E_2 be the curves in $\mathbb{R}^3$ defined as

$$E_2(u) = \frac{1}{|X_v(u, 0)|} X_v(u, 0) \quad \text{and} \quad E_1(v) = \frac{1}{|X_u(0, v)|} X_u(0, v).$$

According to these definitions, one has that

$$E_1(v) \in T_{\beta(v)} S \quad \text{and} \quad E_2(u) \in T_{\alpha(u)} S,$$

and, so, we may consider the following two real functions h_1 and h_2 of one real variable, which can be computed, from (3.10), in terms of f:

$$h_1(v) = \sigma_{\beta(v)}(E_1(v), E_1(v)) = \frac{f_{uu}}{(1 + f_u^2)\sqrt{1 + f_u^2 + f_v^2}}(0, v),$$

$$h_2(u) = \sigma_{\alpha(u)}(E_2(u), E_2(u)) = \frac{f_{vv}}{(1 + f_v^2)\sqrt{1 + f_u^2 + f_v^2}}(u, 0).$$

Therefore, by hypothesis **C**,

$$h_2(0) = \sigma_p(e_2, e_2) = k_2(p) \geq k_2(\alpha(u)) \geq \sigma_{\alpha(u)}(E_2(u), E_2(u)) = h_2(u)$$

and, analogously, using **B**,

$$h_1(0) = \sigma_p(e_1, e_1) = k_1(p) \leq k_1(\beta(v)) \leq \sigma_{\beta(v)}(E_1(v), E_1(v)) = h_1(v).$$

That is, the function h_1 has a local minimum at 0 and h_2 has a local maximum at the same point. Thus,

$$(3.13) \qquad h_2''(0) \le 0 \le h_1''(0).$$

Differentiating with respect to u directly in the expression defining h_2, we arrive at

$$
\begin{aligned}
h_2'(u) &= [-(1+f_v^2)^{-2}2f_v f_{vu}(1+f_u^2+f_v^2)^{-\frac{1}{2}}f_{vv} \\
&\quad - \tfrac{1}{2}(1+f_v^2)^{-1}(1+f_u^2+f_v^2)^{-\frac{3}{2}}(2f_u f_{uu}+2f_v f_{vu})f_{vv} \\
&\quad + (1+f_v^2)^{-1}(1+f_u^2+f_v^2)^{-\frac{1}{2}}f_{vvu}](u,0),
\end{aligned}
$$

and, hence, differentiating again at $u = 0$, we obtain that

$$h_2''(0) = -f_{uu}^2(0,0)f_{vv}(0,0) + f_{vvuu}(0,0).$$

In an analogous way, we get

$$h_1''(0) = -f_{vv}^2(0,0)f_{uu}(0,0) + f_{uuvv}(0,0).$$

Thus, from (3.13), one obtains, applying the Schwarz theorem of elementary calculus,

$$f_{uu}(0,0)f_{vv}(0,0)(f_{vv}(0,0) - f_{uu}(0,0)) \le 0.$$

Substituting here the values that we have computed from these derivatives, we finally have

$$K(p)(k_2(p) - k_1(p)) \le 0.$$

Since **A** asserts that $K(p) > 0$, it follows that p is umbilical. $\qquad\square$

Corollary 3.48 (Jellett-Liebmann's theorem). *A compact connected surface with positive Gauss curvature everywhere and constant mean curvature is necessarily a sphere.*

Proof. Let $c \in \mathbb{R}$ such that $H \equiv c$. Let us see that c is necessarily a non-null constant. Indeed, if $c = 0$, then $k_1 = -k_2$ and $K = -k_1^2 \le 0$, which gives a contradiction to the hypotheses. This requires that the surface S be orientable. In fact, it suffices to apply Exercise 3.9. After having checked the orientability of the surface, we consider the principal curvature functions $k_1 \le k_2$ corresponding to a chosen orientation of S. By Proposition 3.43, they are continuous on S. Since S is compact, there is a point $a \in S$ where k_1 attains its minimum. But $k_2 = 2H - k_1 = 2c - k_1$, and so k_2 attains at this same point its maximum value, say a, on S. On the other hand, the Gauss curvature of S is positive everywhere. From the Hilbert Theorem 3.47, the point a is umbilical. Therefore, for any other point $p \in S$

$$k_2(p) \le k_2(a) = k_1(a) \le k_1(p).$$

Consequently $k_1(p) = k_2(p)$, that is, S is a totally umbilical surface. Now we can finish by applying Corollary 3.31. $\qquad\square$

Corollary 3.49 (Hilbert-Liebmann's theorem). *The only compact connected surfaces with constant Gauss curvature are spheres.*

Proof. If $c \in \mathbb{R}$ is the constant value taken by the Gauss curvature K of the surface S, applying Exercise 3.42, we have $c > 0$. Thus the mean curvature H of S relative to any local orientation cannot vanish anywhere, as (3.6) shows. Similar reasoning to what we have used in the proof of previous results assures us that the surface S must be orientable. We have again two global principal curvature functions k_1 and k_2 on the surface and we can repeat step by step the proof of the Jellett-Liebmann Corollary 3.48, taking into account that the maximum of k_2 has to be reached at the same point where k_1 reaches its minimum, because the product of both functions is nothing more than $K \equiv c$. $\qquad\square$

Exercise 3.50. ↑ See that a compact connected surface with positive Gauss curvature everywhere and such that the quotient H/K is constant has to be a sphere.

Exercise 3.51. Prove that a compact connected surface with positive Gauss curvature everywhere having a constant principal curvature is necessarily a sphere.

Exercises

(1) ↑ Let S be a surface and $p \in S$ a point. Show that there are infinitely many points in the normal line of S at p for which the corresponding distance functions have a local minimum at p.

(2) ↑ Let S be a compact surface contained in a closed ball of $\mathbb{R}^3$ with radius $r > 0$. Prove that there exists at least a point $p \in S$ such that $K(p) \geq 1/r^2$ and $|H(p)| \geq 1/r$.

(3) ↑ Suppose that a compact connected surface S has the following property: for each $p_0 \in \mathbb{R}^3 - A$, where A is a countable set, the square of the distance function at the point p_0 has at most two critical points. Show that this surface is a sphere.

(4) ↑ Let S be an orientable surface and $N : S \to \mathbb{S}^2$ a Gauss map. The *support function of S based on the point* $p_0 \in \mathbb{R}^3$ is defined as:

$$f(p) = \langle p - p_0, N(p) \rangle, \qquad \forall p \in S.$$

Show that, if S is connected, f has constant sign if and only if S is star-shaped with respect to the point p_0; see Exercise (11) at the end of Chapter 2. Prove also that, if f is constant and there exists a point

$p \in S$ with non-zero Gauss curvature, then S is an open subset of a sphere centred at p_0.

(5) If $N : S \to \mathbb{S}^2$ is the Gauss map of an oriented surface, check that the Jacobian of N and the Gauss curvature function coincide at each point. Note that we consider, on the unit sphere $\mathbb{S}^2$, the orientation corresponding to the outer normal field. (See Exercise 3.17 and Remark 8.3.)

(6) Prove that the Gauss map of a compact oriented surface is a local diffeomorphism if and only if its Gauss curvature is positive everywhere. (See Exercise (13) at the end of Chapter 2.)

(7) Let $P \subset \mathbb{R}^3$ be a plane passing through the origin and $\alpha : I \to P$ a regular curve which is a homeomorphism onto its image. If $a \in \mathbb{R}^3$ is a unit vector normal to P and $X : I \times \mathbb{R} \to \mathbb{R}^3$ is defined by

$$X(s, t) = \alpha(s) + t\, a,$$

show that $S = X(I \times \mathbb{R})$ is a surface with identically zero Gauss curvature. Compute its mean curvature at each point. Deduce that, if C is a simple curve in the plane P, the set

$$S = \{p + ta \,|\, p \in C \text{ and } t \in \mathbb{R}\}$$

is a surface with identically zero Gauss curvature everywhere. It is usually called the *right cylinder with directrix C*. (See Subsection 7.7.4.)

(8) Use the methods of Section 3.6 to compute the Gauss and mean curvatures of the torus of revolution described in Example 2.17. Find its elliptic, hyperbolic, and parabolic points.

(9) ↑ Prove that the only surfaces of revolution with identically zero mean curvature which are closed as subsets of $\mathbb{R}^3$ are *catenoids*, i.e., surfaces of revolution generated by a catenary, and planes.

(10) Study the surfaces of revolution with constant Gauss curvature.

(11) ↑ Determine the umbilical points of the ellipsoid of the equation

$$\frac{x^2}{a^2} + \frac{y^2}{b^2} + \frac{z^2}{c^2} = 1,$$

where $0 < a < b < c$.

(12) ↑ Let X be a parametrization of a surface S such that $F = \langle X_u, X_v \rangle = 0$—*orthogonal parametrization*. Prove that the Gauss curvature K of S verifies

$$K \circ X = -\frac{1}{2\sqrt{EG}} \left[\left(\frac{E_v}{\sqrt{EG}} \right)_v + \left(\frac{G_u}{\sqrt{EG}} \right)_u \right].$$

(13) Let S be an orientable connected surface with constant principal curvatures. If S has some elliptic point, then it is an open subset of a sphere.

(14) ↑ Consider a surface S and a dilatation $\phi : \mathbb{R}^3 \to \mathbb{R}^3$. Compare the second fundamental forms and the curvatures of S and of the image surface $S' = \phi(S)$.

(15) ↑ Repeat the previous exercise assuming that ϕ is no longer a Euclidean dilatation but the inversion $\phi : \mathbb{R}^3 - \{0\} \to \mathbb{R}^3 - \{0\}$ given by

$$\phi(p) = \frac{p}{|p|^2}, \qquad p \in \mathbb{R}^3 - \{0\}.$$

(16) Let S be an orientable surface and $N : S \to \mathbb{S}^2$ a Gauss map. Define $F_r : S \to \mathbb{R}^3$ by

$$F_r(p) = p + rN(p), \qquad \forall p \in S,$$

where $r > 0$. Suppose that $S_r = F_r(S)$ is a surface and that $F_r : S \to S_r$ is a diffeomorphism (see Remark 4.29). Show the following.
- If $p \in S$ and $e_1, e_2 \in T_pS$ are principal directions, then $(dF_r)_p(e_i) = (1 - rk_i(p))e_i$, $1 - rk_i(p) > 0$, $i = 1, 2$, and $1 - 2rH(p) + r^2K(p) > 0$.
- $T_pS = T_{F_r(p)}S_r$ for each $p \in S$ and, hence, there is a Gauss map N' for S_r such that $N' \circ F_r = N$.
- The—affine—normal line of S at $p \in S$ coincides with the—affine— normal line of S_r at $F_r(p)$.
- If $e_1, e_2 \in T_pS$ are principal directions at $p \in S$ with principal curvatures $k_1(p)$ and $k_2(p)$, respectively, then e_1 and e_2 are also principal directions of S_r at $F_r(p)$ with principal directions

$$k_i'(F_r(p)) = \frac{k_i(p)}{1 - rk_i(p)}, \qquad i = 1, 2.$$

- Prove that the Gauss curvature K' and the mean curvature H' of S_r are given by

$$K' \circ F_r = \frac{K}{1 - 2rH + r^2K} \quad \text{and} \quad H' \circ F_r = \frac{H - rK}{1 - 2rH + r^2K}.$$

- If S has constant mean curvature $H \equiv 1/2r$, show that $K(p) \neq 0$ for all $p \in S$ and that S_r has constant Gauss curvature $K' \equiv 1/r^2$.

(17) ↑ *Comparison of two surfaces at a point.* Let S_1 and S_2 be two orientable surfaces which are tangent at a common point $p \in S_1 \cap S_2$, and let N_1 and N_2 be corresponding Gauss maps such that $N_1(p) = N_2(p)$. We express S_1 and S_2, near the point p, as two graphs of functions f_1 and f_2 defined on the common tangent plane. We say that S_1 is over S_2 at p if there exists a neighbourhood of the origin of this plane where $f_1 \geq f_2$.
- If S_1 is over S_2 at p, show that $\sigma_1(v, v) \geq \sigma_2(v, v)$ for each $v \in T_pS_1 = T_pS_2$. In particular, $H_1(p) \geq H_2(p)$.
- If $\sigma_1(v, v) > \sigma_2(v, v)$ for each $v \neq 0$, prove that S_1 is over S_2 at p.

(Compare with Exercise (16) at the end of Chapter 1.)

(18) ↑ Let S be a graph over a disc of $\mathbb{R}^2$ with radius $r > 0$. If S has mean curvature $H \geq a$ at each point for some real number a, show that $ar \leq 1$.

(19) Prove that, if an orientable surface has constant principal curvatures, then either each of its points is umbilical or the Gauss curvature is non-positive everywhere. ◆

Hints for solving the exercises

Exercise 3.9: It is enough to define $G : S \to \mathbb{S}^2 \subset \mathbb{R}^3$, taking each point $p \in S$ to the only unit vector $G(p)$ normal to $T_p S$ obeying $\mathcal{P}$. This map is well-defined by the first hypothesis about $\mathcal{P}$ and is differentiable because, by the second hypothesis, it coincides locally with a differentiable field.

Exercise 3.10: Let A be the set of $c \in [a, b]$ such that, on $[a, c]$, one can find a continuous map N_c taking values in $\mathbb{R}^3$ with $N_c(a) = n$, $|N_c(t)| = 1$, and $N_c(t) \perp T_{\alpha(t)} S$ for all $t \in [a, c]$. Since $a \in A$, A is a non-empty set. Furthermore, it is an open subset of $[a, b]$. In fact, if $c \in A$, denote by V an open neighbourhood of $\alpha(c)$ in S where a Gauss map $N : V \to \mathbb{S}^2$ is defined. This can be done because of Lemma 3.2. By the continuity of α, if $c < b$, there exists $\varepsilon > 0$ such that $\alpha((c - 2\varepsilon, c + 2\varepsilon)) \subset V$. Moreover, changing N to its opposite, if necessary, we may assume that $N(\alpha(c)) = N_c(c)$. Define $N_{c+\varepsilon} : [a, c + \varepsilon] \to \mathbb{R}^3$ by

$$N_{c+\varepsilon}(t) = \begin{cases} N_c(t) & \text{if } t \in [a, c], \\ (N \circ \alpha)(t) & \text{if } t \in [c, c + \varepsilon], \end{cases}$$

which is continuous because of the choice of N. Thus $[a, c + \varepsilon] \subset A$, and in conclusion A is an open subset. Let us see that it is closed too. To show this, let $\{c_n\}_{n \in \mathbb{N}}$ be a sequence in A converging to a number $c \in [a, b]$. We take an open neighbourhood V of $\alpha(c)$ as in the previous case. We know that there is a $k \in \mathbb{N}$ such that, if $n \geq k$, then $\alpha([c_n, c]) \subset V$. Eventually changing the sign of the Gauss map N that we dispose of on V, suppose that $N_{c_k}(c_k) = N(\alpha(c_k))$. Then we define $N_c : [a, c] \to \mathbb{R}^3$ as

$$N_c(t) = \begin{cases} N_{c_k}(t) & \text{if } t \in [a, c_k], \\ (N \circ \alpha)(t) & \text{if } t \in [c_k, c], \end{cases}$$

which is continuous by the choice of N on V. Hence $c \in A$ and A is a closed subset of $[a, b]$. Since $[a, b]$ is connected, we conclude that $A = [a, b]$ and, so, there exists a continuous map $N : [a, b] \to \mathbb{R}^3$ with the required conditions. If there were another N' with the same conditions, then $\langle N, N' \rangle : [a, b] \to \mathbb{R}$

would be continuous and would take only the values 1 or -1. Since this product takes the value 1 at a, we infer that $N' = N$.

Suppose now that the curve α is closed, i.e., that $\alpha(a) = \alpha(b)$ and suppose also that S is orientable. Take a Gauss map N on S with $N(\alpha(a)) = n$, and let N_α be the normal field along α with initial condition $N_\alpha(a) = n$. Then, by uniqueness, $N \circ \alpha = N_\alpha$. Thus

$$N_\alpha(b) = N(\alpha(b)) = N(\alpha(a)) = n.$$

To show the converse, we may assume that S is connected and, hence, arc-connected, by working on one of its components. Choose $p_0 \in S$ and a unit $n \in \mathbb{R}^3$ normal to S at p_0. Define $N : S \to \mathbb{S}^2$ as

$$N(p) = N_\alpha(1),$$

where $\alpha : [0, 1] \to S$ is a continuous curve such that $\alpha(0) = p_0$ and $\alpha(1) = p$ and where N_α is the normal field along α with initial condition n. This definition does not depend on the chosen curve α. Indeed, if $\beta : [0, 1] \to S$ is another curve of this type and N_β is the corresponding field, and if we had that $N_\beta(1) = -N_\alpha(1)$, then $N_\gamma = (-N_\beta)^{-1} * N_\alpha$ would be the continuous normal field along $\gamma = \beta^{-1} * \alpha$, which is a closed curve with $\gamma(0) = p_0$. But

$$N_\gamma(0) = N_\alpha(0) = n \quad \text{and} \quad N_\gamma(1) = -(N_\beta)^{-1}(1) = -N_\beta(0) = -n,$$

which is impossible by our hypotheses. It remains to see that the so-defined map N is differentiable. Suppose that V is an open connected subset of S where, according to Lemma 3.2, we have defined a Gauss map G. Then $G \circ \alpha$ is a continuous normal field along α, for each continuous curve α whose trace is contained in V. The reasoning we used to show that N was a well-defined map shows now that either $G = N_{|V}$ or $G = -N_{|V}$, and we finish.

Exercise 3.13: Both surfaces S_1 and S_2 are open subsets of S, by Proposition 2.77. Let N_1 and N_2 be Gauss maps for S_1 and S_2, respectively. By Lemma 3.3, N_1 and N_2 either coincide or are opposite on $S_1 \cap S_2$. Suppose the first, reversing the sign of some of them if necessary. Then it suffices to define $N : S \to \mathbb{S}^2$ by $N_{|S_1} = N_1$ and $N_{|S_2} = N_2$.

Exercise 3.14: Let N_1 and N_2 be Gauss maps for S_1 and S_2, respectively, and suppose that S is orientable. Choose a Gauss map N for S. Applying Lemma 3.3, we know that $N_{|S_1}$ and N_1 are either equal or opposite. Change N by $-N$ if necessary so that $N_{|S_1} = N_1$. Then $N_{1|S_1 \cap S_2} = N_{|S_1 \cap S_2}$. Since S_2 is also connected, using Lemma 3.3 again, we have that $N_{|S_2}$ and N_2 either coincide or are opposite. Hence

$$N_{1|S_1 \cap S_2} = N_{|S_1 \cap S_2} = N_{2|S_1 \cap S_2} \quad \text{or} \quad N_{1|S_1 \cap S_2} = N_{|S_1 \cap S_2} = -N_{2|S_1 \cap S_2},$$

which is impossible since $S_1 \cap S_2 = A \cup B$, $N_{1|A} = N_{2|A}$, and $N_{1|B} = -N_{2|B}$.

To see that this is the case for the Moebius strip, using the notation of Section 3.2 where it was described, we set $S_1 = F([0, \pi/2) \times (-1, 1) \cup (3\pi/2, 2\pi] \times (-1, 1))$ and $S_2 = F((\pi/4, 7\pi/4) \times (-1, 1))$. Both of the two open subsets are orientable because they are covered by a parametrization. The remaining properties can be easily checked.

Exercise 3.15: N_1 is well-defined in this way as a consequence of the fact that, if $p \in S_1$ and $\{a, b\}$, $\{c, d\}$ are two bases of $T_p S_1$ and M is the matrix of the change of basis, then one has

$$\det((df)_p(c), (df)_p(d), N_2(f(p))) = \det M \det((df)_p(a), (df)_p(b), N_2(f(p))).$$

To see that N_1 is differentiable, note that, if $X : U \to S$ is a parametrization defined on a connected subset of $\mathbb{R}^2$, then, since $\{X_u, X_v\}$ is a basis of the corresponding tangent plane at each point of U and since we have

$$(df)_X(X_u) = (f \circ X)_u \quad \text{and} \quad (df)_X(X_v) = (f \circ X)_v$$

and since $(df)_X$ is an isomorphism, we deduce that

$$\det((df)_X(X_u), (df)_X(X_v), N_2 \circ f \circ X)$$

is a non-vanishing continuous function on U. We change, if necessary, the parametrization X by $Y(u, v) = X(v, u)$ so that this sign is always positive. Then, according to the definition,

$$N_1 \circ X = \frac{X_u \wedge X_v}{|X_u \wedge X_v|}$$

and so N_1 is differentiable.

Exercise 3.23: Definitions (3.4) and (3.5) say that, under these hypotheses, the second fundamental form of the surface corresponding to any Gauss map N is identically zero. Thus, the differential of N vanishes at any point of the surface. We conclude applying Proposition 2.63 and Exercise 2.70.

Exercise 3.24: The quadric S is the inverse image of 0 by means of the function f defined on the whole of $\mathbb{R}^3$ by

$$f(p) = \langle Bp, p \rangle + 2\langle b, p \rangle + c, \qquad p \in \mathbb{R}^3.$$

It follows from this that

$$(\nabla f)_p = Bp + b, \qquad \forall p \in \mathbb{R}^3,$$

which does not vanish on S because S is regular; see Example 2.16. It only remains to apply Example 3.8 together with the definition of the second fundamental form and to remember that S is an ellipsoid if and only if the matrix B is definite.

Exercise 3.32: The tangent plane of S along that curve is constant and so any local Gauss map is constant when restricted to the points of the curve. Therefore, the tangent vector at each point of the curve belongs to the kernel of the differential of the Gauss map and, so, this map has an identically null determinant.

Exercise 3.33: Let I be an open interval of $\mathbb{R}$ and $\alpha : I \to S$ a segment of a straight line p.b.a.l. contained in an open subset of S where we have defined a Gauss map N. Since $\langle (N \circ \alpha)(s), \alpha'(s) \rangle = 0$ for each $s \in I$, we have

$$\sigma_{\alpha(s)}(\alpha'(s), \alpha'(s)) = -\langle (dN)_{\alpha(s)}(\alpha'(s)), \alpha'(s) \rangle = \langle (N \circ \alpha)(s), \alpha''(s) \rangle = 0,$$

because the acceleration of the line vanishes. Then, the quadratic form $\sigma_{\alpha(s)}$ is not definite for any $s \in I$.

Exercise 3.40: Choose any plane. By the compacity, there is a global extreme on the surface for the height function relative to this plane. Applying Propositions 3.35 and 3.37, the second fundamental form of the surface at this point is positive semi-definite and, hence, its determinant is non-negative.

Exercise 3.41: A point $p \in S$ is elliptic when its Gauss curvature is positive, that is, when the second fundamental form at this point σ_p is definite. Choosing a suitable local orientation, we may assert that p is elliptic if and only if there is a local Gauss map N relative to whether the second fundamental form is positive definite, or, equivalently, there exists $C \in \mathbb{R}^+$ such that $\sigma_p(v, v) \geq C|v|^2$ for each $v \in T_pS$. But the latter, according to Example 3.39 and Proposition 3.35, is equivalent to the fact that p is a local maximum of the square of the distance function to the point $p_0 = p + (1/C)N(p)$.

Exercise 3.42: One picks $p_0 \in \mathbb{R}^3$ arbitrarily and, by compacity, one gets a global maximum on S for the square of the distance function to this point. Applying Exercise 3.41, previously solved, this point is elliptic.

Exercise 3.50: Since $K > 0$, from (3.6), $H^2 > 0$. Using Exercise 3.9, S is orientable and, for a given orientation, there is some $c \in \mathbb{R}$ such that $H = cK$. Then $k_1 + k_2 = 2ck_1k_2$, and so $k_1(1 - 2ck_2) = -k_2$ and, since $k_2 \geq k_1 > 0$, one obtains

$$k_1 = \frac{k_2}{2ck_2 - 1}.$$

That is, k_1 is a decreasing function of k_2 and, hence, k_1 will attain its minimum at the points where k_2 attains its maximum. Now we conclude by taking into account the Hilbert Theorem 3.47 and Corollaries 3.48 and 3.49.

Exercise (1): Given $p \in S$, we have

$$|\sigma_p(v,v)| \leq \max\{|k_1(p)|, |k_2(p)|\}|v|^2$$

for all $v \in T_p S$. If we represent by $C \in \mathbb{R}$ the above maximum, for $\lambda \in \mathbb{R}$ near enough to zero, one has that $1 - |\lambda|C > 0$. On the other hand, reexamining Example 3.39, we know that p is a critical point of the function f_λ which measures the square of the distance to the point $a_\lambda = p + \lambda N(p)$, $\lambda \in \mathbb{R}$, of its normal line. For these small λ one has

$$(d^2 f_\lambda)_p(v) = 2[|v|^2 - \lambda \sigma_p(v,v)] \geq 2|v|^2[1 - |\lambda|C],$$

and thus, by Proposition 3.35, p is a local minimum.

Exercise (2): Let $p \in S$ be a point where the square of the distance to a, the centre of the ball of radius r in which S is contained, attains a maximum. Then, according to Proposition 3.35 and Example 3.39, a is on the normal line of S at p and

$$1 \leq \langle a - p, N(p) \rangle k_i(p), \qquad i = 1, 2.$$

Bearing in mind that $\langle a - p, N(p) \rangle^2 = |a - p|^2 \leq r^2$, it follows from the two required inequalities by multiplying or adding the two inequalities.

Exercise (3): Let $p \in S$ be any point. This point is critical for f_λ, the square of the distance function to $a_\lambda = p + \lambda N(p)$, for all $\lambda \in \mathbb{R}$. Since S is compact, there exist global extremes for this function. By hypothesis, p must be one of these points for all $\lambda \in \mathbb{R}$, except for maybe a countable number of them. Therefore $(d^2 f_\lambda)_p$ is a semi-definite form, according to Proposition 3.35, for all these λ. Using Example 3.39, one has that $1 - \lambda k_i$, $i = 1, 2$, have the same sign. That is,

$$0 \leq (1 - \lambda k_1(p))(1 - \lambda k_2(p)) = 1 - 2\lambda H(p) + \lambda^2 K(p),$$

which is valid for all $\lambda \in \mathbb{R}$ except for maybe a countable subset. Then, the discriminant of this second degree polynomial has to be non-positive. Therefore $H(p)^2 - K(p) \leq 0$ and, from (3.6), the point p is umbilical. Since this is true for each $p \in S$, one finishes by applying Corollary 3.31.

Exercise (4): The first part is clear if we recall the definition of star-shaped; see Exercise (11) at the end of Chapter 2. If the support function based on p_0 is constant and equal to $c \in \mathbb{R}$, its differential, given by

$$(df)_p(v) = \langle p - p_0, (dN)_p(v) \rangle, \qquad \forall v \in T_p S,$$

vanishes at each $p \in S$. Thus

$$k_i(p)\langle p - p_0, e_i \rangle = 0, \qquad i = 1, 2,$$

for each $p \in S$ where $e_1, e_2 \in T_p S$ are principal directions. Consequently

$$K(p) = 0 \quad \text{or} \quad p - p_0 = cN(p),$$

for each $p \in S$. As a consequence, the points of the open set where the Gauss curvature does not vanish are in a sphere with centre p_0 and radius $|c| > 0$. Since K is continuous, S is connected, and there are points where $K \neq 0$ and K can take only the values 0 and $1/c^2$, one has that there are no points with null Gauss curvature on S, and so we conclude.

Exercise (9): We know, since we solved Exercise (3) at the end of Chapter 2, that

$$X(u,v) = x(u)\, a + y(u)\, \cos v\, b + y(u)\, \sin v\, c, \qquad (u,v) \in I \times J,$$

is a parametrization of the surface, where J is any interval of $\mathbb{R}$ of length 2π, the vectors $\{a, b, c\}$ form a positively oriented orthonormal basis, and $\alpha : I \to \mathbb{R}^3$ given by

$$\alpha(u) = x(u)\, b + y(u)\, c, \qquad y > 0,$$

is a local parametrization of the generating curve of S. From these parametrizations we may compute the coefficients of the first and second fundamental forms and the mean curvature H at the points of S covered by them, as in (3.12). We obtain in this way that the mean curvature of S is identically zero if and only if the components of α satisfy

$$y\left(\frac{dx}{du}\frac{d^2y}{du^2} - \frac{dy}{du}\frac{d^2x}{du^2}\right) - \frac{dx}{du}\left[\left(\frac{dx}{du}\right)^2 + \left(\frac{dy}{du}\right)^2\right] = 0.$$

Let us work now on the open subset of I where $dx/du \neq 0$. On it, there exists the inverse function of x and we can reparametrize α so that y is a function of x or, equivalently, so that α is a graph. Dividing the above equality by $(dx/du)^3$, it follows that

$$(*) \qquad\qquad yy'' - 1 - (y')^2 = 0,$$

where $'$ represents derivatives with respect to the new variable x. From this equality it follows, first, that $y'' > 0$ and, second, taking derivatives again, that $yy''' - y'y'' = 0$, or, equivalently, that $(\log y)' = (\log y'')'$. We deduce from this that

$$y'' = Ky, \qquad K > 0.$$

All the solutions to this second order linear equation are of the form

$$y = \lambda \cosh \sqrt{K}x + \mu \sinh \sqrt{K}x, \qquad \lambda, \mu \in \mathbb{R}.$$

But, if we impose that such a solution obeys $(*)$, we have $K\lambda^2 - K\mu^2 = 1$, and so

$$y = \frac{1}{\sqrt{K}}\cosh(\sqrt{K}x + L), \qquad K \in \mathbb{R}^+, L \in \mathbb{R},$$

which is a catenary. Since the generating curve must be connected, either it coincides entirely with a catenary or we have that dx/du vanishes everywhere. In this case, the generating curve is an open half-line perpendicular to the rotation axis and the surface is a plane.

Exercise (11): The gradient of the quadratic form defining the ellipsoid S supplies us a non-null normal vector at each point. That is, there is a non-vanishing differentiable function h defined on the ellipsoid and a Gauss map N such that

$$h(x, y, z)N(x, y, z) = \left(\frac{x}{a^2}, \frac{y}{b^2}, \frac{z}{c^2}\right),$$

for all $p = (x, y, z) \in S$. If $v = (A, B, C) \in T_pS$, one has that

$$(dh)_p(v)N(p) + h(p)(dN)_p(v) = \left(\frac{A}{a^2}, \frac{B}{b^2}, \frac{C}{c^2}\right).$$

The point p is umbilical if and only if the linear map $(dN)_p$ is a multiple of the identity map. This happens if and only if after scalar multiplication by $N(p) \wedge v$ the member of the left-hand side vanishes. That is, $p = (x, y, z) \in S$ is umbilical if and only if

$$\begin{vmatrix} \dfrac{A}{a^2} & \dfrac{B}{b^2} & \dfrac{C}{c^2} \\ A & B & C \\ \dfrac{x}{a^2} & \dfrac{y}{b^2} & \dfrac{z}{c^2} \end{vmatrix} = 0, \quad \text{if} \quad \frac{x}{a^2}A + \frac{y}{b^2}B + \frac{z}{c^2}C = 0.$$

The determinant on the left-hand side is a quadratic form in the A, B, C variables which vanishes on a plane. This forces it to be degenerate, and so, the determinant of the corresponding symmetric matrix must vanish. This implies that $x = 0$ or $y = 0$ or $z = 0$. Examining each possibility, we see that the only valid choice is $y = 0$. From this, we deduce that there are four umbilical points determined by the equations

$$y = 0, \qquad c^2(c^2 - b^2)x^2 = a^2(b^2 - a^2)z^2.$$

Exercise (12): We will use the notation established in Section 3.6. From equalities (3.10) and (3.4), one has, since $F = 0$,

$$(K \circ X)EG = eg - f^2 = \langle X_{uu}, N^X\rangle\langle X_{vv}, N^X\rangle - \langle X_{uv}, N^X\rangle^2$$

$$= \langle X_{uu}, X_{vv}\rangle - \langle X_{uu}^T, X_{vv}^T\rangle - |X_{uv}|^2 + |X_{uv}^T|^2,$$

where T represents the part tangent to the surface. Now, we must express all the summands on the right-hand side in terms of the non-trivial coefficients E and G and of their derivatives. In fact, since $F = 0$, we have

$$\langle X_{uu}, X_{vv} \rangle - |X_{uv}|^2 = \langle X_{uu}, X_v \rangle_v - \langle X_{uv}, X_v \rangle_u$$

$$= -\langle X_u, X_{vu} \rangle_v - \langle X_{uv}, X_v \rangle_u = -\tfrac{1}{2}E_{vv} - \tfrac{1}{2}G_{uu}.$$

Analogously, the fact that $F = \langle X_u, X_v \rangle$ vanishes implies that

$$X_{uu}^T = \frac{1}{2}\frac{E_u}{E}X_u - \frac{1}{2}\frac{E_v}{G}X_v,$$

$$X_{uv}^T = \frac{1}{2}\frac{E_v}{E}X_u + \frac{1}{2}\frac{G_u}{G}X_v,$$

$$X_{vv}^T = -\frac{1}{2}\frac{G_u}{E}X_u + \frac{1}{2}\frac{G_v}{G}X_v.$$

We have, then, all the necessary data to compute the right-hand side of the first equality. One can see that

$$(K \circ X)EG = -\frac{1}{2}E_{vv} - \frac{1}{2}G_uu + \frac{1}{4}\frac{E_v^2}{E} + \frac{1}{4}\frac{G_u^2}{G} + \frac{1}{4}\frac{E_uG_u}{E} + \frac{1}{4}\frac{E_vG_v}{G},$$

and now a straightforward calculation leads to the required equation.

Exercise (14): A dilatation ϕ with centre $a \in \mathbb{R}^3$ and ratio $\mu \in \mathbb{R}^*$ is given by

$$\phi(p) = a + \mu(p - a), \qquad \forall p \in \mathbb{R}^3.$$

Thus, if $p \in S$ and $v \in T_pS$, according to Example 2.58, its differential is

$$(d\phi)_p(v) = \mu v,$$

and so $T_{\phi(p)}S' = (d\phi)_p(T_pS) = T_pS$, for each $p \in S$. Therefore, if N is a local Gauss map for S, $N' = N \circ \phi^{-1}$ is a local Gauss map for S'. Applying the chain rule, we obtain

$$(dN')_{\phi(p)} = \frac{1}{\mu}(dN)_p, \qquad p \in S,$$

and hence $\sigma'_{\phi(p)} = (1/\mu)\sigma_p$. It follows that $H' \circ \phi = (1/\mu)H$ and $K' \circ \phi = (1/\mu^2)K$.

Exercise (15): The differential of this inversion with centre at the origin is given by

$$(d\phi)_p(v) = \frac{1}{|p|^2}v - 2\frac{\langle p, v \rangle}{|p|^4}p, \qquad p \in S, \quad v \in T_pS.$$

If N is a local Gauss map for S, one can see that the map N' defined by the equality

$$(N' \circ \phi)(p) = N(p) - 2\frac{\langle N(p), p \rangle}{|p|^2}p, \qquad p \in S,$$

is a local Gauss map for S', because it gives a unit field perpendicular to each $T_{\phi(p)}S' = (d\phi)_p(T_pS)$. From this, we can deduce a relation between $(dN')_{\phi(p)} \circ (d\phi)_p$ and $(dN)_p$. It is not difficult to check now that, if $p \in S$ and $e_i \in T_pS$, $i = 1, 2$, is a principal direction associated to the principal curvature $k_i(p)$, then $e_i' = (d\phi)_p(e_i)$ is a principal direction of S' and its associated principal curvature $k_i'(\phi(p))$ is related to the previous one as

$$k_i'(\phi(p)) = |p|^2 k_i(p) + 2\langle N(p), p \rangle.$$

Therefore

$$(H' \circ \phi)(p) = |p|^2 H(p) + 2\langle N(p), p \rangle,$$

$$(K' \circ \phi)(p) = |p|^4 K(p) + 4|p|^2 H(p) + 4\langle N(p), p \rangle^2.$$

Exercise (17):

- The function $h = f_1 - f_2$ measures the difference of heights relative to the common tangent plane at $p \in S_1 \cap S_2$, which can be assumed to be given by the equation $z = 0$. This function has, by hypothesis, a minimum at the origin. Hence, its Hessian matrix in the sense of elementary calculus is positive semi-definite. But, if we examine the proof of the Hilbert Theorem 3.47, the Hessian matrices of f_1 and f_2 at the origin are the matrices of the second fundamental forms of S_1 and of S_2 at p relative to the basis of the tangent plane formed by the vectors $(1, 0, 0)$ and $(0, 1, 0)$. We finish by seeing that the Hessian matrix of h is nothing more than the difference of those of f_1 and f_2.

- In this case, the hypothesis implies that $h = f_1 - f_2$ has at the origin a critical point with positive definite Hessian matrix. Therefore, it has a local minimum there.

Exercise (18): If $a \leq 0$, the result is clear. Suppose that $a > 0$. Take $0 < R < r$, and let $\mathbb{S}_R^2$ be a sphere with radius R, which has constant mean curvature $1/R$ with respect to its inner normal. Assume that S is a graph over an open disc with radius r centred at the origin of the plane $z = 0$. Drop the sphere $\mathbb{S}_R^2$ onto S in the two possible ways and obtain, using Exercise (20) at the end of Chapter 2, two points of tangency between S and $\mathbb{S}_R^2$. At least at one of them, say p, the normal of S coincides with the inner normal of $\mathbb{S}_R^2$. Now use Exercise (17) and so obtain

$$a \leq H(p) \leq \frac{1}{R}.$$

Then $aR \leq 1$ for all $R < r$.

Chapter 4

Separation and Orientability

4.1. Introduction

The geometry of a surface near a point does not inform us at all about its shape near other distant points. The local geometrical behaviour is governed by such things as the Gauss map, the second fundamental form, and its associated functions: Gauss and mean curvatures and principal curvatures. In spite of this, at the end of Chapter 3, in the Jellett and Liebmann results of Corollaries 3.48 and 3.49, we saw that a particular behaviour of some of the curvatures, together with a *totality* condition, such as compacity, which prevents the surface to be an open subset of a larger surface, can determine the whole surface geometrically and topologically.

To tackle the global study of a surface in Euclidean space with adequate tools, we will start in this chapter and continue in the following chapter by showing some results of an analytical nature and a topological nature concerning mainly *compact* surfaces. Let us remark that compacity, or the weaker hypothesis of closedness, is a natural condition to impose in order to avoid that a surface be contained as a subset in a larger one.

Our first goal in this chapter will be to prove that a connected compact surface separates Euclidean space into exactly two domains. That is, we want to show a three-dimensional version of the famous Jordan curve theorem—whose history and proof will be commented on in the introduction of Chapter 9—by giving a proof which is also valid in the two-dimensional case. This result is usually known as the JORDAN-BROWER theorem, in honour of CAMILLE JORDAN (1838-1922), who introduced it the first time for curves, as a more or less trivial assertion, and of the Dutch mathematician L. E. J. BROWER, one of the initiators of algebraic topology, who proved it in a general version in 1912. We will provide a proof based on an elementary version of the $\mathbb{Z}_2$-degree theory. By the way, this proof will enable us to consider in detail the problem of the intersection of lines and surfaces. It will follow a proof of the fact that any compact surface in $\mathbb{R}^3$ is orientable, another result due to BROWER for abstract surfaces, although, with our definition of a surface inside of Euclidean space, the proof by H. SAMELSON in 1969 is much better known, as well as the existence of tubular neighbourhoods which will be profusely exploited in what follows.

4.2. Local separation

The first objective of this chapter is to see that, in the presence of a compact—or even closed—surface, the remaining points of the space are clearly divided into two classes. As a first approach, let us check that any surface satisfies a local version of this property.

Lemma 4.1. *Let $p \in S$ be a point in a surface. There exists a connected open neighbourhood W of p in $\mathbb{R}^3$ such that $W \cap S$ is connected and $W - S$ has exactly two connected components, either one having $W \cap S$ as its boundary in W.*

Proof. We know that the surface S, near any of its points $p \in S$, can be described as an inverse image of a regular value; see Exercise 2.25. That is, there exists an open neighbourhood O of p in $\mathbb{R}^3$ and a differentiable function $f : O \to \mathbb{R}$ having 0 as a regular value and such that $O \cap S = f^{-1}(\{0\})$. Hence, p is not a critical point for f. Then $(df)_p \neq 0$ and we may assume, for instance, that $f_z(p) \neq 0$. Define a differentiable map $G : O \to \mathbb{R}^3$ by

$$G(x, y, z) = (x, y, f(x, y, z)), \qquad (x, y, z) \in O.$$

It can easily see that $G(p) = (p_1, p_2, 0)$ and that $(dG)_p : \mathbb{R}^3 \to \mathbb{R}^3$ is regular. By the inverse function theorem of calculus, there are an open neighbourhood $W \subset O$ of p in $\mathbb{R}^3$ and an open ball B centred at $(p_1, p_2, 0)$ such that $G(W) = B$ and $G : W \to B$ is a diffeomorphism. Therefore, $W \cap S$ is an open neighbourhood of p in S diffeomorphic to $G(W \cap S) = B \cap P$, where P is the plane of $\mathbb{R}^3$ of equation $z = 0$. The required statement follows immediately.   □

Remark 4.2. Lemma 4.1 above implies that, up to a diffeomorphism between open subsets of $\mathbb{R}^3$, any point of a surface possesses a neighbourhood which can be imagined as the equatorial disc of a Euclidean ball. The following is a consequence of this.

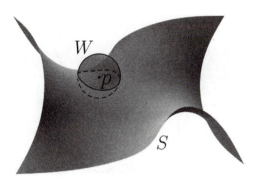

Figure 4.1. *Local separation*

A surface separates Euclidean space locally into two components. For this reason, we can distinguish, locally, two sides on the surface.

Remark 4.3. Another more or less immediate consequence of Lemma 4.1 is that, locally, the problem of determining the intersection of a simple curve—or the trace of a regular curve—with a surface can be reduced, up to a diffeomorphism, to studying the case where the surface is the plane $z = 0$. For this, we will first define the concept of transversality between curves and surfaces corresponding to what we considered in Example 2.64 for a pair of surfaces—and, for curves, in Exercise (17) at the end of Chapter 1. Let S be a surface and $\alpha : I \to \mathbb{R}^3$ a regular curve defined on an interval I of $\mathbb{R}$ intersecting at a point $p = \alpha(s_0) \in S$, with $s_0 \in I$. We will say that the curve α and the surface S cut *transversely* at p when the tangent vector $\alpha'(s_0)$ of the curve is not tangent to the surface S at $p = \alpha(s_0)$. From this we may deduce a corresponding concept of tranverse intersection at a common point of a surface S and a simple curve C; see Example 2.18. We will say that a surface S and a simple curve C of $\mathbb{R}^3$ are *transverse* when they intersect transversely at any point of $S \cap C$. We first intend to establish the following property:

> *If S is a surface and C a simple curve intersecting transversely in $\mathbb{R}^3$, then the intersection $S \cap C$ is a discrete subset of $\mathbb{R}^3$. Moreover, in each neighbourhood of a point in the intersection, there exist points of the curve on each side of the surface—in the sense that we have just mentioned.*

In fact, since diffeomorphisms of $\mathbb{R}^3$ preserve transversality trivially, we have already pointed out that it is enough to consider the case where $S = P$ is the plane of equation $z = 0$. In this case, if $\alpha : (-\varepsilon, \varepsilon) \to \mathbb{R}^3$ is a regular

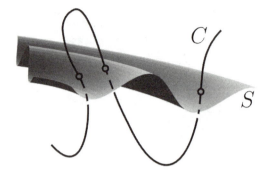

Figure 4.2. *Transverse intersection*

curve with $\alpha(0) = p \in P$ which is transverse to P at p, we have that, if

$$\alpha(t) = (x(t), y(t), z(t)), \qquad t \in (-\varepsilon, \varepsilon),$$

then $z(0) = 0$ and $z'(0) \neq 0$. This implies that there is an interval centred at $t = 0$ in which $z(t)$ vanishes only at $t = 0$ and in which there are points with $z(t) > 0$ and with $z(t) < 0$, that is, points above and under the plane P. This proves the statement above.

A direct consequence of the existence of this class of neighbourhoods for each point of a surface and of elementary topological notions is the fact that the complement in $\mathbb{R}^3$ of a closed connected surface S either is connected or has two connected components.

Proposition 4.4. *Let S be a closed connected surface. Then $\mathbb{R}^3 - S$ has at most two connected components. Moreover, each connected component of $\mathbb{R}^3 - S$ has S as its boundary.*

Proof. Let C be a connected component of $\mathbb{R}^3 - S$. Our first observation is that, since S is closed, $\mathbb{R}^3 - S$ is locally connected, and so, each C is open in $\mathbb{R}^3$. We also have $\text{Bdry}\, C \neq \emptyset$. Otherwise, we would have that $\overline{C} = \text{int}\, C \cup \text{Bdry}\, C = \text{int}\, C = C$, and then C would be simultaneously an open and closed subset of $\mathbb{R}^3$ and, consequently, $C = \mathbb{R}^3$, which is impossible. Now, let us put

$$\mathbb{R}^3 - S = C \cup C',$$

where C' is the union of all the connected components of $\mathbb{R}^3 - S$ different from C, and so an open subset of $\mathbb{R}^3$. It follows that $C \cup S = \mathbb{R}^3 - C'$ is closed and then $\overline{C} \subset C \cup S$. Hence

$$\text{Bdry}\, C = \overline{C} - C \subset S.$$

Therefore $\text{Bdry}\, C$ is a non-empty closed subset of S. Let us now see that it is an open subset of S as well. Indeed, if $p \in \text{Bdry}\, C \subset S$, we know, from Lemma 4.1, that there exists an open neighbourhood W of p in $\mathbb{R}^3$ such that

$W - S$ has exactly two connected components, namely C_1 and C_2, whose common boundary relative to W is $S \cap W$. Since $W - S \subset \mathbb{R}^3 - S$, we have that C_1 and C_2 are contained in different respective connected components of $\mathbb{R}^3 - S$. In particular, each C_i, $i = 1, 2$, is contained in C or, alternatively, does not cut C. If the latter occurs for $i = 1$ and $i = 2$, then we would have $C \cap W = \emptyset$, and consequently $p \notin \overline{C}$. This contradicts the fact that $p \in \text{Bdry}\, C$. Therefore, at least one of the sets C_i is contained in C, for example, $C_1 \subset C$. Then

$$W \cap S = \text{Bdry}\, C_1 \cap W \subset \overline{C_1} \subset \overline{C},$$

and, thus,

$$W \cap S \subset \text{Bdry}\, C.$$

This proves that the point p is interior to $\text{Bdry}\, C$ and, so, that $\text{Bdry}\, C$ is open in S. Since we are assuming that the surface S is connected, we have $\text{Bdry}\, C = S$, as mentioned in the last statement that we desire to show. But, furthermore, the above reasoning also proves that, if C is a connected component of $\mathbb{R}^3 - S$ and W is the neighbourhood given by Lemma 4.1 for an arbitrary point $p \in S$, then C contains some of the two connected components C_1 or C_2 of $W - S$. Since these components are disjoint, there cannot exist more than two. $\qquad \square$

4.3. Surfaces, straight lines, and planes

In this section, we will take an additional step in order to find out about the form in which a closed surface embeds in $\mathbb{R}^3$. Indeed, we will see that the complement of such a surface is never connected. To do this, we will be interested in some aspects of relative positions of straight lines and surfaces and of planes and surfaces in Euclidean space.

First, since straight lines are the easiest examples of simple curves, what we said in Remark 4.3 can be directly applied to studying the transverse intersection of a straight line and a surface . In fact, if a line and a surface meet transversely, then the intersection is a discrete set and, as near as one wants to get to each point of this intersection, one can find points of the line on each side of the surface—in the local sense of Lemma 4.1.

Let us see that there are situations where we can guarantee the existence of straight lines cutting a given surface transversely. For this, we need to control the set of regular values of differentiable maps between surfaces, which we will define now by analogy with Example 2.13.

Definition 4.5. Represent by S_1 and S_2, indiscriminately, surfaces of $\mathbb{R}^3$ or open subsets of $\mathbb{R}^2$ and let $f : S_1 \to S_2$ be a differentiable map between them. A point $q \in S_2$ is said to be a *regular value* of f when, for each

Figure 4.3. *Transverse straight line*

$p \in f^{-1}(\{q\})$, we have that $(df)_p$ is an isomorphism between $T_p S_1$ and $T_q S_2$, or, equivalently—according to Exercise (13) at the end of Chapter 2—when

$$|\operatorname{Jac} f|(p) \neq 0, \qquad \forall p \in f^{-1}(\{q\}).$$

For example, all the points of S_2 which do not belong to the image of f are regular values in the sense that we have just defined. The following result assures, among other things, that *most* of the points of S_2 are regular values of the differentiable map f.

Theorem 4.6 (Weak version of Sard's theorem). *Let $f : S_1 \to S_2$ be a differentiable map, where S_1 and S_2 represent surfaces or open subsets of $\mathbb{R}^2$ indiscriminately. Then the set of regular values of f is dense in S_2. In particular, there always exist regular values of f.*

Proof. The proof of the stronger assertion, usually known as Sard's theorem, will be given in Appendix 4.6 at the end of this chapter. □

Remark 4.7. The same proof that we will give for this result proves that, given a finite—or even countable—number of maps defined either on surfaces or on open subsets of $\mathbb{R}^2$, $S_1, \ldots, S_n$, taking values in the same surface or open subset of $\mathbb{R}^2$, say S, the set of the points of S which are simultaneously regular values for $f_1, \ldots, f_n$ is dense in S.

Example 4.8 (Existence of transverse half-lines). Let S be a surface and $p_0 \notin S$ an exterior point. In Example 2.79, we studied the central projection map $f : S \to \mathbb{S}^2$ given by

$$f(p) = \frac{p - p_0}{|p - p_0|}, \qquad \forall p \in S.$$

We computed its differential and saw that, if $p \in S$, $(df)_p$ is an isomorphism if and only if the straight line passing through p_0 and p is transverse to the surface at p. Therefore, a unit vector $a \in \mathbb{S}^2$ is a regular value of this f when the half-line starting at p_0 in the direction of a cuts the surface transversely. Thus, the following assertion is a consequence of the Sard Theorem 4.6:

> *From each point of $\mathbb{R}^3 - S$ one can draw a half-line intersecting S transversely with direction arbitrarily close to a*

> *given direction. Taking Remark 4.3 into account, the in-*
> *tersection of these half-lines and S are discrete sets, and*
> *so they are finite sets if S is compact.*

Example 4.9 (Transverse planes). Let S be an arbitrary surface and P a plane of $\mathbb{R}^3$, perpendicular to a unit vector $a \in \mathbb{S}^2$. We may cover, according to Lemma 3.2, the surface with countably many open orientable connected subsets. We consider on each of them the two possible Gauss maps; see Lemma 3.3. We get in this way a countable family of differentiable maps all of them taking values in the sphere $\mathbb{S}^2$. From Remark 4.7, there is a unit vector $b \in \mathbb{S}^2$, arbitrarily close to a, which is a regular value for all these maps. This means that all the points of the surface S having their tangent plane perpendicular to the vector b are regular points of these local Gauss maps. In particular, by the inverse function Theorem 2.75, these points form a discrete subset of S and, so, they are at most countable. As a consequence, we can choose a plane P', arbitrarily close to P, in the pencil of the planes perpendicular to the vector b, which avoids this countable set. Therefore, at the points of $S \cap P'$, the tangent plane of S is not the direction plane of P'. That is, S and P' intersect transversely. We can summarize all of this in the following statement:

> *Given a surface S and a plane P in Euclidean space $\mathbb{R}^3$,*
> *one can always find a plane P' arbitrarily close to P—in*
> *the sense, for instance, of the unit normal vectors and the*
> *distances to the origin—cutting S transversely.*

We will go more deeply into transversality of lines and surfaces by showing that, in a suitable sense, transversality is an *open* condition. In fact, we want to prove that, if a line and a surface intersect transversely at a point, then, when we move the line just slightly, transversality is preserved. In order to say explicitly what *just slightly* means, we will now discuss sequences and limits of straight lines in Euclidean space. Indeed, we say that a sequence of straight lines $\{R_n\}_{n \in \mathbb{N}}$ of $\mathbb{R}^3$ converges to a straight line R when there are affine parametrizations $\alpha_n, \alpha : \mathbb{R} \to \mathbb{R}^3$ for each R_n and for R such that

$$\lim_{n \to \infty} \alpha_n(t) = \alpha(t), \qquad \forall t \in \mathbb{R}.$$

This definition and subsequent properties extend immediately to sequences of half-lines and to sequences of segments. The following exercises, which can be easily solved, emphasize the main properties of limits of straight lines.

Exercise 4.10. Consider a sequence $\{R_n\}_{n \in \mathbb{N}}$ of straight lines in $\mathbb{R}^3$ and a straight line R. If $p_n \in R_n$ and $v_n \in \mathbb{R}^3$ is a direction vector of R_n, for all $n \in \mathbb{N}$, and $p \in R$ and $v \in \mathbb{R}^3$ is a direction vector of R, prove that $\lim_{n \to \infty} R_n = R$ if

$$\lim_{n \to \infty} p_n = p \quad \text{and} \quad \lim_{n \to \infty} v_n = v.$$

Exercise 4.11. ↑ Let $\{R_n\}_{n \in \mathbb{N}}$ be a sequence of lines of $\mathbb{R}^3$ converging to a line R. If $p_n \in R_n$ and $v_n \in \mathbb{R}^3$ is a direction vector of R_n, for all $n \in \mathbb{N}$, and the limits

$$\lim_{n \to \infty} p_n = p \quad \text{and} \quad \lim_{n \to \infty} v_n = v \neq 0$$

exist, show that $p \in R$ and that v is a direction vector of R.

Exercise 4.12. ↑ If $\{R_n\}_{n \in \mathbb{N}}$ is a sequence of lines all of which intersect a given compact set of $\mathbb{R}^3$, prove that there exists a partial sequence converging to some line.

Exercise 4.13. ↑ Let $\{p_n\}_{n \in \mathbb{N}}$ be a sequence of points of a surface S converging to a point $p \in S$, and let $\{R_n\}_{n \in \mathbb{N}}$ be a sequence of lines converging to a line R. If each R_n is tangent to S at the point p_n, prove that R is tangent to S at the point p.

Now let us see, as promised, that transversality of lines and surfaces is an open condition. This fact will imply that, if the surface is compact, the number of points of a transverse intersection is preserved when one moves the line just slightly.

Proposition 4.14. *Let R^+ be a half-line, whose origin is in $\mathbb{R}^3 - S$, which intersects a compact surface S transversely. If $\{R_n^+\}_{n \in \mathbb{N}}$ is a sequence of half-lines starting at points of $\mathbb{R}^3 - S$ converging to R^+, then there exists a number $N \in \mathbb{N}$ such that, if $n \geq N$, R_n^+ cuts S transversely and the number of points of the sets $R_n^+ \cap S$ and $R^+ \cap S$ coincide.*

Proof. Transversality of the intersection and compacity of S guarantee that $S \cap R^+$ is a finite set, namely

$$S \cap R^+ = \{p_1, \dots, p_k\}.$$

Suppose that the assertion about the sequence $\{R_n^+\}_{n \in \mathbb{N}}$ is false. Then we can assume, passing to a subsequence of the original one, that the sequence $\{R_n^+\}_{n \in \mathbb{N}}$ of half-lines starting from points of $\mathbb{R}^3 - S$ and converging to R^+ is formed by half-lines satisfying at least one of the following conditions:

> **A:** R_n^+ is tangent to S at some point.

> **B:** The number of points of $R_n^+ \cap S$ is greater than the number of points of $R^+ \cap S$.

> **C:** The number of points of $R_n^+ \cap S$ is smaller than the number of points of $R^+ \cap S$.

Let us check that any of the above alternatives leads us to a contradiction. If **A** occurred, there would be a sequence of points $p_n \in S \cap R_n$ such that R_n would be tangent to S at each p_n, for all $n \in \mathbb{N}$. As S is compact, we might suppose that this sequence converges to a point $p \in S$, which also belongs to R by Exercise 4.11. Then, using Exercise 4.13, the straight line R would be tangent to S at p. This is a contradiction to the transversality hypothesis. Instead, if **B** occurred, since any sequence of points of the form $p_n \in S \cap R_n$ has to converge, by the compacity of S and Exercise 4.11, for some of the $p_1, \ldots, p_k \in S \cap R$, we could find two of these sequences with different terms converging to the same limit point in $S \cap R$. We will call such a point a point of *double intersection* of S and R. We will see later that this implies that S and R are tangent at this point and so we would again arrive at a contradiction. Finally, if alternative **C** were true, then some of the points $p_1, \ldots, p_k$ of $S \cap R$ would not be limit points of a sequence $p_n \in S \cap R_n$. Then we would find a neighbourhood of p in S which would not intersect any of the R_n. We will call such a point a point of *first contact* between S and R. We will also see below that this would require R to be tangent to S at this point. In conclusion, none of the three possibilities **A**, **B**, or **C** can be true, provided we have proved the following result. □

Lemma 4.15. *Let S be a surface and R a straight line of $\mathbb{R}^3$ and denote by p a point in $S \cap R$. We say that p is a point of* first contact *between S and R if there is a sequence of lines $\{R_n\}_{n \in \mathbb{N}}$ converging to R and a neighbourhood V of p in S such that $R_n \cap V = \emptyset$ for all $n \in \mathbb{N}$. We say that p is a point of* double intersection *of S and R if there are two sequences $\{p_n\}_{n \in \mathbb{N}}$ and $\{q_n\}_{n \in \mathbb{N}}$ of points of S converging to p such that $p_n \neq q_n$ for all $n \in \mathbb{N}$ and the sequence of lines R_n determined by the pairs (p_n, q_n) converges to R. In both of these two cases, R is tangent to S at the point p.*

Proof. Suppose that p satisfies the first case. Using Lemma 4.1, we may find an open connected neighbourhood W of p in $\mathbb{R}^3$ such that $W \cap S \subset V$ and $W - S$ has exactly two connected components C_1 and C_2. Let $B \subset W$ be an open ball of $\mathbb{R}^3$ centred at p. Since each $R_n \cap B$ is connected, R_n cannot intersect $C_1 \cap B$ and $C_2 \cap B$ at the same time. We extract a subsequence formed by straight lines which cut only one of the two open sets, say $C_1 \cap B$. By continuity, the limit R cannot intersect $C_2 \cap B$. Hence, R touches S at p and it is locally on a side of S. Then R is tangent to S at p from Remark 4.3.

Instead, if we were in the second case, we might assume, using rigid motions, that p is the origin and R is the z-axis in $\mathbb{R}^3$. If R were not tangent to S at p, by reasoning similar to what we used in Corollary 2.76, a suitable neighbourhood of p in S could be described as the graph of a function f defined on a disc D centred at the origin of the plane $\mathbb{R}^2$, such that $f(0,0) = 0$ and such that 0 is a regular value. Let p'_n and q'_n be the points of D which are projections over $z = 0$ of p_n and q_n converging to p. Then p_n' and q_n' would converge to the origin and we might take as a direction vector of the line R_n joining p_n and q_n the vector

$$v_n = \frac{q_n - p_n}{|q_n' - p_n'|} = \left(\frac{q_n' - p_n'}{|q_n' - p_n'|}, \frac{f(q_n') - f(p_n')}{|q_n' - p_n'|} \right).$$

Taking a subsequence, we can imagine that the sequence of the projections v_n' of $\{v_n\}_{n \in \mathbb{N}}$ converges to a vector $v' \in \mathbb{R}^3$, which should have unit length. Then, the sequence of z-components converges to $(df)_{(0,0)}(v')$ by the mean value theorem of calculus. But this vector, from Exercise 4.11, is a direction vector of R, that is, of the z-axis. This is a contradiction, showing that R is tangent to S at p.   □

4.4. The Jordan-Brower separation theorem

When, with regard to the definition of a surface, we mentioned what is understood by a simple curve in Euclidean space—see Example 2.18—we asserted that such a curve must be topologically a circle, provided that it is compact and connected. In fact, we will give the proof of this fact in Chapter 9. Once we have accepted this result, we may intuitively expect that, as happens with the circle, any compact connected simple curve divides the plane into exactly two connected components. This assertion is the subject of the Jordan curve theorem. With respect to surfaces, the situation is not exactly the same, because, even in the compact case, they do not have a fixed topological structure: there are spheres, tori, surfaces with two *holes*, etc. However, we will show that an analogue to the Jordan curve theorem remains true. It is widely known as the Jordan-Brower theorem. Compact connected surfaces, whatever their topologies are, separate Euclidean space into exactly two pieces. More generally, the same can be said for surfaces that are closed as subsets of $\mathbb{R}^3$. The following proof for the case of surfaces is also valid for curves, as we will point out in Chapter 9, although in this case it could be remarkably simplified.

Theorem 4.16 (Jordan-Brower's separation theorem). *Let S be a connected surface that is closed as a subset of $\mathbb{R}^3$. Then $\mathbb{R}^3 - S$ has exactly two connected components whose common boundary is S.*

Proof. We will give the proof in the case when the surface S is compact (for the closed non-compact case, see Exercise (6) at the end of this chapter). Bearing Proposition 4.4 in mind, it suffices to prove that $\mathbb{R}^3 - S$ is not connected. Now, since S is supposed to be compact, as shown in Example 4.8, from each point in $\mathbb{R}^3 - S$ one may draw a half-line which intersects S transversely and, hence, at a finite set of points. Call Ω and Ω', respectively, the subsets of $\mathbb{R}^3 - S$ from which one can draw a half-line cutting

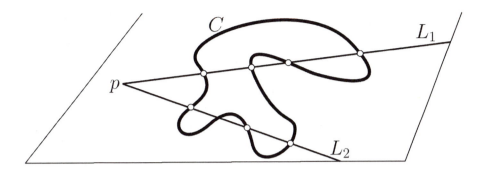

Figure 4.4. *Angular region and simple curve*

transversely at an even or an odd number of points. Obviously we have that $\mathbb{R}^3 - S = \Omega \cup \Omega'$ from the above. A decisive consequence of Proposition 4.14 is the fact that both Ω and Ω' are open subsets of $\mathbb{R}^3$. Since S is compact, one can easily find half-lines which do not intersect it. Thus, Ω' is a non-empty set. But Ω is not either, because, if $p \in S$ is a point of the surface where any height function attains its minimum, it is straightforward to find a half-line contained in the normal line of S at p which intersects S only at p. Therefore, we have expressed the complement $\mathbb{R}^3 - S$ as a union of two non-empty open sets of $\mathbb{R}^3$. Let us now see that they are disjoint. Suppose, on the contrary, that there are a point $x \notin S$ and two half-lines L_1 and L_2 starting at x, cutting S transversely and such that $L_1 \cap S$ is a set with an odd number of points and $L_2 \cap S$ is a set with an even number of points. Let P be the plane determined by the two half-lines. By Example 4.9, we have a plane P', close to P, intersecting S transversely. Hence, by Theorem 2.67, their intersection is a simple curve C, which is necessarily a closed, and so compact, subset of S. Again by Proposition 4.14, we find in P' two half-lines L_1' and L_2' starting at a point x' which still cut S transversely at an odd number and an even number of points, respectively. Then $(L_1' \cup L_2') \cap S$ has an odd number of points. But $(L_1' \cup L_2') \cap S = C \cap (L_1' \cup L_2')$ and we finally arrive at a contradiction to the following result about the intersection of simple curves and angles in the plane. $\qquad \square$

Lemma 4.17. *Let $\Gamma = L_1 \cup L_2$ be the union of two different half-lines of $\mathbb{R}^3$ with common origin at a point p, and let C be a compact simple curve in the plane P determined by Γ, which does not contain p and which intersects L_1 and L_2 transversely. Then $C \cap \Gamma$ has an even number of points.*

Proof. Γ divides the plane P into exactly two connected components, say O_1 and O_2. On the other hand, we know that each connected component C' of C is homeomorphic to a circle; see Example 2.18 and Theorem 9.10. We move along C' with the orientation determined by any homeomorphism of this type. By transversality, $C' \cap \Gamma$ has a finite number of points, and each time that C' cuts Γ, it passes from one component O_i, $i = 1, 2$, to the other one. This implies that, if we start to walk along C' from a point of O_1, each time that we come back to the same component O_1, we have crossed over Γ an even number of times. In particular, this can be applied to a complete turnaround C'. Thus $C' \cap \Gamma$ has an even number of points. Since C has finitely many connected components, we deduce that the same happens for $C \cap \Gamma$. $\qquad\qquad\qquad\qquad\qquad\qquad\qquad\qquad\qquad\qquad\qquad\qquad\quad\square$

Remark 4.18. Since Euclidean space $\mathbb{R}^3$ is locally connected and S is a closed subset of it, the connected components of $\mathbb{R}^3 - S$ are open subsets of $\mathbb{R}^3$. On the other hand, since Euclidean space is locally arc-connected and $\mathbb{R}^3 - S$ is an open subset, its connected and arc-connected components coincide. Therefore, Theorem 4.16 also shows that a closed connected surface separates $\mathbb{R}^3$ into two arc-connected components.

Remark 4.19. The proof of the Jordan-Brower Theorem 4.16 also provides a geometrical criterion for deciding whether a given point in $\mathbb{R}^3 - S$ is in either domain determined by S. This criterion consists of counting the parity of the number of points in the intersection of the surface and any half-line starting at the point and cutting the surface transversely.

Remark 4.20 (Inner domain determined by a compact connected surface). If B is a ball containing the compact surface S, then its complement $\mathbb{R}^3 - B$ is a connected subset of $\mathbb{R}^3 - S$. Hence, it belongs to one of the two connected components C_1 and C_2. Suppose that $\mathbb{R}^3 - B \subset C_2$. Then C_2 is not bounded and $C_1 \cup S = \mathbb{R}^3 - C_2 \subset B$ and, so, C_1 is bounded. Therefore, only one of the two connected components in which a compact connected surface divides the space is bounded. We will call it the *inner domain*—or simply the *domain*—determined by S and we will usually denote it by Ω. Clearly, $\overline{\Omega} = \Omega \cup \mathrm{Bdry}\,\Omega = \Omega \cup S$ is, from the above, a compact set. It is also clear that any half-line drawn from a point outside the ball B does not intersect the surface. Taking Remark 4.19 into account, the points of the *outer domain* $\mathbb{R}^3 - \overline{\Omega}$ are characterized as the origins of transverse half-lines intersecting the surface at an *even* number of points. Consequently, the points of the inner domain Ω are characterized as the initial points of transverse half-lines cutting the surface an *odd* number of times.

As a first remarkable consequence of the separation property stated in the Jordan-Brower theorem, we have the following result, originally due to Brower in 1912, already mentioned when we defined the orientation of surfaces in Remark 3.12.

Theorem 4.21 (Brower-Samelson's theorem). *Any compact surface in Euclidean space is orientable.*

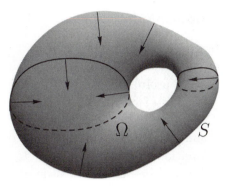

Figure 4.5. *Inner domain and inner normal field*

Proof. It is enough to prove the result for each connected component of the surface. We will assume, then, that S is a compact connected surface. Consider the inner domain Ω determined by the Jordan-Brower Theorem 4.16. If $v \in \mathbb{R}^3$ is a unit vector normal to the surface at a point $p \in S$, we will say that v points *inward*, or v is *inner*, when there is a neighbourhood of p in the half-line starting at p with direction v which is contained in Ω. Let us see that this property of being *inner*, referring to non-null vectors normal to the surface, obeys the conditions required in Exercise 3.9. In fact, if $\alpha : \mathbb{R} \to \mathbb{R}^3$ is the straight line

$$\alpha(t) = p + tv,$$

we have that α cuts S at $\alpha(0) = p$ transversely, because $\alpha'(0) = v \notin T_p S$. Hence—use Remark 4.3—there exists $\varepsilon > 0$ such that $\alpha(-\varepsilon, \varepsilon) \cap S = \{p\}$ and there are points of the image of α on both sides of the surface. Consequently, either $\alpha(0, \varepsilon) \subset \Omega$ or $\alpha(-\varepsilon, 0) \subset \Omega$. In other words, either v is inner or $-v$ is inner, and the two things never occur simultaneously. On the other hand, given a unit vector v normal to S at p, there are a connected neighbourhood V of p in S and a Gauss map N defined on V such that $N(p) = v$. Choosing this V as $W \cap S$ with W as given in Lemma 4.1, one has that either all the $N(q)$ or all the $-N(q)$, with $q \in V$, are simultaneously inner to the surface. Then, we can apply Exercise 3.9 as mentioned, and the surface has to be orientable. $\qquad\square$

Remark 4.22. It follows from the above that, on any compact surface S, we dispose of a Gauss map $N : S \to \mathbb{S}^2$ such that the half-line starting at each $p \in S$ in the direction of $N(p)$ goes into the inner domain Ω determined by the surface—at least, a small enough piece. This N will be called the *inner unit normal* of the surface S. Henceforth, unless otherwise specified, we will always consider the orientation supplied by the inner normal fields of its connected components. Note that this distinction between inner and outer normal fields coincides with what we did for spheres; see Example 3.6.

4.5. Tubular neighbourhoods

We will proceed by showing that compact surfaces in Euclidean space—for the case of simple plane curves, see Chapter 9—have special neighbourhoods, whose existence will be very useful for proving their so-called *global* properties. An example of this kind of result already appeared in the Liebmann and Jellett results in Corollaries 3.49 and 3.48. By the term *global* we mean that we have taken out some information on the whole of the surface—for instance, topological information—from data relative to the diverse classes of curvature or another local object—geometrical data—which indicates how the surface bends near a point.

Let S be a surface. Since $\mathbb{R}^3$ is a metric space, the neighbourhoods easiest to construct for S are the so-called metric neighbourhoods: for each $\delta > 0$, we represent by $B_\delta(S)$ the set of the points of the space whose distance to S is less than δ, that is,

$$B_\delta(S) = \{p \in \mathbb{R}^3 \mid \text{dist}\,(p, S) < \delta\},$$

where $\text{dist}\,(p, S) = \inf_{q \in S} |p - q|$. It is clear that $B_\delta(S)$ is an open neighbourhood of the surface S for each $\delta > 0$. A first observation is that these neighbourhoods consist of segments of normal lines with length 2δ centred at the points of the surface, at least when it is closed as a subset of $\mathbb{R}^3$.

Lemma 4.23. *If S is a surface closed in $\mathbb{R}^3$, the set $B_\delta(S)$ consisting of the points of $\mathbb{R}^3$ whose distance to S is less than $\delta > 0$ coincides with the union $N_\delta(S) = \bigcup_{p \in S} N_\delta(p)$ of all the open segments of normal lines of S with radius δ and centre at the points $p \in S$.*

Proof. In fact, if $p \in S$ is a point of the surface and δ a positive real number, denote by $N_\delta(p)$ the segment with centre p and radius δ in the normal line of S at p. If $q \in N_\delta(p)$ with $p \in S$, then it is immediate that $\text{dist}\,(q, S) \leq |q - p| < \delta$ and, so, $q \in B_\delta(S)$. Conversely, if $q \in B_\delta(S)$, consider on S the distance function to the point q. Suppose that *the surface S is closed* as a subset of $\mathbb{R}^3$. Then, from Exercise 2.73, this function attains its minimum at a point $p \in S$ and, from Example 2.69, q is on the normal line of S at p. Since we also have that $|p - q| = \text{dist}\,(q, S) < \delta$, we conclude that $q \in N_\delta(p)$. $\qquad \square$

This geometrical description of the metric neighbourhoods of a closed surface suggests that, if we get all the involved segments of normal lines not to intersect each other, then we will able to assert that these neighbourhoods are topological products of the surface and an open interval of $\mathbb{R}$. It seems that it could be done if the radius is small enough. In any case, we should be able to control, at the same time, all the normal lines of the surface. For

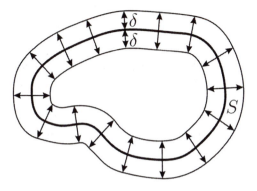

Figure 4.6. *Tubular neighbourhood $N_\delta(S)$*

an orientable surface S, not necessarily closed, this can be done by means of the map $F : S \times \mathbb{R} \to \mathbb{R}^3$ given by

$$(4.1) \qquad F(p, t) = p + tN(p), \qquad \forall (p, t) \in S \times \mathbb{R},$$

where $N : S \to \mathbb{S}^2 \subset \mathbb{R}^3$ is a Gauss map of S. This map F is clearly differentiable, in the sense of Exercises (14)–(19) at the end of Chapter 2, and takes each pair formed by a point p of the surface and a real number t to the point at distance t on the normal line of S at p, on the side of the surface to which $N(p)$ points. Hence, it is clear that

$$F(S \times (-\delta, \delta)) = N_\delta(S) = \bigcup_{p \in S} N_\delta(p), \qquad \forall \delta > 0.$$

Now, note the fact that the normal segments of radius δ do not touch each other is equivalent to the map F being injective on $S \times (-\delta, \delta)$. In fact, we will require something more of the neighbourhoods that we are looking for.

Definition 4.24 (Tubular neighbourhood). The union $N_\delta(S)$ of all the normal segments of radius $\delta > 0$ centred at the points of an orientable surface S is called a tubular neighbourhood of radius $\delta > 0$ if it is open as a subset of $\mathbb{R}^3$ and the map $F : S \times (-\delta, \delta) \to N_\delta(S)$ defined in (4.1) is a diffeomorphism.

We want to see that, under suitable hypotheses, these special neighbourhoods exist.

Lemma 4.25. *For each point $p \in S$ of a surface S there exist an open orientable neighbourhood V and a real number $\delta > 0$ such that the set $N_\delta(V)$ is a tubular neighbourhood of V.*

Proof. The differential of the map $F : S \times \mathbb{R} \to \mathbb{R}^3$ at a pair $(p, t) \in S \times \mathbb{R}$ is given by

$$(4.2) \qquad (dF)_{(p,t)}(v, 0) = v + t(dN)_p(v), \qquad v \in T_p S,$$
$$(dF)_{(p,t)}(0, 1) = N(p).$$

Thus, if we put $t = 0$, it follows that

$$(dF)_{(p,0)}(v, 0) = v \quad \text{and} \quad (dF)_{(p,0)}(0, 1) = N(p)$$

for each $v \in T_p S$. Then $(dF)_{(p,0)}$ is a linear isomorphism from $T_p S \times \mathbb{R}$ to $\mathbb{R}^3$. By the inverse function theorem—see Exercise (19) at the end of Chapter 2—there exist a neighbourhood V of p in S and a number $\delta > 0$ such that

$$F_{|V \times (-\delta, \delta)} : V \times (-\delta, \delta) \longrightarrow F(V \times (-\delta, \delta)) = N_\delta(V)$$

is a diffeomorphism. We conclude using the fact that any surface is locally orientable; see Lemma 3.2. $\qquad \square$

Assume now that the surface S is orientable and that $\mathcal{R} \subset S$ is a relatively compact open subset of S. Using the corresponding cover of $\overline{\mathcal{R}}$ of open sets given in Lemma 4.25, we may extract a subcover and choose, as a common radius, the smaller of the tubular neighbourhoods of the different sets of the cover. This idea leads us to prove the following existence result.

Theorem 4.26 (Existence of tubular neighbourhoods). *Let S be an orientable surface and $\mathcal{R} \subset S$ a relatively compact open subset. Then there exists a number $\varepsilon > 0$ such that the set $N_\varepsilon(\mathcal{R})$ consisting of all the segments of normal lines centred at the points of $\mathcal{R}$ with radius ε is a tubular neighbourhood of the surface $\mathcal{R}$, that is, it is an open subset of $\mathbb{R}^3$ and the map $F : \mathcal{R} \times (-\varepsilon, \varepsilon) \to N_\varepsilon(\mathcal{R})$ given by*

$$F(p, t) = p + tN(p), \qquad (p, t) \in \mathcal{R} \times (-\varepsilon, \varepsilon),$$

is a diffeomorphism. In particular, when S is compact, there exists a number $\varepsilon > 0$ such that the metric neighbourhood $B_\varepsilon(S) = N_\varepsilon(S)$ is a tubular neighbourhood of S.

Proof. By the compacity of $\overline{\mathcal{R}}$, it can be covered by finitely many open subsets of S, each of them having a corresponding tubular neighbourhood, as proved in Lemma 4.25. Let $\delta > 0$ be the smallest radius among them. Note that, in this situation, the map F, defined in (4.1), restricted to $\mathcal{R} \times (-\delta, \delta)$ is a local diffeomorphism. Let us see that, moreover, we can find a number $\varepsilon \in (0, \delta)$ such that the map F restricted to $\mathcal{R} \times (-\varepsilon, \varepsilon)$ is injective, i.e., such that the segments $N_\varepsilon(p)$ of normal lines, with $p \in \mathcal{R}$, do not intersect each other. Suppose that this is not the case. Then, for each $n \in \mathbb{N}$, we would find points $p_n, q_n \in S$ with $p_n \neq q_n$ and

$$N_{1/n}(p_n) \cap N_{1/n}(q_n) \neq \emptyset.$$

Since $\overline{\mathcal{R}}$ is compact, there are subsequences of $\{p_n\}_{n\in\mathbb{N}}$ and of $\{q_n\}_{n\in\mathbb{N}}$ which converge in $\overline{\mathcal{R}}$. We may, by restricting ourselves to them, assume that there are points $p, q \in \overline{\mathcal{R}}$ such that

$$\lim_{n\to\infty} p_n = p \quad \text{and} \quad \lim_{n\to\infty} q_n = q.$$

On the other hand, let $r_n \in N_{1/n}(p_n) \cap N_{1/n}(q_n)$. Then

$$|p_n - q_n| \le |p_n - r_n| + |r_n - q_n| < \frac{1}{n} + \frac{1}{n} = \frac{2}{n}$$

and the two limits p and q coincide. Now we apply Lemma 4.25 to the point $p = q \in S$ and obtain an open neighbourhood V of $p = q$ in S and a real number $\rho > 0$ such that $N_\rho(V)$ is a tubular neighbourhood. But there is also an $N \in \mathbb{N}$ such that

$$n \ge N \implies \left\{ \begin{array}{l} p_n, q_n \in V, \\ 1/n < \rho. \end{array} \right.$$

Therefore, we have a contradiction because

$$N_{1/n}(p_n) \cap N_{1/n}(q_n) \subset N_\rho(p_n) \cap N_\rho(q_n) = \emptyset,$$

since $N_\rho(V)$ is a tubular neighbourhood and so F restricted to $V \times (-\rho, \rho)$ must be injective. In conclusion, we can find a positive number ε such that

$$F : \mathcal{R} \times (-\varepsilon, \varepsilon) \longrightarrow N_\varepsilon(\mathcal{R})$$

is injective and is a local diffeomorphism. Since it is obviously surjective, we conclude the proof. $\qquad\square$

Remark 4.27. One of the consequences which can be deduced from the existence of tubular neighbourhoods for compact surfaces is that the sets of points at a distance from the surface less than a certain $\varepsilon > 0$—just the number supplied by Theorem 4.26—are topological products of the surface and an open interval of $\mathbb{R}$. In particular, the metric neighbourhoods of a compact surface with small enough radius have the same homotopy type as the surface.

Remark 4.28. If S is a compact surface, Ω is the inner domain, and $N_\varepsilon(S)$, with $\varepsilon > 0$, is a tubular neighbourhood of S, then the set

$$V_\varepsilon(S) = F(S \times (0, \varepsilon)) \subset N_\varepsilon(S)$$

is connected and does not intersect S. If the orientation that we are using for S is what the inner normal field gives, then it is clear that $V_\varepsilon(S) \subset \Omega$.

Remark 4.29 (Parallel surfaces). Consider again that $N_\varepsilon(S)$, for some $\varepsilon > 0$, is a tubular neighbourhood of a compact surface S with a fix Gauss map N. Since the map $F : S \times (-\varepsilon, \varepsilon) \to N_\varepsilon(S)$ defined in (4.1) is a

Figure 4.7. *Parallel surfaces*

diffeomorphism, for all $t \in (-\varepsilon, \varepsilon)$, the map $F_t : S \to N_\varepsilon(S) \subset \mathbb{R}^3$ defined by

$$F_t(p) = F(p, t) = p + tN(p)$$

is a homeomorphism onto its image. On the other hand, if $p \in S$ and $e_1, e_2 \in T_pS$ are principal directions of S at p, we have

$$(dF_t)_p(e_i) = e_i + t(dN)_p(e_i) = (1 - tk_i(p))e_i, \qquad i = 1, 2.$$

Now then, from (4.2), we see that F_t has an injective differential at each point, for $t \in (-\varepsilon, \varepsilon)$. Hence, using Exercise (9) at the end of Chapter 2, the image $F_t(S)$, which we will represent by S_t, is a surface and $F_t : S \to S_t$ is a diffeomorphism for each t, $|t| < \varepsilon$. We will call the surface S_t the *parallel surface* of S at—oriented—distance t. In Exercise (16) at the end of Chapter 3, the reader already found some useful geometrical properties of these surfaces.

Remark 4.30. When S is a non-compact surface which is a closed subset of Euclidean space, there also exist tubular neighbourhoods, although they do not necessarily have a *constant* radius. In fact, it can be shown that there is a positive continuous function $\varepsilon : S \to \mathbb{R}$ and that the map

$$F : E(S, \varepsilon) \subset S \times \mathbb{R} \longrightarrow \mathbb{R}^3$$

defined in (4.1) is a diffeomorphism onto its image, where $E(S, \varepsilon)$ denotes the open subset of $S \times \mathbb{R}$ given by

$$E(S, \varepsilon) = \{(p, t) \in S \times \mathbb{R} \mid |t| < \varepsilon(p)\}.$$

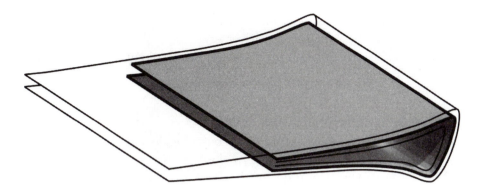

Figure 4.8. *Tubular neighbourhood of a closed surface*

Exercises

(1) ↑ Let S be an orientable surface and $N_\delta(S)$ a tubular neighbourhood, with $\delta > 0$. Represent by π and r, respectively, the first and second components of the map

$$F^{-1} : N_\delta(S) \longrightarrow S \times (-\delta, \delta).$$

They are called *projection* and the *oriented distance*, respectively. Study their properties and compute their differentials.

(2) ↑ Let S be a compact connected surface, Ω its inner domain, and p a point of S. Show that there is a sphere $\mathbb{S}_r^2$, $r > 0$, contained in $\overline{\Omega}$, tangent to S at p with $S \cap \mathbb{S}_r^2 = \{p\}$.

(3) Suppose that S is a surface and $N_\rho(S)$ a tubular neighbourhood. Prove that, if $f : S \to (-\rho, \rho) \subset \mathbb{R}$ is a differentiable function, then $S_f = \{x \in N_\rho(S) \,|\, r(x) = (f \circ \pi)(x)\}$ is a surface by checking that 0 is a regular value of $r - f \circ \pi$.

(4) ↑ Show that, if S is a compact surface, there are points of S whose second fundamental form relative to the inner normal field is positive definite.

(5) ↑ Let S be a compact surface, and let $P \subset S$ be the subset of S formed by the points with non-negative Gauss curvature. Prove that the restriction of the Gauss map of S to P is surjective. Is it also true if one takes the set of points with positive Gauss curvature?

(6) *The Jordan-Brower and Brower-Samelson theorems for closed surfaces.* Let S be a connected surface closed as a subset of $\mathbb{R}^3$, and let $B = B_\varepsilon(p_0) \subset \mathbb{R}^3 - S$ be an open ball of radius $\varepsilon > 0$. Show the following.

a: For each $p_1 \in \mathbb{R}^3 - S$ one can find a point $p_2 \in B$ such that the segment $[p_1, p_2]$ cuts S transversely.

b: If p_3 is another point of B such that $[p_1, p_3]$ cuts S transversely as well, then, in

$$([p_1, p_2] \cup [p_1, p_3]) \cap S$$

there is an even number of points.

c: $\mathbb{R}^3 - S$ is not connected. As a consequence, show that the Brower separation theorem extends to closed surfaces.

Now, observe that the proof of the Brower-Samelson Theorem 4.21 is valid in this case without any changes.

(7) ↑ Let S be a compact connected surface and Ω the inner domain determined by S. Suppose that a straight line R intersects $\overline{\Omega}$ only at a point p. Prove that $p \in S$. Using the restriction to R of the oriented distance function in a tubular neighbourhood of S, show that R is tangent to S at p.

(8) ↑ Let S be a compact connected surface with positive mean curvature everywhere and Ω the inner domain determined by S. Take $q \in \Omega$. Show that, if $p \in S$ is the point of S closest to q, then $q = p + tN(p)$, where $N(p)$ is the inner normal and $0 \leq t \leq 1/k_2(p)$.

(9) ↑ Let S be a compact connected surface and Ω its inner domain. Prove that, if $x \notin \overline{\Omega}$, then there are at least two straight lines tangent to S passing through x.

(10) ↑ Prove that any compact surface can be expressed as the inverse image of a regular value of a differentiable function defined on $\mathbb{R}^3$.

(11) ↑ Prove that any surface has Lebesgue measure zero in $\mathbb{R}^3$.

Hints for solving the exercises

Exercise 4.11: There exist, by definition, affine parametrizations $\alpha_n(t) = q_n + tw_n$, $t \in \mathbb{R}$, of each straight line R_n, $n \in \mathbb{N}$, such that

$$\lim_{n \to \infty} \alpha_n(t) = \alpha(t), \qquad t \in \mathbb{R},$$

where $\alpha(t) = q + tw$. Hence $q \in R$ is the limit of $\{q_n\}_{n \in \mathbb{N}}$ and $w \neq 0$ is the limit of $\{w_n\}_{n \in \mathbb{N}}$ and it is a direction vector of R. Since $p_n \in R_n$ and v_n is a direction vector of R_n, one has

$$p_n = q_n + t_n w_n \quad \text{and} \quad v_n = \lambda_n w_n, \qquad n \in \mathbb{N},$$

for some $t_n \in \mathbb{R}$. It follows from these expressions that

$$t_n = \frac{\langle p_n - q_n, w_n \rangle}{|w_n|^2} \quad \text{and} \quad \lambda_n = \frac{\langle v_n, w_n \rangle}{|w_n|^2}, \qquad n \in \mathbb{N},$$

and, thus, the sequences $\{t_n\}_{n\in\mathbb{N}}$ and $\{\lambda_n\}_{n\in\mathbb{N}}$ are convergent to real numbers t and λ. Therefore, taking limits in the first two equalities, we get

$$p = q + tw \quad \text{and} \quad v = \lambda w,$$

and so $p \in R$ and v is a direction vector of R.

Exercise 4.12: Let K be a compact subset of $\mathbb{R}^3$ which intersects all the lines of the sequence $\{R_n\}_{n\in\mathbb{N}}$. We choose, for each $n \in \mathbb{N}$, a point $p_n \in R_n \cap K$ and a direction unit vector v_n of R_n. By the compacity of K and of the unit sphere in $\mathbb{R}^3$, there exist subsequences of the sequences $\{p_n\}_{n\in\mathbb{N}}$ and $\{v_n\}_{n\in\mathbb{N}}$, respectively, which we will still denote by the same symbols, converging, respectively, to $p \in K$ and $v \in \mathbb{S}^2$. Then

$$\lim_{n\to\infty} (p_n + tv_n) = p + tv, \qquad \forall t \in \mathbb{R}.$$

Consequently, $\{R_n\}_{n\in\mathbb{N}}$ converges to the straight line passing through p in the direction of v.

Exercise 4.13: By definition, one can find a sequence $\{v_n\}_{n\in\mathbb{N}}$ of direction vectors of the lines R_n converging to a direction vector v of R. Moreover, we can suppose, forgetting finitely many points if necessary, that the whole of the sequence $\{p_n\}_{n\in\mathbb{N}}$ is contained in an open neighbourhood of p where we dispose of a Gauss map N. Then, it is enough to take limits in the equality $\langle v_n, N(p_n) \rangle = 0$, which is valid for each $n \in \mathbb{N}$.

Exercise (1): If $N_\delta(S)$ is a tubular neighbourhood of S, each point $x \in N_\delta(S)$ is in a unique normal segment of radius δ. Represent by $\pi(x) \in S$ the point of S on which this normal segment lies and by $r(x) \in (-\delta, \delta)$ the distance from the point x to its projection on S, if x is on the side of S towards $N(\pi(x))$ points, and, otherwise, the opposite of this distance. These two maps are the components of F^{-1}. Both are differentiable, because one can take into account that

$$\pi = p_1 \circ F^{-1}_{|S \times (-\delta,\delta)} \quad \text{and} \quad r = p_2 \circ F^{-1}_{|S \times (-\delta,\delta)},$$

where p_1 and p_2 are the projections of the product $S \times (-\delta, \delta)$ onto either factors. The following statement summarizes the most relevant properties of π and r.

Let S be an oriented surface, $N_\delta(S)$, $\delta > 0$, a tubular neighbourhood of S, and $\pi : N_\delta(S) \to S$, $r : N_\delta(S) \to (-\delta, \delta)$ the corresponding projection and the oriented distance. Then the following hold.

A: $r^{-1}(\{0\}) = S$.

B: If $0 < \rho < \delta$, then $N_\rho(S)$ is another tubular neighbourhood of S.

C: Reversing the orientation of S, the projection π remains unchanged and the oriented distance r changes its sign.

D: If $p \in S$, $e_1, e_2 \in T_p S$ are principal directions of S at p, and $t \in (-\delta, \delta)$, one has that $1 - tk_i(p) > 0$ and

$$(d\pi)_{p+tN(p)}(e_i) = \frac{1}{1 - tk_i(p)} e_i, \quad i = 1, 2, \quad (d\pi)_{p+tN(p)}(N(p)) = 0,$$
$$(dr)_{p+tN(p)}(e_i) = 0, \quad i = 1, 2, \quad (dr)_{p+tN(p)}(N(p)) = 1.$$

Proof. It is clear that $r^{-1}(\{0\}) = S$ from definition (4.1). Furthermore, the two assertions in **C** about the changes that the projection and the oriented distance suffer when we reverse the orientation of the surface are immediate from the corresponding definitions. The same happens for **B**. Now, take a curve $\alpha : I \to S$, where I is an interval of $\mathbb{R}$ containing the origin and such that $\alpha(0) = p$ and $\alpha'(0) = e_i$, $i = 1, 2$. We define another curve β in $\mathbb{R}^3$ by

$$\beta(s) = F(\alpha(s), t) = \alpha(s) + tN(\alpha(s)), \quad s \in I.$$

Then this new curve β belongs to the open set $N_\delta(S)$ and

$$\beta(0) = p + tN(p) \quad \text{and} \quad \beta'(0) = (1 - tk_i(p))e_i = (dF)_{(p,t)}(e_i, 0),$$
$$\pi(\beta(s)) = \alpha(s) \quad \text{and} \quad r(\beta(s)) = t, \quad s \in I.$$

Thus, since F is a diffeomorphism when restricted to $S \times (-\delta, \delta)$, we have that $(dF)_{(p,t)}(e_i, 0)$ is a non-null vector, that is, $1 - tk_i(p) \neq 0$. But $(-\delta, \delta)$ is connected and $1 - tk_i(p)$ is positive for $t = 0$. Thus $1 - tk_i(p) > 0$ on $S \times (-\delta, \delta)$. Moreover

$$(d\pi)_{p+tN(p)}([1 - tk_i(p)]e_i) = \left.\frac{d}{ds}\right|_{s=0} \pi(\beta(s)) = \alpha'(0) = e_i.$$

By the linearity of the differential and the fact that $1 - tk_i(p) \neq 0$, we obtain

$$(d\pi)_{p+tN(p)}(e_i) = \frac{1}{1 - tk_i(p)} e_i, \quad i = 1, 2.$$

Furthermore

$$(dr)_{p+tN(p)}(e_i) = \left.\frac{d}{ds}\right|_{s=0} r(\beta(s)) = 0.$$

Similarly, if we consider the segment γ of the normal line in the tubular neighbourhood $N_\delta(S)$ given by

$$\gamma(s) = F(p, t + s) = p + (t + s)N(p), \quad s \in I,$$

we clearly have that

$$\gamma(0) = p + tN(p) \quad \text{and} \quad \gamma'(0) = N(p),$$
$$\pi(\gamma(s)) = p \quad \text{and} \quad r(\gamma(s)) = s + t, \quad s \in I.$$

Taking derivatives at $s = 0$ in the two last equalities, we get

$$(d\pi)_{p+tN(p)}(N(p)) = 0 \quad \text{and} \quad (dr)_{p+tN(p)}(N(p)) = 1,$$

as required. □

Exercise (2): If $t > 0$ is small, the point $p_t = p + tN(p)$, where N is the inner normal of S, belongs to Ω and, according to Exercise (1) at the end of Chapter 3, p is an isolated local minimum of the square of the distance function to p_t. Then there exists an open neighbourhood V of p in S such that V is outside the open ball B_t with centre p_t and radius t and such that $V \cap \mathbb{S}_t^2 = \{p\}$, where $\mathbb{S}_t^2$ is the corresponding sphere. Since $\mathbb{S}_t^2$ tends to the point p as t goes to zero and $S - V$ is a compact set which does not contain p, we have $S \cap B_t = \emptyset$ for small t. Since B_t is connected, either $B_t \subset \Omega$ or $B_t \subset \mathbb{R}^3 - \overline{\Omega}$. The correct choice is the first one, because $p_t \in B_t \cap \Omega$. Thus, $\mathbb{S}_t^2 \subset \overline{B_t} \subset \overline{\Omega}$ and $\mathbb{S}_t^2 \cap S = \{p\}$.

Exercise (4): Let us show that any point p of S where the square of the distance function to a fix point $a \in \mathbb{R}^3$ reaches a maximum must have positive definite second fundamental form, with respect to the inner normal field. Indeed, all the points of the half-line starting at a and passing through p which are beyond p cannot belong to S. That is,

$$p + t(p - a) \notin S, \qquad \forall t > 0.$$

Since this set is not bounded, one has that $p + t(p - a) \in \mathbb{R}^3 - \overline{\Omega}$ for each $t > 0$. On the other hand, we know, from Example 3.39, that, for some $\lambda \in \mathbb{R}$,

$$a = p + \lambda N(p) \quad \text{and} \quad |v|^2 - \lambda\sigma_p(v, v) \leq 0, \qquad v \in T_pS.$$

Hence, we have

$$p + t(p - a) = p - t\lambda N(p) \in \mathbb{R}^3 - \overline{\Omega}, \qquad t > 0.$$

Thus, if N is the inner normal, $t\lambda > 0$ for each $t > 0$ small enough and, so, $\lambda = |p - a| > 0$. Consequently

$$\sigma_p(v, v) \geq \frac{1}{|p - a|}|v|^2, \qquad v \in T_pS.$$

Exercise (5): Let $a \in \mathbb{S}^2$, and let q be a point of S where the height function $h(p) = \langle p, a \rangle$, $p \in S$, relative to a fix plane perpendicular to the vector a, attains its minimum. By Proposition 3.35 and Example 3.36, the second fundamental form of S at q, relative to any choice of normal, is semi-definite and, thus, $K(q) \geq 0$. That is, $q \in P$. Now it suffices to show that $N(q) = a$, where N is the inner normal of S. Since q is a critical point for h, part **A** of Example 2.59 implies that $N(q) = \pm a$. On the other hand, since S is over the plane passing through q perpendicularly to a, Ω is over this plane as well. Then, for $t > 0$ small, since $q + tN(q) \in \Omega$, we have $h(q + tN(q)) = h(q) + t\langle N(q), a \rangle \geq h(q)$. It follows that $\langle N(q), a \rangle \geq 0$ and, so, $N(q) = a$.

The final answer is negative. One can see this by studying the region of the elliptic points in the torus of revolution, as proposed in Exercise (8) at the end of Chapter 3.

Exercise (7): If $p \in \Omega$, since Ω is an open subset of $\mathbb{R}^3$, $R \cap \Omega$ would be open in R and contained in $R \cap \overline{\Omega} = \{p\}$. Hence $p \in \overline{\Omega} - \Omega = S$. Let $\varepsilon > 0$ be a number such that $N_\varepsilon(S)$ is a tubular neighbourhood of S and $v \in \mathbb{R}^3$ a direction vector of the straight line R. For t close to 0, one has that $p + tv \in N_\varepsilon(S)$. Since R touches $\overline{\Omega}$ only at p, taking Exercise (1) into account, we have $r(p + tv) < 0$ for each $t \neq 0$ small. That is, at $t = 0$ there is a maximum of the function $t \mapsto r(p + tv)$. Therefore, $(dr)_p(v) = 0$, and, using Exercise (1) again, we deduce that $\langle N(p), v \rangle = 0$.

Exercise (8): Taking Example 3.39 into account, we know that $q = p + tN(p)$ for some $t \in \mathbb{R}$ and that $1 - tk_2(p) \geq 0$. Since $k_2(p) \geq H(p) > 0$, it follows that $t \leq 1/k_2(p)$. It only remain to show that $t \geq 0$. Since p is the point of S closest to $q \in \Omega$, the segment $]p, q]$ does not intersect S. This same segment is connected and does cut Ω. Hence, $]p, q] \subset \Omega$. If $t \leq 0$, then $p + \lambda N(p) \in \Omega$ for $\lambda \in [t, 0)$, which is not true according to Remark 4.22. Then $t \geq 0$.

Exercise (9): Consider the support function f based on the point x relative to the inner normal of S, as was defined in Exercise (4) of Chapter 3. Since S is compact, we can find two different points $p, q \in S$ for which the distance to the point x reaches a maximum and a minimum, respectively. If we argue in a way similar to Exercise (8) above, we can see, using that $x \notin \overline{\Omega}$, that $f(p) = -|p - x| < 0$ and that $f(q) = |q - x| > 0$. Since S is connected, there must exist a point $r \in S$ where $0 = f(r) = \langle r - x, N(r) \rangle$. Furthermore, one can draw a continuous curve joining p and q in S avoiding the point r. In this way, we obtain another point, different from r, where f vanishes. Now, it is enough to prove that any point where the support function vanishes supplies a point from which we may draw a tangent line passing through x.

Exercise (10): We take $\varepsilon > 0$ so that $N_\varepsilon(S)$ is a tubular neighbourhood and a differentiable function $h : \mathbb{R} \to \mathbb{R}$ such that $h(t) = -\varepsilon$ for each $t \leq -\varepsilon$ and $h(t) = \varepsilon$ for each $t \geq \varepsilon$, which is increasing on the interval $(-\varepsilon, \varepsilon)$. Then we define $f : \mathbb{R}^3 \to \mathbb{R}$ by

$$f(x) = \begin{cases} (h \circ r)(x) & \text{if } x \in N_\varepsilon(S), \\ \varepsilon & \text{if } x \in \Omega - N_\varepsilon(S), \\ -\varepsilon & \text{otherwise}, \end{cases}$$

where r is the corresponding oriented distance on the tubular neighbourhood, studied in Exercise (1) above, and Ω is the domain determined by S. This definition and the choice of h force f to be differentiable. Moreover, it is clear that $f^{-1}(\{0\}) \subset N_\varepsilon(S)$ and so

$$f^{-1}(\{0\}) = r^{-1}(h^{-1}(\{0\})) = r^{-1}(\{0\}) = S.$$

On the other hand, if $p \in S$,

$$(df)_p = h'(0)(dr)_p \neq 0,$$

because h is increasing and $(dr)_p(N_p) = 1$.

Exercise (11): Cover the surface by countably many open subsets like those provided by Lemma 4.1. We finish by showing that any of them has Lebesgue measure zero. This is immediate if we remember that having measure zero is a property invariant under diffeomorphisms and that a plane in $\mathbb{R}^3$ has measure zero.

4.6. Appendix: Proof of Sard's theorem

The version of Sard's theorem that we stated as Theorem 4.6 is a direct consequence of a stronger result concerning differentiable maps between surfaces that, to be established, requires the notion of the *measure zero* set in a surface. We will base our study on the corresponding concept of Lebesgue measure zero for subsets of $\mathbb{R}^2$. Recall that, as a consequence of the formula for change of variables for the Lebesgue integral on $\mathbb{R}^2$, a diffeomorphism between open subsets of $\mathbb{R}^2$ takes measure zero subsets into subsets with this same property. This is why the following definition makes sense.

Definition 4.31 (Measure zero set). A subset A of a surface S in Euclidean space $\mathbb{R}^3$ is said to have measure zero if, for each parametrization $X : U \to S$ of the surface, the set $X^{-1}(A)$ has Lebesgue measure (area) zero in $\mathbb{R}^2$; compare this with Remark 5.9.

The following properties follow directly from the previous definition and from the corresponding ones of the Lebesgue measure on $\mathbb{R}^2$.

Proposition 4.32 (Properties of the measure zero sets). *Let S be a surface in Euclidean space and $A \subset S$ a subset. The following assertion are true:*

A: *A has measure zero if and only if there exists a family $\{X_i\}_{i \in I}$ of parametrizations of S whose images cover S and such that $X_i^{-1}(A)$ has Lebesgue measure zero in $\mathbb{R}^2$ for each $i \in I$.*

B: *If A has measure zero, then it has empty interior.*

C: *If A is a union of countably many measure zero subsets of S, then A has measure zero.*

We may now give Sard's theorem in a stronger and more common version than what was needed in Theorem 4.6. In fact, that result can be deduced from the following statement, along with property **B** of Proposition 4.32.

Theorem 4.33 (Sard's theorem). *Let $f : S_1 \to S_2$ be a differentiable map, where S_1 and S_2 represent indistinctly either surfaces or open subsets of $\mathbb{R}^2$. Then the complement of the set of regular values of f has measure zero in S_2.*

As in other places where the proof is somewhat longer than usual, we will divide the proof of this Sard theorem into several intermediate steps.

Proposition 4.34. *Let O be a relatively compact open subset of $\mathbb{R}^2$ and $\phi : \overline{O} \to \mathbb{R}^2$ a differentiable map. Represent by $\mathcal{N}$ the set*

$$\mathcal{N} = \{x \in O \,|\, (\operatorname{Jac} \phi)(x) = 0\}.$$

Then, if K is a compact subset of $\mathcal{N}$, for each $\varepsilon > 0$, there exists an open subset V_ε of $\mathbb{R}^2$ such that

$$K \subset V_\varepsilon \subset O \quad \text{and} \quad \operatorname{area} \phi(V_\varepsilon) < \varepsilon.$$

Thus $\phi(\mathcal{N})$ has Lebesgue measure zero in $\mathbb{R}^2$.

Proof. We will separate the proof into different steps.

Assertion 1. *If $x, y \in O$ are two points of the open set O such that the segment $[x, y]$ is contained in O, then, for each $a \in \mathbb{R}^2$, there exists $t \in [0, 1]$ such that*

$$\langle \phi(y) - \phi(x), a \rangle = \langle (d\phi)_{x+t(y-x)}(y - x), a \rangle.$$

In fact, if $[x, y] \subset O$ and a is a vector in $\mathbb{R}^2$, we may define a function f on $[0, 1]$ by

$$f(t) = \langle \phi(x + t(y - x)), a \rangle,$$

which is derivable. Now it is enough to apply the Cauchy mean value theorem to this function and to take the chain rule into account to compute its first derivative.

Assertion 2. *For each $\varepsilon > 0$ there exists $\delta > 0$ such that, if C is a square in $\mathbb{R}^2$ included in O with $C \cap \mathcal{N} \neq \emptyset$ and $\operatorname{diam} C < \delta$, one has $\operatorname{area} \phi(C) \leq \varepsilon \operatorname{area}(C)$.*

Given an arbitrary $\varepsilon > 0$, set $\varepsilon' = \varepsilon/c > 0$ where $c = (2\sqrt{2})^2 M^2$ and $M = 1 + \max_{x \in \overline{O}} |(d\phi)_x| > 0$. Since the map

$$x \in \overline{O} \longmapsto (d\phi)_x \in M_{2\times 2}(\mathbb{R}) \equiv \mathbb{R}^4$$

is continuous on the compact set $\overline{O}$, then it is uniformly continuous. Hence, for the chosen $\varepsilon' > 0$ there exists $\delta > 0$ such that

$$\left. \begin{array}{l} x, y \in \overline{O}, \\ |x - y| < \delta \end{array} \right\} \implies |(d\phi)_x - (d\phi)_y| < \varepsilon'.$$

Once this δ is determined, let C be a square of $\mathbb{R}^2$ such that $C \subset O$ with $\mathrm{diam}\,(C) < \delta$ and such that there is at least a point $a \in C \cap \mathcal{N}$, that is, a point $a \in C$ with $(\mathrm{Jac}\,\phi)(a) = 0$. Then the linear map $(d\phi)_a$ is not regular and, so, its image is not the whole of $\mathbb{R}^2$. Then, we may find a unit vector $v \in \mathbb{R}^2$ perpendicular to this image. Let $\{v, w\}$ be an orthonormal basis of $\mathbb{R}^2$. If $x \in C$, since C is convex, from Assertion 1 above, we have

$$\langle \phi(x) - \phi(a), w \rangle = \langle (d\phi)_{a+t(x-a)}(x - a), w \rangle,$$

where $t \in [0, 1]$. Using the Schwarz inequality,

$$|\langle \phi(x) - \phi(a), w \rangle| \le |(d\phi)_{a+t(x-a)}||x - a| \le M\,\mathrm{diam}\,C,$$

for each $x \in C$. On the other hand, again by Assertion 1 and by the choice of the vector v,

$$\langle \phi(x) - \phi(a), v \rangle = \langle [(d\phi)_{a+t(x-a)} - (d\phi)_a](x - a), v \rangle,$$

for some $t \in [0, 1]$. From this, the Schwarz inequality gives

$$|\langle \phi(x) - \phi(a), v \rangle| \le |(d\phi)_{a+t(x-a)} - (d\phi)_a||x - a| \le \varepsilon'\mathrm{diam}\,C,$$

because $|a + t(x - a) - a| = |t||x - a| \le |x - a| \le \mathrm{diam}\,C < \delta$. Since the the length of the side of a square is exactly its diameter divided by $\sqrt{2}$, we arrive at, for each $x \in C$,

$$\begin{aligned} |\langle \phi(x) - \phi(a), w \rangle| &\le \sqrt{2}\,M\,\mathrm{side}\,C, \\ |\langle \phi(x) - \phi(a), v \rangle| &\le \sqrt{2}\,\varepsilon'\,\mathrm{side}\,C. \end{aligned}$$

That is, $\phi(C)$ is contained in a rectangle whose sides have lengths

$$2\sqrt{2}M\,\mathrm{side}\,C \quad \text{and} \quad 2\sqrt{2}\varepsilon'\,\mathrm{side}\,C.$$

Then, we obtain

$$\mathrm{area}\,\phi(C) \le (2\sqrt{2})^2 M^2 \varepsilon'\,(\mathrm{side}\,C)^2 = \varepsilon\,\mathrm{area}\,(C),$$

as we wanted.

Now we can prove Proposition 4.32. Given $\varepsilon > 0$, put $\varepsilon' = \varepsilon/\mathrm{vol}\,(O) > 0$ and call $\delta > 0$ the positive number provided by Assertion 2 above. Now take a partition of the compact set K with norm less than this δ, by means of squares $C_1, \ldots, C_n \subset O$. Then, we define the open set V_ε as the interior of $C_1 \cup \cdots \cup C_n$. Clearly $K \subset V_\varepsilon \subset O$ and

$$\text{area } \phi(V_\varepsilon) \leq \text{area } \phi \left(\bigcup_{i=1}^{n} C_i \right) = \text{area } \bigcup_{i=1}^{n} \phi(C_i).$$

Since the diameter of each square C_i, $i = 1, \ldots, n$, is less than δ and since each C_i cuts K and, so, $\mathcal{N}$, we have $\text{area } \phi(C_i) \leq \varepsilon' \text{ area } (C_i)$. In this way, we get

$$\text{area } \phi(V_\varepsilon) \leq \varepsilon' \sum_{i=1}^{n} \text{area } (C_i) \leq \varepsilon' \text{ area } (O) = \varepsilon,$$

which was the required condition. Hence K has Lebesgue measure zero in $\mathbb{R}^2$. To see that the same holds for $\mathcal{N}$, it suffices to choose an open subset W of $\mathbb{R}^2$ with $\overline{O} \subset W$ and such that ϕ is defined on $\overline{W}$. Now we apply the above changing O to W and K to $\overline{\mathcal{N}}$. $\square$

Proof of Sard's theorem. Let us show that the image through f of the set

$$\mathcal{N} = \{ p \in S_1 \mid (\text{Jac } f)(p) = 0 \}$$

has measure zero in S_2. Let $X_2 : U_2 \to S_2$ be any parametrization of S_2 defined on an open subset U_2 of $\mathbb{R}^2$. Since f is continuous, $V_1 = f^{-1}(X_2(U_2))$ is open in S_1. Then, we may cover the surface V_1 with countably many parametrizations $X_1^n : U_1^n \to S_1$ of S_1, $n \in \mathbb{N}$, defined on relatively compact open subsets of $\mathbb{R}^2$ which can be extended to differentiable maps on each $\overline{U_1^n}$. We denote by ϕ_n the compositions

$$\phi_n = X_2^{-1} \circ f \circ X_1^n : U_1^n \subset \mathbb{R}^2 \to \mathbb{R}^2,$$

which are differentiable and obey the conditions that we need for applying Proposition 4.32. Therefore, the set

$$\mathcal{N}_n = \{ x \in U_1^n \mid (\text{Jac } \phi_n)(x) = 0 \}, \qquad n \in \mathbb{N}$$

has the property that $\phi_n(\mathcal{N}_n)$ has measure zero and, so,

$$\text{area } \bigcup_{n \in \mathbb{N}} \phi_n(\mathcal{N}_n) = 0.$$

Now, it is enough to observe that we dispose of the equality

$$\bigcup_{n \in \mathbb{N}} \phi_n(\mathcal{N}_n) = X_2^{-1}(f(\mathcal{N}))$$

to conclude that $f(\mathcal{N})$ has measure zero in S_2, according to Definition 4.31. $\square$

Chapter 5

Integration on Surfaces

5.1. Introduction

In this chapter, we will proceed by giving some more tools adequate for the study of global geometry that we develop later. At this time, they will not have a topological nature, as in Chapter 4, but a mainly analytical flavour.

Perhaps the most famous result that we will prove in this chapter is the *divergence theorem*, whose statement requires, in some sense, the JORDAN-BROWER separation theorem for a compact surface in $\mathbb{R}^3$. If S is a compact connected surface in Euclidean space and Ω is the inner domain determined by S, the divergence theorem will enable us to relate the integrals of certain functions defined on $\overline{\Omega}$ with integrals on the surface S. In this way, it is a generalization of the fundamental theorem of calculus. Even the names of the concepts involved in this theorem—fields, divergence, etc.—show its physical origin. Indeed, the first formulation of this theorem appears in a postscript of a letter dated in 1850 and addressed by LORD KELVIN to STOKES, professor at the University of Cambridge. LORD KELVIN himself gave two different proofs of the result. There is also another one due to MAXWELL in this same period. But all the above presuppose the development on the surface of a certain *integral calculus*.

Indeed, we will begin with the task of endowing surfaces with an integral calculus and we will do it in a non-standard, though natural, way. In fact, to construct the differential calculus on surfaces—definition of a differentiable function, differential of a function, etc.—in Chapter 2, we used differential calculus on the plane and we translated it on surfaces by means of parametrizations, although we were always careful that the definitions that we gave did not depend on the choice of the precise parametrization.

A similar method could be followed to built integral calculus on surfaces: use any theory of integration already known on the plane $\mathbb{R}^2$, for instance, Lebesgue integration, and try to transpose it on the surface by means of parametrizations. This is the usual way to introduce an integral calculus on surfaces.

Instead, in this book, the integral of a function on a compact surface—or on a relatively compact open subset of an orientable surface—will not be defined from the integral of functions on the plane through parametrizations of the surface and the so-called partitions of unity, but we will use the fact that our surfaces are not abstract manifolds since they are embedded in $\mathbb{R}^3$. We will define the integral of a function on a compact surface from the integral on any of its tubular neighbourhoods—which is a relatively compact open subset of $\mathbb{R}^3$—of the function obtained by extending the original one as a constant along each normal segment. That is, we will exploit the following result that we established for surfaces: even though a relatively compact subset of a surface does not have to be diffeomorphic to an open subset of the plane, it has a tubular neighbourhood which is diffeomorphic to an open subset of Euclidean space $\mathbb{R}^3$.

We will develop the usual properties of Lebesgue integration, give a proof of the *divergence theorem*, and, later, we will finish with the *Brower fixed point theorem*, as an example of an application of the first result. In Chapter 6, we will see that integration is really a powerful tool for getting theorems of a global nature in the theory of surfaces.

5.2. Integrable functions and integration on $S \times \mathbb{R}$

We will use the term *region*, in an orientable surface S, for any relatively compact open subset of S. For example, if S is a surface closed as a subset of Euclidean space, from Exercise (6) of Chapter 4 generalizing the Brower-Samelson theorem, each bounded open subset is a region of S in the sense that we have just defined. Note that, if S is compact, the whole of S is a region of itself. We will usually reserve the symbol $\mathcal{R}$ to represent such subsets of orientable surfaces.

Take a region $\mathcal{R}$ in an orientable surface S, and let O be an open subset of $\mathbb{R}^3$. Suppose that $\phi : O \to \mathcal{R} \times (a, b)$ is a diffeomorphism, where $a < b$ are two real numbers. We will introduce the *absolute value of the Jacobian*, by analogy with the usual definition for differentiable maps between open subsets of Euclidean spaces—see Exercise (13) of Chapter 2 for the case of surfaces—as the function $|\operatorname{Jac} \phi| : O \to \mathbb{R}$ given by

$$|\operatorname{Jac} \phi|(x) = |\det(d\phi)_x|,$$

for each $x \in O$, where the absolute value of the determinant of the linear map $(d\phi)_x : \mathbb{R}^3 \to T_{\phi^1(x)} S \times \mathbb{R}$, with $\phi^1 : O \to S$ the first component of ϕ, must be understood in the sense of the following exercises.

Exercise 5.1. Let V_1 and V_2 be two Euclidean vector spaces with the same dimension, and let $f : V_1 \to V_2$ be a linear map between them. If M is the matrix of the map f relative to two orthonormal bases of V_1 and V_2, prove that $|\det M|$ does not depend on the chosen bases. This non-negative real number is called the absolute value of the determinant of f and we will denote it by $|\det f|$. Show that $|\det f| > 0$ if and only if f is an isomorphism.

Exercise 5.2. Let $f : V_1 \to V_2$ and $g : V_2 \to V_3$ be two linear maps between Euclidean vector spaces of the same dimension. Prove that

$$|\det(g \circ f)| = |\det g| \, |\det f|,$$

that is, the absolute value of the determinant is multiplicative with respect to the composition of maps.

In general, we can see that this absolute value of the Jacobian can be defined, in a completely analogous way, for differentiable maps between objects such that open subsets of Euclidean space and products of surfaces by intervals of $\mathbb{R}$ are still multiplicative with respect to the composition. Explicitly, if $\phi : O_1 \to O_2$ and $\psi : O_2 \to O_3$ are two differentiable maps between either open subsets of $\mathbb{R}^3$ or open subsets of products of surfaces and $\mathbb{R}$, we have, by the chain rule and Exercise 5.2 above, that

$$|\mathrm{Jac} \, (\psi \circ \phi)| = [|\mathrm{Jac} \, \psi| \circ \phi] |\mathrm{Jac} \, \phi|.$$

Moreover, by Exercise 5.1, such a map $\phi : O_1 \to O_2$ is a diffeomorphism if and only if $|\mathrm{Jac} \, \phi| > 0$.

We will take advantage now of the fact that, since O is an open, and so measurable, subset of $\mathbb{R}^3$, we know how to integrate functions on O, in order to define the integral of functions on the product $\mathcal{R} \times (a, b)$ with $a < b$. The idea is to work as if O and $\mathcal{R} \times (a, b)$ were both open subsets of $\mathbb{R}^3$ and to transfer integrals from each to the other by imagining that the formula for the change of variables is still valid for any diffeomorphism $\phi : O \to \mathcal{R} \times (a, b)$.

Definition 5.3 (Integral on $S \times \mathbb{R}$). Once we have fixed a diffeomorphism $\phi : O \to \mathcal{R} \times (a, b)$ defined on an open subset O of $\mathbb{R}^3$, we will say that a function $h : \mathcal{R} \times (a, b) \to \mathbb{R}$ is integrable on $\mathcal{R} \times (a, b)$ when $(h \circ \phi)|\mathrm{Jac} \, \phi|$ is integrable on O in the Lebesgue sense. In this case, the real number given by

$$(5.1) \qquad \int_O (h \circ \phi)|\mathrm{Jac} \, \phi|$$

will be called the integral of h on $\mathcal{R} \times (a, b)$, and it will be denoted by any of the symbols

$$\int_{\mathcal{R} \times (a,b)} h \quad \text{or} \quad \int_{\mathcal{R} \times (a,b)} h(p, t)\, dp\, dt.$$

First of all, let us check that this is a good definition in the sense that it depends on neither the open set O nor the diffeomorphism $\phi : O \to \mathcal{R} \times (a, b)$. We will decide later whether or not we can dispose of open sets and diffeomorphisms of this type. In fact, if O_1 and O_2 are two open subsets of $\mathbb{R}^3$ and $\phi_i : O_i \to \mathcal{R} \times (a, b)$, $i = 1, 2$, are two diffeomorphisms, then the composite map $\psi = \phi_2^{-1} \circ \phi_1 : O_1 \to O_2$ is a diffeomorphism between open subsets of $\mathbb{R}^3$ satisfying

$$|\mathrm{Jac}\, \phi_1| = |(\mathrm{Jac}\, \phi_2) \circ \psi||\mathrm{Jac}\, \psi|,$$

on O_1. Thus, if $h : \mathcal{R} \times (a, b) \to \mathbb{R}$ is a function such that $(h \circ \phi_2)|\mathrm{Jac}\, \phi_2|$ is integrable on O_2, the formula for the change of variables for the Lebesgue integral says us that

$$(h \circ \phi_1)|\mathrm{Jac}\, \phi_1| = [(h \circ \phi_2) \circ \psi][|(\mathrm{Jac}\, \phi_2)| \circ \psi]|\mathrm{Jac}\, \psi|$$

is integrable on O_1 and that

$$\int_{O_2} (h \circ \phi_2)|\mathrm{Jac}\, \phi_2| = \int_{O_1} [(h \circ \phi_2) \circ \psi][|(\mathrm{Jac}\, \phi_2)| \circ \psi]|\mathrm{Jac}\, \psi|.$$

That is, in this case, we have

$$\int_{O_2} (h \circ \phi_2)|\mathrm{Jac}\, \phi_2| = \int_{O_1} (h \circ \phi_1)|\mathrm{Jac}\, \phi_1|.$$

Second, to make sense out of the previous definition, we need to dispose, for each region $\mathcal{R}$ on the surface, of some diffeomorphism $\phi : O \to \mathcal{R} \times (a, b)$, where O is an open subset of three-space. But note that, indeed, we can take as O any tubular neighbourhood $N_\varepsilon(\mathcal{R})$ supplied by Theorem 4.26. In this way, $\phi : (I_\mathcal{R} \times h) \circ F^{-1} : O \to \mathcal{R} \times (a, b)$, where $h : (-\varepsilon, \varepsilon) \to (a, b)$ is any diffeomorphism, is the diffeomorphism that we are looking for.

From this definition, it is more or less immediate that the integral of functions on subsets of $S \times \mathbb{R}$ of the form $\mathcal{R} \times (a, b)$ keeps all the elementary properties—linearity, monotony, etc.—of Lebesgue integration. In the following result, we gather the most significant of them.

Proposition 5.4 (Properties of the integration on $S \times \mathbb{R}$). *Let $\mathcal{R}$ be a region of an orientable surface and (a, b) a non-empty open interval of $\mathbb{R}$. The following standard properties of the Lebesgue integral remain valid for the integration of functions $\mathcal{R} \times (a, b)$:*

    A: *Any continuous function on $\overline{\mathcal{R}} \times [a, b]$ is integrable on $\mathcal{R} \times (a, b)$.*

    B: *The integral of functions is linear and monotone.*

C: *There are convergence theorems, and theorems for the continuous and differentiable dependence of parameters, under the same conditions as for the Lebesgue integral on $\mathbb{R}^3$; to recall them, see Remark 5.22.*

D: *Let $\mathcal{R}' \subset \mathcal{R}$ be a subregion of $\mathcal{R}$ and suppose that the function h is integrable on $\mathcal{R} \times (a,b)$ and vanishes on $(\mathcal{R} - \mathcal{R}') \times (a,b)$. Then*

$$\int_{\mathcal{R}\times(a,b)} h = \int_{\mathcal{R}'\times(a,b)} h.$$

E: *If $\mathcal{R}'$ is a region of another orientable surface S', (a',b') is another open interval of $\mathbb{R}$, and $\psi : \mathcal{R} \times (a,b) \to \mathcal{R}' \times (a',b')$ is a diffeomorphism, then*

$$\int_{\mathcal{R}'\times(a',b')} h = \int_{\mathcal{R}\times(a,b)} (h \circ \psi)|\operatorname{Jac}\psi|,$$

for each integrable function h on $\mathcal{R}' \times (a',b')$.

Proof. The only assertion which perhaps is not a direct consequence of the definition is the last one. It follows easily from the fact, already proved, that the definition of the integral on $\mathcal{R} \times (a,b)$ depends on neither the open set O nor the diffeomorphism $\phi : O \to \mathcal{R} \times (a,b)$ and from the multiplicativity of the absolute value of the Jacobian. $\qquad\square$

5.3. Integrable functions and integration on surfaces

Consider now a function $f : \mathcal{R} \to \mathbb{R}$ defined on a region $\mathcal{R}$ of an orientable surface S. We may think of f as a function on the product $\mathcal{R} \times (a,b)$, for any real numbers $a < b$, as

$$f(p,t) = f(p), \qquad \forall p \in \mathcal{R} \subset S \quad \forall t \in (a,b).$$

For this class of functions, we have that

$$\frac{1}{b-a} \int_{\mathcal{R}\times(a,b)} f(p)\, dp\, dt$$

is independent of (a,b), provided that this integral exists. The reason is that, if we take another numbers $c < d$, we may define a diffeomorphism $\phi : S \times (a,b) \to S \times (c,d)$ by $\phi = \operatorname{id}_S \times h$, where $h : (a,b) \to (c,d)$ is an affine diffeomorphism. Now, one applies property **E** above. This enables us to establish the following definition.

Definition 5.5 (Integral on S). Let $\mathcal{R}$ be a region of an orientable surface and $f : \mathcal{R} \to \mathbb{R}$ a function. We will say that f is integrable on $\mathcal{R}$ when the function $(p, t) \in \mathcal{R} \times (0, 1) \mapsto f(p)$ is integrable on $\mathcal{R} \times (0, 1)$. The real number

(5.2)
$$\int_{\mathcal{R} \times (0,1)} f(p) \, dp \, dt$$

will be called the integral of f on $\mathcal{R}$ and it will be represented by any of the symbols

$$\int_{\mathcal{R}} f, \quad \int_{\mathcal{R}} f(p) \, dp, \quad \text{or} \quad \int_{\mathcal{R}} f \, dA.$$

The following properties are inherited, through definition (5.2), from the analogous ones of either the integral that we defined on $S \times \mathbb{R}$ in (5.1) or the Lebesgue integral on $\mathbb{R}^3$.

Proposition 5.6 (Properties of the integration on S). *Let $\mathcal{R}$ be a region of an orientable surface and $f, g : \mathcal{R} \to \mathbb{R}$ two functions defined on $\mathcal{R}$. The following assertions are true:*

A: *If f is continuous on $\overline{\mathcal{R}}$, then it is integrable on $\mathcal{R}$.*

B: *If f and g are integrable and $\lambda \in \mathbb{R}$, then $f + g$ and λf are integrable and moreover*

$$\int_{\mathcal{R}} f + g = \int_{\mathcal{R}} f + \int_{\mathcal{R}} g \quad \text{and} \quad \int_{\mathcal{R}} \lambda f = \lambda \int_{\mathcal{R}} f.$$

C: *If f and g are integrable and $f \leq g$, then*

$$\int_{\mathcal{R}} f \leq \int_{\mathcal{R}} g.$$

If, moreover, f and g are continuous, the equality holds if and only if $f = g$.

D: *Let $\mathcal{R}' \subset \mathcal{R}$ be a subregion of $\mathcal{R}$ and suppose that f vanishes on $\mathcal{R} - \mathcal{R}'$. Then*

$$\int_{\mathcal{R}} f = \int_{\mathcal{R}'} f.$$

E: *If $\mathcal{R}' \subset \mathcal{R}$ is a subregion of $\mathcal{R}$ such that $\mathcal{R} - \mathcal{R}'$ is a finite union of curves and points of S, then*

$$\int_{\mathcal{R}'} f = \int_{\mathcal{R}} f,$$

for each f integrable.

Proof. It is worthwhile to say something about the proof of **E** only. We have that $\mathcal{R}' \times (0, 1) \subset \mathcal{R} \times (0, 1)$ and that the difference of the two sets is a finite union of products of the form $C \times (0, 1)$ and $\{p\} \times (0, 1)$, where C is a curve of S and p is a point. But, since $\phi : O \to \mathcal{R} \times (0, 1)$ is a diffeomorphism, it is enough to realize that $\phi^{-1}(C \times (0, 1))$ is a surface and that $\phi^{-1}(\{p\} \times (0,1))$ is a curve, both in the open subset O of $\mathbb{R}^3$. Note now that both curves and surfaces have Lebesgue measure zero in $\mathbb{R}^3$, as seen in Exercise (11) of Chapter 4. $\qquad\square$

Remark 5.7 (Area of a region). Since the constant function 1 is continuous, properties **A** and **C** above assure us that the integral of the constant function 1 exists and that it is positive on each region $\mathcal{R}$ of an orientable S. This number will be called the *area* of the region $\mathcal{R}$. It will be denoted by $A(\mathcal{R})$ and so we may write down

$$A(\mathcal{R}) = \int_{\mathcal{R}} 1.$$

At the beginning of this chapter, we alluded to the standard way of presenting a theory of integration on surfaces. One bases it on a known theory for $\mathbb{R}^2$ and transfers it by means of parametrizations. The following statement relates this procedure to our definition of integral and, by the way, provides a useful manner for computing integrals when we dispose of an explicit parametrization which covers the surface *almost* entirely.

Proposition 5.8. *Let $\mathcal{R}$ be a region of an orientable surface and $X : U \to S$ a parametrization defined on an open subset $U \subset \mathbb{R}^2$. Suppose that $X(U) \subset \mathcal{R}$ and that the difference $\mathcal{R} - X(U)$ consists of finitely many points and curves. Then*

$$\int_{\mathcal{R}} f = \int_U (f \circ X)(u,v)|X_u \wedge X_v|(u,v) \, du \, dv,$$

for each integrable function $f : \mathcal{R} \to \mathbb{R}$.

Proof. Note that $X(U)$ is a region of S included in $\mathcal{R}$ and that, by part **E** of Proposition 5.6, we may put

$$\int_{\mathcal{R}} f = \int_{X(U)} f$$

for each function f integrable on $\mathcal{R}$. Definition (5.2) of the integral on the region $X(U)$ gives

$$\int_{X(U)} f = \int_{X(U) \times (0,1)} f(p) \, dp \, dt.$$

But, the map $\phi : U \times (0,1) \to X(U) \times (0,1)$ given by $\phi = X \times \mathrm{id}_{(0,1)}$, i.e., by

$$\phi(u,v,t) = (X(u,v), t), \qquad (u,v) \in U, \quad t \in (0,1),$$

is clearly a diffeomorphism defined on the open subset $U \times (0,1)$ of $\mathbb{R}^3$. Furthermore

$$\phi_u = (X_u, 0), \quad \phi_v = (X_v, 0), \quad \text{and} \quad \phi_t = (0,1)$$

and, hence, the corresponding absolute value of the Jacobian is

$$|\mathrm{Jac}\,\phi|(u,v,t) = |X_u \wedge X_v|(u,v), \qquad \forall (u,v) \in U \quad \forall t \in (0,1).$$

Thus, we may use this diffeomorphism ϕ in definition (5.1) and obtain

$$\int_{X(U)\times(0,1)} f(p)\,dp\,dt = \int_{U\times(0,1)} (f \circ \phi)|\mathrm{Jac}\,\phi| = \int_U (f \circ X)|X_u \wedge X_v|,$$

where we have applied the Fubini theorem for the Lebesgue integral to compute the integral on the product $U \times (0,1) \subset \mathbb{R}^3$. $\qquad\square$

Remark 5.9 (Measure zero sets in an orientable surface). Let S be an orientable surface and $\mathcal{R}$ a region of S. A set $A \subset \mathcal{R}$ will be said to have measure zero when

$$\int_{\mathcal{R}} \chi_A = 0,$$

where χ_A is the characteristic function of A in $\mathcal{R}$. For an arbitrary subset A of S, the notion of measure zero can be defined as follows: A has measure zero when $A \cap \mathcal{R}$ has measure zero for all regions $\mathcal{R}$ of S. From Proposition 5.8 above, we deduce, among other things, the consequence that this concept of measure zero coincides with what we introduced in Definition 4.31, based on parametrizations and on the Lebesgue theory on $\mathbb{R}^2$. Now, it is clear, from the definition we have just proposed, that, if a function is defined on a region $\mathcal{R}$ of S except maybe on a measure zero subset, we may question if it is integrable on $\mathcal{R}$ or not.

Example 5.10. Denote by S the torus of revolution of Example 2.17 whose minimum and maximum distances to the revolution axis are, respectively, $a - r$ and $a + r$, with $a > r > 0$. Then the map $X : (0,2\pi) \times (0,2\pi) \to \mathbb{R}^3$ given by

$$X(u,v) = ((a + r\cos u)\cos v, (a + r\cos u)\sin v, r\sin u)$$

is a parametrization—see Exercise (3) of Chapter 2—of S covering it entirely except for two circles with a common point. Then we may apply Proposition 5.8 to this specific parametrization to integrate any integrable function on the whole of the torus S. For instance, take the constant function 1. We have

$$A(S) = \int_S 1 = \int_0^{2\pi} \int_0^{2\pi} |X_u \wedge X_v|(u,v)\,du\,dv.$$

But

$$X_u(u,v) = (-r\sin u\cos v, -r\sin u\sin v, r\cos u),$$
$$X_v(u,v) = (-(a + r\cos u)\sin v, (a + r\cos u)\cos v, 0),$$

and consequently

$$|X_u \wedge X_v|^2 = |X_u|^2|X_v|^2 - \langle X_u, X_v\rangle^2 = r^2(a + r\cos u)^2.$$

Finally the area of the torus is

$$A(S) = \int_0^{2\pi} \int_0^{2\pi} r(a + r \cos u) \, du \, dv = 4\pi^2 ar.$$

Exercise 5.11. Calculate the integral of the Gauss curvature function on the torus of revolution of Example 5.10 above by using Proposition 5.8.

Exercise 5.12. Show that the map $X : (0, \pi) \times (0, 2\pi) \to \mathbb{R}^3$ defined by

$$X(u, v) = (a_1, a_2, a_3) + r(\sin u \cos v, \sin u \sin v, \cos u)$$

is a parametrization of the sphere $\mathbb{S}^2(r)$ with centre $a = (a_1, a_2, a_3) \in \mathbb{R}^3$ and radius $r > 0$. Use this, together with Proposition 5.8, to prove that the area of this sphere is $4\pi r^2$.

Now let $V_1, \ldots, V_k$ be a finite cover of the region $\mathcal{R}$ by open subsets of the orientable surface S. We may define on $\mathcal{R}$ an integer valued function $n : \mathcal{R} \to \mathbb{N}$ by taking each point $p \in \mathcal{R}$ into the number of open sets of the cover that p belongs to. Then $1 \leq n(p) \leq k$ for each $p \in \mathcal{R}$. Using this function, we define $\chi_1, \ldots, \chi_k$, as many real functions as the open sets of the cover have, all of them on $\mathcal{R}$, by

$$\chi_i(p) = \begin{cases} 0 & \text{if } p \notin V_i, \\ \dfrac{1}{n(p)} & \text{if } p \in V_i, \end{cases}$$

for all $i = 1, \ldots, k$ and all $p \in \mathcal{R}$. It is clear that, from this definition, the sum $\chi_1 + \cdots + \chi_k$ is the constant function 1 on $\mathcal{R}$. This family $\chi_1, \ldots, \chi_k$ of functions—which are not of course differentiable—is usually called a *partition of the unity* on $\mathcal{R}$ associated to the cover $V_1, \ldots, V_k$. We will use it to see that, as mentioned before, there is an alternative definition for the integral of the functions on surfaces, taking as a base Lebesgue theory on $\mathbb{R}^2$ and not on $\mathbb{R}^3$.

Proposition 5.13. *Let $\mathcal{R}$ be a region of an orientable surface S and let $f : \mathcal{R} \to \mathbb{R}$ be an integrable function on S. Then*

$$\int_{\mathcal{R}} f = \sum_{i=1}^k \int_{U_i} ((\chi_i f) \circ X^i) |X_u^i \wedge X_v^i|,$$

where each $X^i : U_i \to S$, $i = 1, \ldots, k$, is a parametrization of S defined on an open subset U_i of $\mathbb{R}^2$, such that $\mathcal{R} \subset X^1(U_1) \cup \cdots \cup X^k(U_k)$ and $\chi_1, \ldots, \chi_k$ is the partition of the unity on $\mathcal{R}$ associated to this cover.

Proof. Consider the partition of the unity $\chi_1, \ldots, \chi_k$ associated to the finite cover $X^1(U_1), \ldots, X^k(U_k)$ of the region $\mathcal{R}$. The properties listed in

Proposition 5.6 give us

$$\int_{\mathcal{R}} f = \int_{\mathcal{R}} \left(\sum_{i=1}^{k} \chi_i \right) f = \sum_{i=1}^{k} \int_{\mathcal{R}} \chi_i \, f = \sum_{i=1}^{k} \int_{X^i(U_i)} \chi_i \, f.$$

It suffices to now apply Proposition 5.8 to each of the summands.   □

5.4. Formula for the change of variables

One of the fundamental features of Lebesgue integration on Euclidean spaces is the formula for the change of variables. It enables us to compute integrals by passing from an integration domain to another one more convenient. In the context of surfaces we have an analogous result.

Theorem 5.14 (Change of variables formula). *Let $\phi : \mathcal{R}_1 \to \mathcal{R}_2$ be a diffeomorphism between two regions of two orientable surfaces, and let $f : \mathcal{R}_2 \to \mathbb{R}$ be an integrable function. Then the function $(f \circ \phi)|\mathrm{Jac}\,\phi|$ is integrable on $\mathcal{R}_1$ and moreover*

$$\int_{\mathcal{R}_2} f = \int_{\mathcal{R}_1} (f \circ \phi)|\mathrm{Jac}\,\phi|.$$

Proof. We can dispose of a diffeomorphism ψ between the products $\mathcal{R}_1 \times (0,1)$ and $\mathcal{R}_2 \times (0,1)$ by setting $\psi = \phi \times \mathrm{id}_{(0,1)}$. This diffeomorphism obviously satisfies

$$|\mathrm{Jac}\,\psi|(p,t) = |\mathrm{Jac}\,\phi|(p), \qquad \forall (p,t) \in \mathcal{R}_1 \times (0,1).$$

Using this equality, property **E** of the integration on $S \times \mathbb{R}$ listed in Proposition 5.4, and definition (5.1), we complete the proof.   □

Example 5.15 (Invariance under rigid motions). Let S be an orientable surface, $\mathcal{R}$ a region of S, and $\phi : \mathbb{R}^3 \to \mathbb{R}^3$ the rigid motion given by $\phi(x) = Ax + b$ with $A \in O(3)$ and $b \in \mathbb{R}^3$. We know that $\phi(S)$ is another orientable surface and that $\phi(\mathcal{R})$ is a region of it. Then, one can view $\phi : \mathcal{R} \to \phi(\mathcal{R})$ as a diffeomorphism between the two regions. If f is an integrable function on $\phi(\mathcal{R})$, one has, by the change of variables formula,

$$\int_{\phi(\mathcal{R})} f = \int_{\mathcal{R}} f \circ \phi,$$

because, if $p \in \mathcal{R}$ and $e_1, e_2 \in T_p S$ form an orthonormal basis,

$$|\mathrm{Jac}\,\phi|(p) = |(d\phi)_p(e_1) \wedge (d\phi)_p(e_2)| = |Ae_1 \wedge Ae_2| = |\det A| = 1.$$

In particular, if we choose $f \equiv 1$, we deduce that $A(\phi(\mathcal{R})) = A(\mathcal{R})$, that is, *the area of a compact surface is invariant under rigid motions.*

Example 5.16 (Odd functions on the sphere). Any rigid motion $\phi : \mathbb{R}^3 \to \mathbb{R}^3$ fixing the origin fixes the unit sphere $\mathbb{S}^2$ as well. Applying the invariance of the previous example to this particular situation, we have

$$\int_{\mathbb{S}^2} f = \int_{\mathbb{S}^2} f \circ \phi.$$

For example, if $\phi = -I_{\mathbb{R}^3}$ is the antipodal map, we obtain

$$\int_{\mathbb{S}^2} f(p)\,dp = \int_{\mathbb{S}^2} f(-p)\,dp$$

for each function f integrable on the sphere. Then, if f is an odd function, that is, $f(-p) = -f(p)$ for each $p \in \mathbb{S}^2$, one has

$$\int_{\mathbb{S}^2} f(p)\,dp = 0.$$

Relevant examples of odd functions on the sphere are the height functions $f(p) = \langle p, a \rangle$ with $a \in \mathbb{R}^3$, or, in general, any restriction to the sphere of an odd degree homogeneous polynomial.

Example 5.17 (Area of a parallel surface). We pointed out in Remark 4.29 that, if $t \neq 0$ is small enough, the map $F_t : S \to \mathbb{R}^3$, given by $F_t(p) = p + tN(p)$, where S is a compact surface and N is its Gauss map, is a diffeomorphism onto its image $S_t = F_t(S)$, the parallel surface at distance t. Moreover

$$(\mathrm{Jac}\, F_t)(p) = |(dF_t)_p(e_1) \wedge (dF_t)_p(e_2)| = 1 - 2tH(p) + t^2 K(p).$$

Applying the change of variables formula to the constant function 1 on S_t, we obtain the following expression for the area of the parallel surface at distance t:

$$A(S_t) = A(S) - 2t \int_S H(p)\,dp + t^2 \int_S K(p)\,dp,$$

where H and K are the mean and Gauss curvatures of S, respectively.

5.5. The Fubini theorem and other properties

We will carry on with the development of some other properties of the integral of functions on orientable surfaces, defined in Section 5.3. They will be used in the proofs of the global theorems that we will deal with in the text that follows. Now, in this section, we pretend to show the corresponding version of the very useful Fubini theorem of calculus. Later, we will recall some properties about continuous and differentiable dependence of parameters and show how to use them with some important examples.

Theorem 5.18 (Fubini's theorem). *Let $\mathcal{R}$ be a region of an orientable surface, $a < b$ two real numbers, and $h : \mathcal{R} \times (a, b) \to \mathbb{R}$ an integrable function on the product $\mathcal{R} \times (a, b)$. Then, for almost all $t \in (a, b)$, the function $p \in \mathcal{R} \mapsto h(p, t)$ is integrable on $\mathcal{R}$ and, for almost all $p \in \mathcal{R}$ the function $t \in (a, b) \mapsto h(p, t)$ is integrable on (a, b). Moreover, the functions*

$$p \in \mathcal{R} \mapsto \int_a^b h(p, t)\, dt \quad and \quad t \in (a, b) \mapsto \int_S h(p, t)\, dp$$

are integrable—see Remark 5.9—on $\mathcal{R}$ and on (a, b), respectively. Finally

$$\int_{\mathcal{R} \times (a,b)} h(p, t)\, dp\, dt = \int_{\mathcal{R}} \left(\int_a^b h(p, t)\, dt \right) dp = \int_a^b \left(\int_{\mathcal{R}} h(p, t)\, dp \right) dt.$$

Proof. Since $\mathcal{R}$ is relatively compact, we may cover it by a finite number $X^1(U_1), \ldots, X^k(U_k)$ of images of parametrizations defined on open subsets U_i, with $i = 1, \ldots, k$, of $\mathbb{R}^2$. Let $\{\chi_1, \ldots, \chi_k\}$ be the corresponding partition of the unity. Then each product $h\chi_i$ is integrable on $\mathcal{R} \times (a, b)$ and

$$\int_{\mathcal{R} \times (a,b)} h(p, t)\, dp\, dt = \sum_{i=1}^k \int_{\mathcal{R} \times (a,b)} h(p, t) \chi_i(p)\, dp\, dt.$$

By property **D** in Proposition 5.4 of the integration on $S \times \mathbb{R}$, one obtains

$$\int_{\mathcal{R} \times (a,b)} h(p, t)\, dp\, dt = \sum_{i=1}^k \int_{X^i(U_i) \times (a,b)} h(p, t) \chi_i(p)\, dp\, dt.$$

Reasoning as in the proof of Proposition 5.8, we see that each product $X^i \times \mathrm{id}_{(a,b)} : U_i \times (a, b) \to X^i(U_i) \times (a, b)$ is a diffeomorphism with absolute value of the Jacobian equal to $|X_u^i \wedge X_v^i|$, for $i = 1, \ldots, k$. Hence, we may use it in definition (5.1), and we have

$$(5.3) \qquad \int_{\mathcal{R} \times (a,b)} h = \sum_{i=1}^k \int_{U_i \times (a,b)} (\chi_i \circ X^i)(h \circ (X^i \times \mathrm{id}_{(a,b)})) |X_u^i \wedge X_v^i|.$$

Now, we apply the Fubini theorem of calculus on each integral on the right-hand side. This theorem assures us that the integrands of the right-hand side are integrable with respect to each pair of variables for almost every fixed value of the third variable. Furthermore

$$\int_{U_i \times (a,b)} (\chi_i \circ X^i)(h \circ (X^i \times \mathrm{id}_{(a,b)})) |X_u^i \wedge X_v^i|$$

$$= \int_{U_i} (\chi_i \circ X^i) |X_u^i \wedge X_v^i| \left(\int_a^b h(X^i, t)\, dt \right) du\, dv.$$

This expression, together with (5.3) and Proposition 5.8, gives us the first required equality. Instead, if we apply the Fubini theorem to each summand of (5.3) in the second possible way, we have

$$\int_{U_i \times (a,b)} (\chi_i \circ X^i)(h \circ (X^i \times \mathrm{id}_{(a,b)}))|X_u^i \wedge X_v^i|$$

$$= \int_a^b \left(\int_{U_i} (\chi_i \circ X^i) h(X^i, t)|X_u^i \wedge X_v^i| \, du \, dv \right) dt.$$

Using Proposition 5.8 again for the expression inside of the parentheses and substituting in (5.3), we get the second equality that we were looking for. $\square$

Example 5.19 (Volume enclosed by a parallel surface). Let S be a compact surface and let $\varepsilon > 0$ be such that $N_\varepsilon(S)$ is a tubular neighbourhood. In Remark 4.29, we introduced parallel surfaces S_t at distance $t \in (0, \varepsilon)$. If Ω_t is the inner domain determined by S_t, we have, taking Remark 4.28 into account, that

$$\mathrm{vol}\,\Omega - \mathrm{vol}\,\Omega_t = \mathrm{vol}\,V_t(S) = \int_{V_t(S)} 1.$$

But the map $F : S \times (0, \varepsilon) \to V_\varepsilon(S)$, given by $F(p, t) = p + tN(p)$, is a diffeomorphism. Therefore, if we use it in definition (5.1) of the integral on $S \times \mathbb{R}$, we obtain

$$\mathrm{vol}\,\Omega - \mathrm{vol}\,\Omega_t = \int_{S \times (0,t)} |\mathrm{Jac}\,F|.$$

But, in (4.2), we computed the differential of the map F. It follows from this that, if $(p, t) \in S \times (0, \varepsilon)$, then

$$|\mathrm{Jac}\,F|(p, t) = \det((dF)_{(p,t)}(e_1, 0), (dF)_{(p,t)}(e_2, 0), (dF)_{(p,t)}(0, 1))$$

$$= 1 - 2tH(p) + t^2 K(p),$$

using, for example, an orthonormal basis of the principal directions at the point p of S. Hence, the Fubini Theorem 5.18 above implies

$$\mathrm{vol}\,\Omega - \mathrm{vol}\,\Omega_t = \int_0^t \left(\int_S (1 - 2tH(p) + t^2 K(p)) \, dp \right) dt$$

$$= tA(S) - t^2 \int_S H + \frac{t^3}{3} \int_S K.$$

Example 5.20 (Integration in polar coordinates). Assume now that the compact surface being considered is the unit sphere $\mathbb{S}^2$. We know that the punctured unit ball $B^3 - \{0\}$ is diffeomorphic to the product $\mathbb{S}^2 \times (0, 1)$, because the map $F : \mathbb{S}^2 \times (0, 1) \to B^3 - \{0\}$ given by

$$F(p, t) = p + tN(p) = (1 - t)p$$

is a diffeomorphism, whose absolute value of the Jacobian was computed in Example 5.19 above and, in this particular case, is

$$|\mathrm{Jac}\, F|(p,t) = 1 - 2tH(p) + t^2 K(p) = (1 - t)^2,$$

because the (inner) mean curvature and the Gauss curvature of the unit sphere are both constant functions 1. Then, by definition (5.1), we have

$$\int_{B^3} f = \int_{B^3 - \{0\}} f = \int_{\mathbb{S}^2 \times (0,1)} f((1 - t)p)(1 - t)^2 \, dp \, dt.$$

Now using the Fubini Theorem 5.18, we obtain

$$\int_{B^3} f = \int_0^1 \rho^2 \left(\int_{\mathbb{S}^2} f(\rho p) \, dp \right) d\rho.$$

This is the so-called *polar coordinates integration formula* for functions defined on the unit disc.

We will complete this section by gathering some results relative to convergence and dependence of parameters. Our integration inherits them directly from Lebesgue theory. Finally we will give an example to show how it can be useful in the study of the interactions between the geometry and the analysis of surfaces.

Remark 5.21 (Dominated and monotone convergence theorems). The main properties of Lebesgue integration pass without difficulty to the integration of functions on surfaces. For instance, the following is true.

> *Let $\mathcal{R}$ be a region of an orientable surface and $f_n : \mathcal{R} \to \mathbb{R}$,*
> *$n \in \mathbb{N}$, a sequence of integrable functions on $\mathcal{R}$ such that*
> *$\{f_n\}_{n \in \mathbb{N}}$ converges pointwise to a function f and such that*
> *$|f_n| \leq g$ for some function g integrable on $\mathcal{R}$. Then f is*
> *integrable and*
> $$\lim_{n \to \infty} \int_{\mathcal{R}} f_n = \int_{\mathcal{R}} f.$$

In fact, this is the *dominated convergence theorem*, which can be obtained by using definitions (5.1) and (5.2) from the analogue in the Lebesgue theory on $\mathbb{R}^3$. Similarly, one can get the *monotone convergence theorem*, which says:

> *Let $\mathcal{R}$ be a region of an orientable surface and $f_n : \mathcal{R} \to \mathbb{R}$,*
> *$n \in \mathbb{N}$, a monotone sequence of integrable functions on $\mathcal{R}$*
> *which converges pointwise to a function f such that*
> $$\text{there exists } \lim_{n \to \infty} \int_{\mathcal{R}} f_n.$$

Then the function f is integrable on $\mathcal{R}$ and

$$\lim_{n\to\infty} \int_{\mathcal{R}} f_n = \int_{\mathcal{R}} f.$$

Remark 5.22 (Continuous and differentiable dependence of parameters). We will infer from the dominated convergence theorem, although it could be done directly from the definitions as in the preceding remark, some easy versions, adequate for our needs, of the theorems of continuous and differentiable dependence of integrals whose integrands depend on a real parameter.

> *Let $\mathcal{R}$ be a region of an orientable surface, I an interval of $\mathbb{R}$, and $f : \overline{\mathcal{R}} \times I \to \mathbb{R}$ a continuous function. Taking into account part $\mathbf{E}$ of Proposition 5.4, we define a new function $h : I \to \mathbb{R}$ by*
>
> $$h(\lambda) = \int_{\mathcal{R}} f(p, \lambda)\, dp, \qquad \forall \lambda \in I.$$
>
> *Then h is continuous.*

Let us prove this assertion. If $\lambda \in I$ and $\{\lambda_n\}_{n\in\mathbb{N}}$ is a sequence in I converging to λ, we may imagine that both the sequence and its limits belong to a compact interval $J \subset I$. Since f is continuous, the sequence $f(-, \lambda_n)$ of functions defined on $\mathcal{R}$ converges pointwise to $f(-, \lambda)$. Moreover

$$|f(p, \lambda_n)| \le \max_{(q,\mu)\in\overline{\mathcal{R}}\times J} |f(q, \mu)|.$$

Using Remark 5.21,

$$\lim_{n\to\infty} h(\lambda_n) = \lim_{n\to\infty} \int_{\mathcal{R}} f(p, \lambda_n)\, dp = \int_{\mathcal{R}} f(p, \lambda)\, dp = h(\lambda).$$

Thus, h is continuous at the point λ. This same continuous dependence of parameters can be used to obtain results relative to the derivation of an integral when the integrand depends differentiably of a parameter. For example, we have the following.

> *Let $f : \overline{\mathcal{R}} \times I \to \mathbb{R}$ be a continuous function, where $\mathcal{R}$ is a region of an orientable surface and I is an open interval of $\mathbb{R}$. Suppose that, for each $p \in \overline{\mathcal{R}}$, the function*
>
> $$\lambda \in I \longmapsto f(p, \lambda)$$
>
> *is differentiable and represent its derivative by $(\partial f/\partial\lambda)(p, \lambda)$. If the function*
>
> $$\frac{\partial f}{\partial \lambda} : \overline{\mathcal{R}} \times I \longrightarrow \mathbb{R}$$
>
> *is continuous, then the function $h : I \to \mathbb{R}$ given by*
>
> $$h(\lambda) = \int_{\mathcal{R}} f(p, \lambda)\, dp$$

is of class C^1 and its derivative is

$$h'(\lambda) = \int_{\mathcal{R}} \frac{\partial f}{\partial \lambda}(p, \lambda) \, dp.$$

Let us check that, in fact, this statement is true. For $\lambda \in I$ we choose a compact interval $J \subset I$ and a sequence $\{\lambda_n\}_{n \in \mathbb{R}}$ contained in J converging to λ. Then

$$\frac{h(\lambda_n) - h(\lambda)}{\lambda_n - \lambda} = \int_{\mathcal{R}} \frac{f(p, \lambda_n) - f(p, \lambda)}{\lambda_n - \lambda} \, dp.$$

Since f is differentiable with respect to the λ-variable, the integrand on the right-hand side converges pointwise to $(\partial f / \partial \lambda)(p, \lambda)$. Furthermore

$$\left| \frac{f(p, \lambda_n) - f(p, \lambda)}{\lambda_n - \lambda} \right| = \left| \frac{\partial f}{\partial \lambda}(p, \theta_n) \right| \leq \max_{\mathcal{R} \times J} \left| \frac{\partial f}{\partial \lambda} \right|,$$

where we have applied the Cauchy mean value theorem—θ_n represents a point between λ and λ_n—and the continuity of $\partial f / \partial \lambda$. By the dominated convergence theorem established in Remark 5.21, it follows that

$$\lim_{n \to \infty} \frac{h(\lambda_n) - h(\lambda)}{\lambda_n - \lambda} = \int_{\mathcal{R}} \frac{\partial f}{\partial \lambda}(p, \lambda) \, dp,$$

and the integral on the right-hand side is continuous in λ by the continuous dependence of parameters that we have just proved.

Remark 5.23 (An application to the sphere). Let $O \subset \mathbb{R}^3$ be an open subset of Euclidean space containing the origin and $f : O \to \mathbb{R}$ a differentiable function. For $r > 0$ small enough, the closed ball centred at 0 with radius r is included in O and, so, $\mathbb{S}^2(r) \subset O$ as well. Thus, we may define a function ϕ by

$$r \longmapsto \phi(r) = \frac{1}{4\pi r^2} \int_{\mathbb{S}^2(r)} f(p) \, dp,$$

for $r > 0$ near zero. Let r and r_0 be two positive small numbers. The dilatation with centre the origin and ratio r/r_0 provides a diffeomorphism between $\mathbb{S}^2(r_0)$ and $\mathbb{S}^2(r)$. Then, applying the change of variables formula of Theorem 5.14, we have

$$\int_{\mathbb{S}^2(r)} f(p) \, dp = \frac{r^2}{r_0^2} \int_{\mathbb{S}^2(r_0)} f(\frac{r}{r_0}p) \, dp.$$

Therefore

$$\phi(r) = \frac{1}{4\pi r_0^2} \int_{\mathbb{S}^2(r_0)} f(\frac{r}{r_0}p) \, dp,$$

which shows, by the dependence theorems, that ϕ is a continuous and derivable function and that it extends to zero. Applying also the continuous dependence result and taking into account that the area of $\mathbb{S}^2(r_0)$ is $4\pi r_0^2$, one has

$$\lim_{r \to 0} \frac{1}{4\pi r^2} \int_{\mathbb{S}^2(r)} f(p) \, dp = \phi(0) = f(0).$$

From the same expression we deduce

$$\phi'(r) = \frac{1}{4\pi r_0^2} \int_{\mathbb{S}^2(r_0)} \langle (\nabla f)(\frac{r}{r_0}p), \frac{p}{r_0} \rangle \, dp.$$

From this, the differentiable dependence theorem assures us that ϕ' is also derivable. Moreover

$$\phi'(0) = \frac{1}{4\pi r_0^2} \int_{\mathbb{S}^2(r_0)} \langle (\nabla f)(0), \frac{p}{r_0} \rangle \, dp = 0,$$

where ∇ is the gradient of functions on $\mathbb{R}^3$—see Exercise (5.36) at the end of this chapter—and where the last equality comes from, for instance, Example 5.16. On the other hand, we may define a field X on an open neighbourhood of 0 by the expression $X(p) = (\nabla f)(\frac{r}{r_0}p)$ which is clearly differentiable. Moreover $\operatorname{div} X(p) = \frac{r}{r_0}(\Delta f)(\frac{r}{r_0}p)$. Using the divergence Theorem 5.31 on the ball $B^3(r_0)$, which will be proved later, we obtain

$$\frac{\phi'(r)}{r} = \frac{1}{4\pi r_0^3} \int_{B^2(r_0)} (\Delta f)(\frac{r}{r_0}p) \, dp.$$

Taking the limit as r goes to zero, by the continuous dependence of parameters,

$$\phi''(0) = \frac{1}{4\pi r_0^3}(\Delta f)(0)\operatorname{vol} B^2(r_0) = \frac{(\Delta f)(0)}{3}.$$

Hence, applying the l'Hôpital rule,

$$\lim_{r \to 0} \frac{1}{r^2} \left(\frac{1}{4\pi r^2} \int_{\mathbb{S}^2(r)} f(p) \, dp - f(0) \right) = \frac{(\Delta f)(0)}{6}.$$

With the two limits that we have just computed as an example of the usefulness of the dependence theorems of the integration on surfaces, we can solve Exercise (12) at the end of this chapter.

5.6. Area formula

We will prove, in the following section, the divergence theorem. A lot of different proofs of this result can be found in the literature. We will base our text on a particular case of the so-called *area formula*, an integral formula which generalizes the change of variables formula for the Lebesgue integral, when the involved transformation is no longer a diffeomorphism, but only a differentiable map. This area formula is of interest itself and will enable us to obtain more interesting consequences in the text that follows.

Consider a relatively compact open subset O of $\mathbb{R}^3$, a differentiable map $\phi : O \to \mathbb{R}^3$, and a function $f : O \to \mathbb{R}$. If $x \in \mathbb{R}^3$ is a regular value of the map ϕ—with definition similar to Definition 4.5—the set of the inverse images through ϕ of the point x is discrete, as a direct consequence of the inverse function theorem of calculus. Then, since O is relatively compact, this is a finite subset of O. This is why we may define the following function

$$n(\phi, f) : \mathbb{R}^3 - \phi(\mathcal{N}) \longrightarrow \mathbb{R},$$

where $\mathcal{N} \subset O$ is the set of the points of O which are critical for ϕ, that is,

$$\mathcal{N} = \{x \in O \,|\, (\mathrm{Jac}\,\phi)(x) = 0\},$$

by the expression

$$n(\phi, f)(x) = \sum_{p \in \phi^{-1}(x)} f(p),$$

with the convention that the sum is zero when $x \notin \phi(O)$. The three-dimensional version of the Sard Theorem 4.33 tells us that $\phi(\mathcal{N})$ is a measure zero subset of $\mathbb{R}^3$, and thus the function $n(\phi, f)$ is defined almost everywhere on $\mathbb{R}^3$, that is, except on a measure zero set. Note that, if ϕ were a diffeomorphism onto its image, then $n(\phi, f)$ would be defined on the whole of $\mathbb{R}^3$ and would coincide with the composite map $f \circ \phi^{-1}$ on $\phi(O)$. For this reason, the following result can be thought of as a generalization of the change of variables formula for the Lebesgue integral on $\mathbb{R}^3$.

Theorem 5.24 (Area formula). *Let* $\phi : O \to \mathbb{R}^3$ *be a differentiable map, where O is a relatively compact open subset of $\mathbb{R}^3$ and $f : O \to \mathbb{R}$ a function on O. Suppose that the product $f|\mathrm{Jac}\,\phi|$ is integrable on O. Then the function $n(\phi, f)$ is integrable on $\mathbb{R}^3$ and*

$$\int_{\mathbb{R}^3} n(\phi, f) = \int_O f(x)|\mathrm{Jac}\,\phi|(x)\,dx.$$

Proof. Let us first examine the case where the map ϕ extends differentiably to $\overline{O}$. Let V be a relatively compact open subset of $\mathbb{R}^3$ including $\overline{O}$ such that ϕ is defined on $\overline{V}$ and choose another open subset W of $\mathbb{R}^3$ such that $\overline{O} \subset W \subset \overline{W} \subset V$. Suppose first that all the points of W are not critical for ϕ. Then, for all $p \in W$ there exists, by the inverse function theorem, an open neighbourhood V^p of p in V such that $\phi_{|V^p}$ is a diffeomorphism of V^p onto its image $\phi(V^p)$. Thus, the family $\{V^p\}$ with $p \in W$ is a cover by open sets of the compact $\overline{O}$. Let $\delta > 0$ be the Lebesgue number of this cover and let $P = \{C_1, \ldots, C_n\}$ be a partition by cubes of the compact $\overline{O}$—that is, $\overline{O} \subset \bigcup_{i=1}^n C_i$ and the interiors of the C_i are disjoint—with norm smaller than δ and all of them included in W. Each of these cubes is contained in some open set of the cover. Consequently, $\phi_{|C_i}$ is a diffeomorphism onto its image for each $i = 1, \ldots, n$. On the other hand, since $f|\mathrm{Jac}\,\phi|$ is integrable on O, it is also integrable on each $C_i \cap O$. Therefore, the formula for the change of variables for the Lebesgue integral implies that $f \circ \phi^{-1}$ is integrable on $\phi(C_i \cap O)$ and gives

$$\int_{\phi(C_i \cap O)} f \circ \phi^{-1} = \int_{C_i \cap O} f(x)|\mathrm{Jac}\,\phi|(x)\,dx$$

for all $i = 1, \ldots, n$. Adding these n equalities, we obtain the integral equality of the theorem because $n(\phi, f)$ vanishes out of $\phi(O)$ and, in this particular case, $n(\phi, f) = f \circ \phi^{-1}$.

The case where there are critical points of ϕ can be deduced from the previous case as follows. Consider the set

$$\mathcal{N} = \{p \in V \mid (\mathrm{Jac}\,\phi)(p) = 0\}.$$

Then $K = \mathcal{N} \cap \overline{O}$ is a compact of $\mathcal{N}$. For each $k \in \mathbb{N}$, we take an open set O_k such that

$$K \subset O_k, \quad \mathrm{vol}\,O_k = \mathrm{vol}\,K + \frac{1}{k}, \quad \text{and } O_k \text{ is decreasing.}$$

By Proposition 4.34, there exists another open set V_k such that

$$K \subset V_k \subset \overline{V_k} \subset O_k \quad \text{and} \quad \mathrm{vol}\,\phi(V_k) < \frac{1}{k}.$$

Then, the open set $O - \overline{O_k}$ obeys the conditions of the above particular case, since $W = O - \overline{V_k}$ is an open set containing its adherence and consisting of regular points for ϕ. Hence

$$\int_{\mathbb{R}^3} n(\phi_{|O-\overline{O_k}}, f_{|O-\overline{O_k}}) = \int_{O-\overline{O_k}} f\,|\mathrm{Jac}\,\phi|.$$

The integral on the right-hand side converges to

$$\int_{O-K} f\,|\mathrm{Jac}\,\phi|$$

as $k \to \infty$ because f is integrable on O, $|\mathrm{Jac}\,\phi|$ is bounded on O, and the limit of $\mathrm{vol}\,(\overline{O_k} - K)$ is zero as $k \to \infty$. As a consequence we have that

$$\text{there exists } \lim_{k \to \infty} \int_{\mathbb{R}^3} n(\phi_{|O-\overline{O_k}}, f_{|O-\overline{O_k}}) = \int_{O-K} f\,|\mathrm{Jac}\,\phi|.$$

On the other hand, from the definition of $n(\phi, f)$ and the fact that each regular value of ϕ has finitely many preimages in O, since ϕ is differentiable on the compact $\overline{O}$, we deduce that, for each $x \in \mathbb{R}^3 - \phi(\mathcal{N})$, we can find a $k_x \in \mathbb{N}$ such that

$$n(\phi_{|O-\overline{O_k}}, f_{|O-\overline{O_k}})(x) = n(\phi, f)(x), \qquad k \geq k_x.$$

Therefore

$$n(\phi_{|O-\overline{O_k}}, f_{|O-\overline{O_k}}) \longrightarrow n(\phi, f)$$

pointwise as $k \to \infty$ and moreover, when $f \geq 0$, this convergence is monotone. So, in the case where f is non-negative, by the monotone convergence theorem, we infer that $n(\phi, f)$ is integrable on $\mathbb{R}^3$ and furthermore

$$\int_{\mathbb{R}^3} n(\phi, f) = \int_{O-K} f \, |\mathrm{Jac} \, \phi|.$$

In the case where we do not know anything about the sign of f, since

$$|n(\phi_{|O-\overline{O_k}}, f_{|O-\overline{O_k}})| \leq n(\phi_{|O-\overline{O_k}}, |f|_{|O-\overline{O_k}}) \leq n(\phi, |f|),$$

the dominated convergence theorem gives us the same equality. Now it is enough to realize that

$$\int_{O-K} f \, |\mathrm{Jac} \, \phi| = \int_{O} f \, |\mathrm{Jac} \, \phi|$$

because $|\mathrm{Jac} \, \phi| = 0$ on K.

To prove the general case, it suffices to take a sequence $\{O_k\}_{k \in \mathbb{N}}$ of relatively compact open subsets of O with $\lim_{k \to \infty} \mathrm{vol} \, O_k = \mathrm{vol} \, O$, to apply the above to each term of the sequence, and to check that it is possible to use an adequate convergence theorem. $\square$

Exercise 5.25. ↑ Let $\mathcal{R}$ be a region of an orientable surface S and two real numbers $a < b$, and let $\phi : \mathcal{R} \times (a, b) \to \mathbb{R}^3$ be a differentiable map. If we define

$$\mathcal{N} = \{(p, t) \in \mathcal{R} \times (a, b) \, | \, |\mathrm{Jac} \, \phi|(p, t) = 0\},$$

show that $\phi(\mathcal{N})$ has Lebesgue measure zero in $\mathbb{R}^3$.

Exercise 5.26. ↑ Let S be a compact surface and P a plane of $\mathbb{R}^3$ perpendicular to a unit vector $a \in \mathbb{S}^2$. Prove that almost all the straight lines perpendicular to P intersect S transversely at an even number of points in such a way that the sign of $\langle N, a \rangle$ changes alternatively when we move along the line in a given direction.

The first exercise above allows us to obtain quickly the corresponding versions of the area formula for differentiable maps of $\mathcal{R} \times (a, b)$ to $\mathbb{R}^3$, where $\mathcal{R}$ is a region of an orientable surface, and for differentiable maps between regions of orientable surfaces.

Theorem 5.27 (Area formula for products). *Let $\phi : \mathcal{R} \times (a, b) \to \mathbb{R}^3$ be a differentiable map, where $\mathcal{R}$ is a region of an orientable surface and $a < b$, and let f be a function on $\mathcal{R} \times (a, b)$ such that $f|Jac\,\phi|$ is integrable on $\mathcal{R} \times (a, b)$. Then the function $n(\phi, f)$ given by*

$$n(\phi, f)(x) = \sum_{(p,t) \in \phi^{-1}(x)} f(p, t)$$

is well defined for each $x \in \mathbb{R}^3$ except on a measure zero set, is integrable on $\mathbb{R}^3$, and moreover

$$\int_{\mathbb{R}^3} n(\phi, f) = \int_{\mathcal{R} \times (a,b)} f(p,t) |\mathrm{Jac}\, \phi|(p,t) \, dp \, dt.$$

Proof. The function $n(\phi, f)$ is well defined except on a measure zero subset of $\mathbb{R}^3$ as a consequence of Exercise 5.25 above. To prove the area formula in this case, it suffices to pick a diffeomorphism $\psi : O \to \mathcal{R} \times (a, b)$, where O is a relatively compact open subset of $\mathbb{R}^3$, and to apply Theorem 5.24 to the map $\phi \circ \psi$ and the function $f \circ \psi$. Then

$$\int_O (f \circ \psi) |\mathrm{Jac}\, (\phi \circ \psi)| = \int_{\mathbb{R}^3} n(\phi \circ \psi, f \circ \psi).$$

Now it remains to see that $n(\phi \circ \psi, f \circ \psi) = n(\phi, f)$ and that, by definition (5.1), we have

$$\int_O (f \circ \psi) |\mathrm{Jac}\, (\phi \circ \psi)| = \int_{\mathcal{R} \times (a,b)} f |\mathrm{Jac}\, \phi|.$$

$\square$

On the other hand, there is also an area formula for maps between surfaces. We may give the following version.

Theorem 5.28 (Area formula for surfaces). *Let $\phi : \mathcal{R}_1 \to \mathcal{R}_2$ be a differentiable between regions of two orientable surfaces, and let f be a function on $\mathcal{R}_1$ such that the product $f|\mathrm{Jac}\, \phi|$ is integrable on $\mathcal{R}_1$. Then the function $n(\phi, f)$ given by*

$$n(\phi, f)(q) = \sum_{p \in \phi^{-1}(q)} f(p)$$

is well-defined for each $q \in \mathcal{R}_2$ except on a measure zero set and is integrable on $\mathcal{R}_2$. Furthermore

$$\int_{\mathcal{R}_1} f(p) |\mathrm{Jac}\, \phi|(p) \, dp = \int_{\mathcal{R}_2} n(\phi, f).$$

Proof. To prove this version, recall that, by definition (5.2), one has

$$\int_{\mathcal{R}_1} f |\mathrm{Jac}\, \phi| = \int_{\mathcal{R}_1 \times (0,1)} f(p) |\mathrm{Jac}\, \phi|(p) \, dp \, dt.$$

But, if $p_1 : \mathcal{R}_1 \times (a, b) \to \mathcal{R}_1$ represents the projection onto the first factor, the integral on the right-hand side can be rewritten and so

$$\int_{\mathcal{R}_1} f |\mathrm{Jac}\, \phi| = \int_{\mathcal{R}_1 \times (0,1)} (f \circ p_1) |\mathrm{Jac}\, (\phi \times \mathrm{id}_{(0,1)})|.$$

Now we choose a diffeomorphism $\psi : O \to \mathcal{R}_2 \times (0, 1)$, where O is a relatively compact open subset of $\mathbb{R}^3$. If we denote by $\xi : \mathcal{R}_1 \times (0, 1) \to O$ the composition $\psi^{-1} \circ (\phi \times \mathrm{id}_{(0,1)})$ and use the multiplicativity of the absolute value of the Jacobian, the integral on the right-hand side becomes

$$\int_{\mathcal{R}_1 \times (0,1)} (f \circ p_1) \left(|\mathrm{Jac}\,\psi| \circ \xi \right) |\mathrm{Jac}\,\xi|.$$

Applying the area formula Theorem 5.27 to the map ξ and the function $(f \circ p_1) |\mathrm{Jac}\,\psi| \circ \xi$, we see that

$$\int_{\mathcal{R}_1} f |\mathrm{Jac}\,\phi| = \int_{\mathbb{R}^3} n(\xi, (f \circ p_1) |\mathrm{Jac}\,\psi| \circ \xi).$$

But now it is easy to check that $n(\xi, (f \circ p_1) |\mathrm{Jac}\,\psi| \circ \xi) = n(\xi, f \circ p_1) |\mathrm{Jac}\,\psi|$. Bearing this in mind together with definition (5.1), we arrive at

$$\int_{\mathcal{R}_1} f |\mathrm{Jac}\,\phi| = \int_{\mathcal{R}_2 \times (0,1)} n(\xi, f \circ p_1) \circ \psi^{-1}.$$

The proof can be finished by noting that

$$[n(\xi, f \circ p_1) \circ \psi^{-1}](q, t) = n(\phi, f)(q)$$

for each $q \in \mathcal{R}_2$ and again using definition (5.2) of the integral on a surface. $\square$

An immediate application of these area formulas is the following three-dimensional version of the so-named Chern-Lashoff theorem.

Theorem 5.29 (Chern-Lashoff's theorem). *Let S be a compact surface. Then the following hold.*

> **A:** $\displaystyle\int_S K^+ \geq 4\pi$ *where* $K^+ = \max\{0, K\}$.

> **B:** $\displaystyle\int_S |K| \geq 4\pi$ *and, if equality is attained, then S has non-negative Gauss curvature everywhere.*

Proof. The proof consists simply in considering the compact set $A \subset S$ formed by the points of S with non-negative Gauss curvature and applying the area formula Theorem 5.28 for maps between surfaces to the Gauss map $N : S \to \mathbb{S}^2$ and the function χ_A characteristic of the set A. Then we have

$$\int_S \chi_A |\mathrm{Jac}\,N| = \int_{\mathbb{S}^2} n(N, \chi_A).$$

Now one must check that $\chi_A |\mathrm{Jac}\,N| = K^+ = \max\{0, K\}$ and that, since N is surjective even when restricted to A, as Exercise (5) of Chapter 4 showed, the function $n(N, \chi_A)$ is always greater than or equal to 1 on $\mathbb{S}^2$. This should prove the first inequality. The second inequality and the case of equality follow from the fact that $|K| \geq K^+$. $\square$

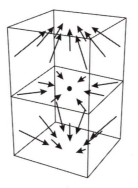

Figure 5.1. *Vector field*

5.7. The divergence theorem

The divergence theorem is also known as the Gauss theorem, the Green theorem, the Ostrogradski theorem, and even the Stokes theorem. In fact, all these mathematicians either used this result in a fundamental way or contributed to stating and to proving it. It is one of the most powerful analytical tools. From the physical point of view, it might appear, in some sense, to be a trivial assertion, because it establishes that the variation of the volume of the inner domain determined by a compact connected surface, when it is deformed through the flow of a vector field, coincides exactly with the total flow of the field crossing along the boundary surface.

Definition 5.30 (Vector field and divergence of a field). If A is a subset of $\mathbb{R}^3$, a—differentiable—vector field on A is a differentiable vector valued function $X : A \to \mathbb{R}^3$. We define the divergence of the field X as the function $\mathrm{div}\, X : A \to \mathbb{R}$ given by $\mathrm{div}\, X(p) = \mathrm{trace}\,(dX)_p$ for each $p \in A$.

Therefore, if $\{e_1, e_2, e_3\}$ is any orthonormal basis of the space $\mathbb{R}^3$, we have $\mathrm{div}\, X(p) = \sum_{i=1}^{3} \langle (dX)_p(e_i), e_i \rangle$. If we make the particular choice $e_1 = (1,0,0)$, $e_2 = (0,1,0)$, and $e_3 = (0,0,1)$, then

$$(dX)_p(e_1) = \left(\frac{\partial X^1}{\partial x}(p), \frac{\partial X^2}{\partial x}(p), \frac{\partial X^3}{\partial x}(p) \right),$$
$$(dX)_p(e_2) = \left(\frac{\partial X^1}{\partial y}(p), \frac{\partial X^2}{\partial y}(p), \frac{\partial X^3}{\partial y}(p) \right),$$
$$(dX)_p(e_3) = \left(\frac{\partial X^1}{\partial z}(p), \frac{\partial X^2}{\partial z}(p), \frac{\partial X^3}{\partial z}(p) \right),$$

provided that $X = (X^1, X^2, X^3)$, that is,

$$\mathrm{div}\, X = \frac{\partial X^1}{\partial x} + \frac{\partial X^2}{\partial y} + \frac{\partial X^3}{\partial z}$$

and, so, $\mathrm{div}\, X$ is a differentiable function.

Theorem 5.31 (Divergence theorem). *Let S be a compact connected surface and Ω the inner domain determined by S. If $X : \overline{\Omega} \to \mathbb{R}^3$ is a vector field, then*

$$\int_{\Omega} \operatorname{div} X = -\int_{S} \langle X, N \rangle,$$

where $N : S \to \mathbb{S}^2$ is the inner unit normal field.

Proof. Denote by P the plane of $\mathbb{R}^3$ of equation $x = 0$ and consider the orthogonal projection $\pi : S \to P$ given by

$$\pi(p) = p - \langle p, a \rangle a, \qquad p \in S,$$

where a is the vector $(1, 0, 0)$ normal to P. This projection is clearly differentiable and its differential at each point p of S is the projection itself restricted to $T_p S$; see Example 2.58. Hence

$$|\operatorname{Jac} \pi|(p) = |\langle N(p), a \rangle|,$$

where N is a Gauss map of S. Consider the following function $f : S \to \mathbb{R}$ defined by

$$f(p) = (\operatorname{sign} \langle N(p), a \rangle) \langle X, a \rangle = (\operatorname{sign} N^1) X^1,$$

where the superscript 1 means the first component of these maps and the function $\operatorname{sign} N^1$ takes the values 1, 0, −1 depending on whether N^1 is positive, zero, or negative. This function f is integrable on S because it is bounded since X is continuous and S compact. We apply the area formula for surfaces of Theorem 5.28 to obtain

$$(5.4) \qquad \int_{S} X^1 N^1 = \int_{\mathbb{R}^2} n(\pi, f)(0, y, z) \, dy \, dz.$$

Taking Exercise 5.26 and the definition of $n(\pi, f)$ into account, we easily deduce that, for each $(0, y, z)$, regular value of π, we have

$$n(\pi, f)(0, y, z) = -\sum_{i=2j \in \{1, \ldots, k\}} [X^1(x_{i+1}, y, z) - X^1(x_i, y, z)],$$

where (x_i, y, z), $i = 1, \ldots, k$, are the points of intersection of S with the vertical line passing through $(0, y, z)$, arranged according to their heights relative to the plane $x = 0$. Now, if we take the functions X^1 and $\partial X^1 / \partial x$ extended by zero outside Ω, the fundamental theorem of calculus, together with the previous equality, give us

$$\int_{-\infty}^{+\infty} \frac{\partial X^1}{\partial x}(x, y, z) \, dx = -n(\pi, f)(0, y, z).$$

Now integrating on $\mathbb{R}^2$ and using the Fubini theorem for the Lebesgue integral and the equality (5.4), we get

$$-\int_{S} X^1 N^1 = \int_{\Omega} \frac{\partial X^1}{\partial x} \, dx \, dy \, dz.$$

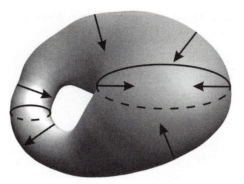

Figure 5.2. *Regular domain*

The same formula can be obtained for the second and third components. Adding the corresponding three equalities, we have the required integral formula. □

Remark 5.32 (Volume enclosed by a compact surface). If S is a compact surface and Ω is its inner domain, consider the identity vector field $X : \overline{\Omega} \to \mathbb{R}^3$ given by $X(x) = x$ for all $x \in \overline{\Omega}$. Its divergence is the constant function 3. Hence, by the divergence theorem

$$\operatorname{vol} \Omega = -\frac{1}{3} \int_S \langle p, N(p) \rangle \, dp,$$

where N is the inner unit normal of S. For example, if we take as the surface S a sphere $\mathbb{S}^2(r)$ of radius $r > 0$ and centre a point $p_0 \in \mathbb{R}^3$, we have $\Omega = B_r(p_0)$ and

$$\operatorname{vol} B_r(p_0) = -\frac{1}{3} \int_{\mathbb{S}^2(r)} \langle p, \frac{1}{r}(p_0 - p) \rangle \, dp.$$

Now using Exercise 5.12 and Example 5.16, we obtain

$$\operatorname{vol} B_r(p_0) = \frac{4}{3} \pi r^3.$$

Remark 5.33 (Divergence theorem on regular domains). We will use the term *regular domain* of $\mathbb{R}^3$ for any bounded connected open subset Ω such that $\operatorname{Bdry} \Omega$ is a compact—not necessarily connected—surface that we will denote by $\partial\Omega$. For example, if S_1 and S_2 are two compact connected disjoint surfaces such that their corresponding inner domains satisfy $\Omega_2 \subset \Omega_1$, then $\Omega = \Omega_1 - \overline{\Omega_2}$ is a regular domain such that $\partial\Omega = S_1 \cup S_2$. A particularly interesting example of this situation is the following: if S is a compact connected surface and S_t is a parallel surface with $t > 0$ small enough, the open set $V_t(S)$ is a regular domain with $\partial V_t(S) = S \cup S_t$ and such that

$\Omega - \overline{\Omega_t} = V_t(S)$, where Ω and Ω_t are the inner domains determined by S and S_t, respectively.

For this kind of domain Ω it also makes sense to talk about its inner unit normal field and about tubular neighbourhoods of $\partial\Omega$ and the fact that the divergence theorem remains valid:

$$\int_\Omega \operatorname{div} X = -\int_{\partial\Omega} \langle X, N \rangle,$$

where the integral on $\partial\Omega$ means the sum of the corresponding integrals on each of its connected components.

Remark 5.34 (Gauss theorem of electrostatics). If S is a compact surface and Ω is its inner domain, the vector field

$$Y(p) = \frac{p-a}{|p-a|^3}$$

is differentiable on $\overline{\Omega}$ if $a \notin \overline{\Omega}$. One easily sees that

$$(dY)_p(v) = \frac{1}{|p-a|^3} v - 3 \frac{\langle p-a, v \rangle}{|p-a|^5}(p-a)$$

for each $v \in \mathbb{R}^3$. Thus, the divergence of Y vanishes everywhere. Consequently

$$\int_S \frac{\langle p-a, N(p) \rangle}{|p-a|^3} \, dp = 0, \qquad \forall a \notin \overline{\Omega}.$$

Suppose now that $a \in \Omega$. Since Ω is open, there exists $\varepsilon > 0$ such that the open ball $B_\varepsilon(a)$ centred at a with radius ε is included in Ω. The set $\Omega' = \Omega - \overline{B_\varepsilon(a)}$ is a bounded connected open subset of the plane. Moreover, its topological boundary is $S \cup \mathbb{S}^2_\varepsilon(a)$. Hence, Ω' obeys the conditions required in Remark 5.33 above to be a regular domain. Since the field Y is differentiable on $\overline{\Omega'}$ and has null divergence, we have

$$0 = \int_S \frac{\langle p-a, N(p) \rangle}{|p-a|^3} \, dp + \int_{\mathbb{S}^2_\varepsilon(a)} \frac{\langle q-a, N(q) \rangle}{|q-a|^3} \, dq.$$

But, in the second integral we have

$$N(q) = \frac{q-a}{|q-a|} = \frac{1}{\varepsilon}(q-a).$$

Then we can explicitly compute this second integral and finally obtain

$$\int_S \frac{\langle a-p, N(p) \rangle}{|a-p|^3} \, dp = 4\pi, \qquad \forall a \in \Omega.$$

Note that this equality is usually known as the *Gauss theorem* of electrostatics.

5.8. Brower fixed point theorem

We will finish this chapter by proving, from the divergence theorem, one of the most famous results of algebraic topology: the Brower fixed point theorem, which we will consider only in the differentiable case. Our proof also works in the two-dimensional case, after suitable modifications that we will point out in Chapter 9.

Exercise 5.35 (Archimedes' principle). Let S be a compact surface and Ω its inner domain. Apply the divergence theorem to the vector fields X and Y defined on $\mathbb{R}^3$ by $X(x) = \langle x, a \rangle b$ and $Y(x) = \langle x, a \rangle (x \wedge b)$, for all $x \in \mathbb{R}^3$, and where $a, b \in \mathbb{R}^3$. We obtain in this way the two equalities

$$\int_S \langle p, a \rangle N(p)\, dp = -(\operatorname{vol}\Omega)a, \qquad a \in \mathbb{R}^3,$$

$$\int_S \langle p, a \rangle [p \wedge N(p)]\, dp = -(\operatorname{vol}\Omega)(G \wedge a), \qquad a \in \mathbb{R}^3,$$

where

$$G = \frac{1}{\operatorname{vol}\Omega} \int_\Omega x\, dx$$

is the centre of gravity of Ω. The first integral equality can be interpreted as follows: *if a force acts at each point of the boundary S of a solid body $\overline{\Omega}$, perpendicularly to S and with a strength equal to the height from this point to a given plane placed over the body, then the body experiences a total upthrust whose strength is equal to its volume.* The second equality refers to the momenta of the involved forces in the previous statement and can be paraphrased as follows: *the resultant effect of all these forces acting on the boundary of the solid body is the same that we would obtain by applying the total upthrust at the centre of gravity of the body.*

Exercise 5.36. We may associate to each differentiable function $f : A \to \mathbb{R}$, defined on a subset A of Euclidean space, a vector field ∇f also defined on A by

$$\langle (\nabla f)(p), v \rangle = (df)_p(v), \qquad \forall v \in \mathbb{R}^3.$$

This field is called the *gradient* of f. Prove that the three components of ∇f are

$$\nabla f = \left(\frac{\partial f}{\partial x}, \frac{\partial f}{\partial y}, \frac{\partial f}{\partial z} \right).$$

Show that, if the gradient of a differentiable function vanishes everywhere, then f is constant on each connected component of A.

Exercise 5.37. If $A \subset \mathbb{R}^3$ and $f, g : A \to \mathbb{R}$ are two differentiable functions, then

$$\operatorname{div}(\nabla f \wedge \nabla g) = 0.$$

Exercise 5.38. If $A \subset \mathbb{R}^3$ and $f, g, h : A \to \mathbb{R}$ are differentiable, then

$$\det(\nabla f, \nabla g, \nabla h) = \operatorname{div}[h(\nabla f \wedge \nabla g)].$$

The following result, which prepares the proof of the fixed point theorem, asserts that, in some sense, a differentiable transformation of the inner domain determined by a compact surface which fixes the surface behaves almost like a preserving orientation diffeomorphism.

Proposition 5.39. *Let Ω be the inner domain determined by a compact connected surface S, and let $\phi : \overline{\Omega} \to \mathbb{R}^3$ be a differentiable map such that $\phi_{|S} = I_S$. Then*

$$\int_\Omega (\operatorname{Jac} \phi)(x)\, dx = \operatorname{vol}(\Omega).$$

Proof. We have $\phi = (\phi^1, \phi^2, \phi^3)$, where ϕ^i, $i = 1, 2, 3$, is the component $\langle \phi, u_i \rangle$ relative to an orthonormal basis $\{u_1, u_2, u_3\}$ of $\mathbb{R}^3$. Using Exercises 5.37 and 5.38 above, we have

$$(\operatorname{Jac} \phi)(x) = \det(d\phi)_x = \det(\nabla \phi^1, \nabla \phi^2, \nabla \phi^3) = \operatorname{div}[\phi^3(\nabla \phi^1 \wedge \nabla \phi^2)].$$

Hence, the divergence Theorem 5.31 gives us

$$\int_\Omega (\operatorname{Jac} \phi)(x)\, dx = -\int_S \phi^3(p) \det((\nabla \phi^1)(p), (\nabla \phi^2)(p), N(p))\, dp.$$

But our hypothesis about ϕ asserts that ϕ restricted to the surface S is the identity map. Thus $\phi^3(p) = \langle p, u_3 \rangle$ if $p \in S$ and, moreover,

$$\langle (\nabla \phi^i)(p), v \rangle = (d\phi^i)_p(v) = \langle v, u_i \rangle$$

for $i = 1, 2, 3$ and $v \in T_p S$. By substituting into the previous equality, we get

$$\int_\Omega (\operatorname{Jac} \phi)(x)\, dx = -\int_S \langle p, u_3 \rangle \det(u_1, u_2, N(p))\, dp.$$

We may conclude by Exercise 5.35 above, where we studied the Archimedes principle. $\qquad \square$

Theorem 5.40. *Let S be a compact connected surface and Ω the corresponding inner domain. There are no differentiable maps $\phi : \overline{\Omega} \to \mathbb{R}^3$ such that $\phi_{|S} = I_S$ and $\phi(\Omega) \subset S$.*

Proof. If such a map ϕ existed, then, for each $x \in \Omega$ and each $v \in \mathbb{R}^3$, we would have

$$(d\phi)_x(v) = \left.\frac{d}{dt}\right|_{t=0} \phi(x + tv) \in T_{\phi(x)} S.$$

Therefore, the image of $(d\phi)_x$ would be two-dimensional at each $x \in \Omega$ and, so, $(\operatorname{Jac} \phi)(x) = \det(d\phi)_x = 0$. In this way, we would arrive at a contradiction to Proposition 5.39 above. $\qquad \square$

Theorem 5.41 (Fixed point theorem). *Let* $\phi : \overline{B_r} \to \overline{B_r} \subset \mathbb{R}^3$ *be a differentiable map, where* $\overline{B_r}$ *is the closed disc of radius* $r > 0$ *centred at the origin. Then* ϕ *has at least a fixed point, that is, there exists* $x \in \overline{B_r}$ *such that* $\phi(x) = x$.

Proof. Suppose that this is not the case. Then, for each $x \in \overline{B_r}$, we denote by $\psi(x)$ the only intersection point of the half-line starting at $\phi(x)$ and passing through x with $\mathbb{S}^2(r)$. The resultant map ψ takes $\overline{B_r}$ into $\mathbb{S}^2(r)$ and satisfies $\psi_{|S} = I_S$. The proof will be finished if we see that ψ is differentiable. But this can be easily done by expressing it in terms of ϕ and using the differentiability of the latter. $\qquad\qquad\square$

Exercises

(1) ↑ Let S be a surface of revolution, and let $\alpha : I \subset \mathbb{R} \to S$ be a curve p.b.a.l. generating the surface. Prove that the area of S is equal to

$$A(S) = 2\pi \int_I \text{dist}\,(\alpha(s), R)\,ds$$

where R is the revolution axis. This result is known as the *Pappus theorem*.

(2) ↑ Let S be a compact surface with non-vanishing Gauss curvature everywhere and injective Gauss map. Show that

$$\int_S K = 4\pi.$$

(3) ↑ Prove that, if $a, b \in \mathbb{R}^3$,

$$\int_{\mathbb{S}^2} \langle p, a \rangle \langle p, b \rangle \, dp = \frac{4}{3}\pi \langle a, b \rangle,$$

and, consequently,

$$\int_{\mathbb{S}^2} \langle Ap, p \rangle \, dp = \frac{4}{3}\pi \, \text{trace}\, A,$$

for each square matrix A of order three.

(4) ↑ Let A be a regular square matrix of order three. Show that the map $\phi : \mathbb{S}^2 \to \mathbb{S}^2$ given by $\phi(p) = Ap/|Ap|$ is a diffeomorphism. Use the change of variables formula to check that

$$\int_{\mathbb{S}^2} \frac{1}{|Ap|^3} \, dp = \frac{4\pi}{|\det A|}.$$

(5) ↑ Let $\phi : \mathbb{S}^2 \to \mathbb{S}^2$ be a map defined by $\phi(p) = (p + a)/|p + a|$, for all $p \in \mathbb{S}^2$, where $a \in \mathbb{R}^3$ is a vector with $|a| < 1$. Show that ϕ is a diffeomorphism and that

$$\int_{\mathbb{S}^2} \frac{1 + \langle p, a \rangle}{|p + a|^3}\, dp = 4\pi.$$

(6) Let ϕ be a dilatation of Euclidean space $\mathbb{R}^3$ with centre $p_0 \in \mathbb{R}^3$ and ratio $\mu \neq 0$. Prove that the areas of a compact surface S and of the image surface $S' = \phi(S)$ are related by $A(S') = \mu^2 A(S)$.

(7) ↑ Demonstrate that the integral of the Gauss curvature on a compact surface coincides with the integral of the Gauss curvature on any of its parallel surfaces.

(8) ↑ Show that, if S is a compact surface,

$$\int_S \langle N, a \rangle = 0,$$

where N is a Gauss map and a is an arbitrary vector.

(9) ↑ Assume that S is a compact surface. Use the divergence theorem and the formula for the change of variables to see that

$$\mathrm{vol}\,\Omega_t = \mathrm{vol}\,\Omega + \frac{t}{3}\int_S (2H\langle p, N\rangle - 1) + \frac{t^2}{3}\int_S (2H - K\langle p, N\rangle) - \frac{t^3}{3}\int_S K,$$

where Ω_t is the inner domain determined by the parallel surface at distance t. (See Example 5.19.)

(10) Let $A \subset \mathbb{R}^3$, let X, Y be two vector fields defined on A, let $f : A \to \mathbb{R}$ be a differentiable map, and let $a, b \in \mathbb{R}$. Prove that

$$\mathrm{div}\,(aX + bY) = a\,\mathrm{div}\,X + b\,\mathrm{div}\,Y \quad \text{and} \quad \mathrm{div}\,(fX) = f\,\mathrm{div}\,X + \langle \nabla f, X \rangle.$$

(11) Let $A \subset \mathbb{R}^3$, and let $f, g : A \to \mathbb{R}$ be two differentiable functions. We define the *Laplacian* of f as the function

$$\Delta f = \frac{\partial^2 f}{\partial x^2} + \frac{\partial^2 f}{\partial y^2} + \frac{\partial^2 f}{\partial z^2}.$$

Prove that $\Delta f = \mathrm{div}\,\nabla f$ and $\Delta(fg) = g\Delta f + f\Delta g + 2\langle \nabla f, \nabla g \rangle$.

(12) Let O be an open subset of Euclidean space $\mathbb{R}^3$, and let $f : O \to \mathbb{R}$ be a differentiable map. Show that the following assertions are equivalent:
   - The function f is *harmonic*, that is, $\Delta f = 0$.
   - For each $x \in O$ and $r > 0$ small enough

$$f(x) = \frac{1}{4\pi r^2} \int_{\mathbb{S}^2_x(r)} f(p)\, dp.$$

- For each $x \in O$ and $r > 0$ small enough

$$\int_{\mathbb{S}_x^2(r)} \langle (\nabla f)(p), p \rangle \, dp = 0,$$

where $\mathbb{S}_x^2(r)$ is the sphere of radius $r > 0$ and centre $x \in O$.

(13) ↑ Let S be a surface and P a plane such that the orthogonal projection onto P restricted to S is a diffeomorphism on an open set $U \subset P$. Prove that $A(U) \leq A(S)$. That is, orthogonal projections on planes decrease area.

(14) ↑ Let S be a compact connected surface which is star-shaped with respect to a point, say, for example, the origin. Show that the central projection $\phi : S \to \mathbb{S}^2$ given by $\phi(p) = p/|p|$, for all $p \in S$, is a diffeomorphism. Thus $S = S(f)$ for some positive function f defined on $\mathbb{S}^2$. (See Exercises (10) and (11) at the end of Chapter 2.)

(15) ↑ *Spheres minimize the integral of the square of the mean curvature.* Prove that, if S is a compact surface, then

$$\int_S H(p)^2 \, dp \geq 4\pi,$$

and equality occurs if and only if S is a sphere.

Hints for solving the exercises

Exercise 5.25: As we argued after Definition 5.3, one can find a diffeomorphism $\psi : O \to \overline{\mathcal{R}} \times (a, b)$, where O is a relatively compact open subset of $\mathbb{R}^3$. Let $\mathcal{M}$ be the set of the points of O where $|\mathrm{Jac}\,(\phi \circ \psi)| = (|\mathrm{Jac}\,\phi| \circ \psi)\,\mathrm{Jac}\,\psi$ vanishes. It is enough to see that this equality implies that $\mathcal{N} = \psi(\mathcal{M})$ and apply a three-dimensional version of the Sard Theorem 4.6 to the map $\phi \circ \psi$.

Exercise 5.26: Let $\phi : S \to P$ be the restriction to S of the orthogonal projection π of $\mathbb{R}^3$ to the plane P. Then, since

$$\pi(x) = x - \langle x, a \rangle a + ca, \qquad \forall x \in \mathbb{R}^3,$$

for some $c \in \mathbb{R}$, we have

$$(d\phi)_p(v) = v - \langle v, a \rangle a, \qquad \forall v \in T_p S \quad \forall p \in S,$$

and, so, an easy computation leads to

$$|\mathrm{Jac}\,\phi| = |\langle N, a \rangle|,$$

where N is a Gauss map of S. Hence, a point $x \in P$ is a regular value of ϕ if all the points $p \in S$ which project onto x satisfy $\langle N(p), a \rangle \neq 0$. Taking the Sard Theorem 4.6 into account, we deduce that, almost every straight line R perpendicular to the plane P satisfies the property that, if $p \in R \cap S$, then $\langle N(p), a \rangle \neq 0$, that is, $a \notin T_pS$ and, so, S and R meet at p transversely.

Suppose now that R is a straight line of this class. From Remark 4.3, we know that $R \cap S$ consists of finitely many points, because S is compact. The open intervals resulting after removing these points from R belong to either Ω or to $\mathbb{R}^3 - \overline{\Omega}$. The fact that Ω is bounded and that, again from Remark 4.3, there are points of Ω and of $\mathbb{R}^3 - \overline{\Omega}$ on each side of the points of $R \cap S$, implies that these intervals lie alternatively in $\mathbb{R}^3 - \overline{\Omega}$ and in Ω, if we move along R in any of the possible senses, for example, corresponding to the vector a. Thus $R \cap S$ has an even number of points. Let $p \in R \cap S$ be a point which is a lower extreme of one of these intervals. Then, for t near zero,

$$p + ta \in \Omega \quad \text{if } t > 0 \qquad \text{and} \qquad p + ta \in \mathbb{R}^3 - \overline{\Omega} \quad \text{if } t < 0.$$

If we work with t small enough so that $p + ta$ is in a tubular neighbourhood of S, the expressions above can be rewritten as

$$r(p + ta) > 0 \quad \text{if } t > 0 \quad \text{and} \quad r(p + ta) < 0 \quad \text{if } t < 0,$$

where r is the oriented distance to S relative to the inner normal (see Exercise (1) of Chapter 4). Therefore

$$0 \leq \frac{d}{dt}\Big|_{t=0} r(p + ta) = (dr)_p(a) = \langle N(p), a \rangle.$$

Hence, since $\langle N(p), a \rangle \neq 0$, we have $\langle N(p), a \rangle > 0$. Analogously one can see that $\langle N, a \rangle < 0$ at the upper extremes of such intervals.

Exercise (1): Use a parametrization $X : I \times (0, 2\pi) \to S$ of the form

$$X(u, v) = x(u)\, a + y(u) \cos v\, b - y(u) \sin v\, c, \qquad \forall u \in I \quad \forall v \in (0, 2\pi),$$

like that one described in Exercise (3) at the end of Chapter 2, where $\alpha(u) = x(u)\, a + y(u)\, b$ for each $u \in I$. Then

$$|X_u \wedge X_v|(u, v) = y(u), \qquad \forall u \in I \quad \forall v \in (0, 2\pi),$$

and one finishes by using Proposition 5.13.

Exercise (2): The Gauss map $N : S \to \mathbb{S}^2$ is a local diffeomorphism because, according to (3.4),

$$0 \neq K(p) = \det(dN)_p, \qquad \forall p \in S,$$

and, so, we may use the inverse function Theorem 2.75. Thus, it is an open map and, so, $N(S)$ is an open subset of $\mathbb{S}^2$, which is also closed since

S is compact. Consequently, N is a diffeomorphism and we may apply the change of variables formula Theorem 5.14 taking into account that $|\mathrm{Jac}\, N| = |K| = K$ (see, for this last equation, Exercise 3.42) and that $A(\mathbb{S}^2) = 4\pi$ (see Exercise 5.12).

Exercise (3): Apply the invariance of the integral on $\mathbb{S}^2$ under rigid motions preserving the origin. If $\langle a, b \rangle = 0$, there is a symmetry $\phi \in O(3)$ such that $\phi(a) = a$ and $\phi(b) = -b$. Then

$$\int_{\mathbb{S}^2} \langle p, a \rangle \langle p, b \rangle \, dp = \int_{\mathbb{S}^2} \langle \phi(p), a \rangle \langle \phi(p), b \rangle \, dp$$

$$= \int_{\mathbb{S}^2} \langle p, \phi(a) \rangle \langle p, \phi(b) \rangle \, dp = -\int_{\mathbb{S}^2} \langle p, a \rangle \langle p, b \rangle \, dp.$$

On the other hand, if $a, b \in \mathbb{R}^3$ are unit vectors, there exists $\phi \in O(3)$ such that $\phi(b) = a$. Hence

$$\int_{\mathbb{S}^2} \langle p, a \rangle^2 \, dp = \int_{\mathbb{S}^2} \langle \phi(p), a \rangle^2 \, dp = \int_{\mathbb{S}^2} \langle p, b \rangle^2 \, dp.$$

Since, furthermore, given a unit $a \in \mathbb{R}^3$, one can choose an orthonormal basis $\{u_1 = a, u_2, u_3\}$ of $\mathbb{R}^3$, one gets

$$3 \int_{\mathbb{S}^2} \langle p, a \rangle^2 \, dp = \sum_{i=1}^{3} \int_{\mathbb{S}^2} \langle p, u_i \rangle^2 \, dp = \int_{\mathbb{S}^2} dp = 4\pi.$$

This suffices for solving the exercise since, if $\{e_1, e_2, e_3\}$ is the standard basis of $\mathbb{R}^3$, one has

$$\langle Ap, p \rangle = \sum_{i,j=1}^{3} A_{ij} \langle p, e_i \rangle \langle p, e_j \rangle.$$

Exercise (4): The map ϕ is differentiable since it is the restriction to $\mathbb{S}^2$ of a differentiable map defined on $\mathbb{R}^3 - \{0\}$. It is a diffeomorphism because one can check immediately that $\psi : \mathbb{S}^2 \to \mathbb{S}^2$, given by

$$\psi(p) = \frac{A^{-1}p}{|A^{-1}p|}, \qquad \forall p \in \mathbb{S}^2,$$

is its inverse map. To conclude, compute the absolute value of the Jacobian of ϕ. Since

$$(d\phi)_p(v) = \frac{Av}{|Ap|} - \frac{\langle Ap, Av \rangle}{|Ap|^3} Ap, \qquad \forall v \in T_pS \quad \forall p \in \mathbb{S}^2,$$

we have, if $p \in \mathbb{S}^2$ and $e_1, e_2 \in T_p\mathbb{S}^2$ is an orthonormal basis,

$$|\mathrm{Jac}\, \phi|(p) = \frac{1}{|Ap|^3} |\det(Ae_1, A_2, Ap)| = \frac{|\det A|}{|Ap|^3},$$

because $\{e_1, e_2, p\}$ is an orthonormal basis of $\mathbb{R}^3$.

Exercise (5): The map ϕ is well defined on $\mathbb{S}^2$ and it is differentiable because it is the restriction of a differentiable map defined on $\mathbb{R}^3 - \{-a\}$. It is a diffeomorphism because $\psi : \mathbb{S}^2 \to \mathbb{S}^2$, given by

$$\psi(p) = \left(\langle a, p \rangle + \sqrt{\langle a, p \rangle^2 + 1 - |a|^2} \right) p - a,$$

is an explicit inverse map for it, which is clearly differentiable (recall that $|a| < 1$). As in Exercise (4) above, it only remains to compute the absolute value of its Jacobian.

Exercise (7): Let S be a compact surface, S_t the parallel surface at distance t, and $F_t : S \to S_t$ the corresponding diffeomorphism studied in Remark 4.29. It suffices to combine the information about $K_t \circ F_t$ supplied by Exercise (16) of Chapter 3 with the calculation of $|\text{Jac} \, F_t|$ that we made in Example 5.17. One can show in this way that $(K_t \circ F_t)|\text{Jac} \, F_t| = K$. Now, use the change of variables Theorem 5.14.

Exercise (8): Apply the divergence Theorem 5.31 on the domain Ω determined by S to the constant vector field equal to the vector a.

Exercise (9): According to Remark 5.32, we have

$$\text{vol} \, \Omega_t = -\frac{1}{3} \int_{S_t} \langle p, N_t(p) \rangle \, dp,$$

where N_t is the inner normal of the surface S_t parallel to S at distance t. Now use the change of variables Theorem 5.14 for the diffeomorphism $F_t : S \to S_t$, taking into account that $N_t \circ F_t = N$ and using the value of $|\text{Jac} \, F_t|$ computed in Example 5.17.

Exercise (13): From Theorem 5.14 for the change of variables, we have that

$$A(U) = \int_U 1 = \int_S |\text{Jac} \, \phi|,$$

where $\phi : S \to U$ is the restriction of the mentioned orthogonal projection. But, in the hint for solving Exercise 5.26, we saw that

$$|\text{Jac} \, \phi| = |\langle N, a \rangle| \leq 1,$$

where N is a Gauss map of S and a is a unit vector perpendicular to P.

Exercise (14): In Example 2.79, we computed the differential of ϕ and, from this, it can be deduced that

$$|\operatorname{Jac}\phi|(p) = \frac{1}{|p|^3}|\langle N(p), p\rangle| = -\frac{1}{|p|^3}\langle N(p), p\rangle,$$

where the last equality, if N is the inner normal, is true because we know that the support function based on the origin does not change its sign, according to Exercise (4) of Chapter 3. By the area formula for surfaces found in Theorem 5.28, one has

$$-\int_S \frac{1}{|p|^3}\langle N(p), p\rangle\, dp = \int_{\mathbb{S}^2} n(\phi, 1),$$

where $n(\phi, 1)(x)$ is the number of inverse images through ϕ of each regular value $x \in \mathbb{S}^2$ of ϕ. Since S is star-shaped with respect to the origin, all the points of $\mathbb{S}^2$ are regular values of ϕ, that is, any half-line starting from the origin cuts S transversely. Moreover, Proposition 4.14 assures us that the function $n(\phi, 1)$ is locally constant and, so, continuous. Since S is compact, $n(\phi, 1) \geq 1$. Therefore

$$4\pi \leq \int_{\mathbb{S}^2} n(\phi, 1) = -\int_S \frac{1}{|p|^3}\langle N(p), p\rangle\, dp = 4\pi,$$

where the latter is due to the Gauss theorem given in Remark 5.34. Hence $n(\phi, 1)$ is the constant function 1. That is, ϕ is injective. On the other hand, ϕ is surjective because its image is closed in $\mathbb{S}^2$, since S is compact, and open in $\mathbb{S}^2$ as a consequence of the inverse function Theorem 2.75.

Exercise (15): Equation (3.6), together with the fact H^2 is non-negative, gives us the inequality $H^2 \geq K^+$. Integrating on S and using the Chern-Lashoff Theorem 5.29, we obtain

$$\int_S H^2 \geq \int_S K^+ \geq 4\pi.$$

If equality holds, Proposition 5.6 gives us that $H^2 = K^+$ and, so, all the elliptic points of S are umbilical. There is at least one such point, say $p_0 \in S$, according to Exercise 3.42. Then, the connected component C containing p_0 of the open subset of S formed by its elliptic points is included in a sphere, by Theorem 3.30. If there exists some point $p \in S$ with $K(p) \leq 0$, since S is connected, we could take a continuous curve $\alpha : [0, 1] \to S$ such that $\alpha(0) = p_0$ and $\alpha(1) = p$. Then, there would exist a first time $x \in (0, 1]$ where $K(\alpha(x)) \leq 0$. Thus $\alpha([0, x)) \subset C$ and, hence, $\alpha([0, x))$ would be included in a sphere and the Gauss curvature along it would be a positive constant. By continuity, the same would occur at $\alpha(x)$ and this would provide a contradiction. Therefore, all the points of S are elliptic and, consequently, umbilical. We finish using Corollary 3.31.

Chapter 6

Global Extrinsic Geometry

6.1. Introduction and historical notes

We already have, after Chapter 3, convincing reasons to accept that the second fundamental form and the corresponding curvatures associated with the Gauss map are efficient tools for controling the shape of a surface. On the other hand, we have developed in Chapter 5 a theory of integration on compact surfaces keeping the usual properties of Lebesgue integration. These are the two main bases for proceeding with a deeper study of the geometry of surfaces in $\mathbb{R}^3$.

From our point of view, and from that of modern differential geometry, greatly influenced by the topology, *deeper* means being able, even in a small measure, to skip from local to global properties. Indeed, the Brower-Samelson theorem, the Jordan-Brower separation theorem, and the Brower fixed point theorem, proved in Chapters 4 and 5, are of course results of a global nature, but they are theorems which fit better into the so-called differential topology. In this chapter, we deal with the class of results and proofs that we think of as the core of—in this case, classical—differential geometry.

Whenever one thinks about global problems in geometry, it is usual to impose some *totality* condition to avoid the case that *pieces* of larger surfaces may appear as possible solutions, that is, surfaces which are open subsets of another larger surface which includes them strictly. One of these conditions is clearly *compacity*. Another totality condition weakening compacity, which appears in most of the texts, is *geodesic completeness*. Since our point of view gives priority to extrinsic geometry, we prefer the *closedness* as a subset of $\mathbb{R}^3$. Thus, we will restrict ourselves in this chapter to studying closed (as subsets of Euclidean space) surfaces and insist on compact surfaces. In this way, we will be able to use, as a fundamental analytical tool, the theory of integration of Chapter 5.

The first problem that we will reflect upon, which repeats some questions that we already posed for curves, is to determine compact surfaces with one or several of their curvatures having the simplest behaviour. For example, which are the compact surfaces with constant Gauss or mean curvature? First, let us consider the Gauss curvature. The local geometry of surfaces with constant Gauss curvature was studied by F. MINDING in 1838, with respect to their relation to non-Euclidean geometries. Relative to the compact case, we know from Chapter 3 that the only example is the sphere, as the HILBERT-LIEBMANN, theorem asserted. This result was first proved by H. LIEBMANN in 1899 and was rediscovered by D. HILBERT in 1901—our proof is a simplification of Hilbert's. Then it is natural to generalize the question and to think of compact surfaces whose Gauss curvature does not change sign. From what we have seen until now, we may deduce that this sign must be positive. We face up to the study of compact surfaces with positive Gauss curvature, the so-called *ovaloids*, and even of the positively curved closed surfaces. Study of this class of surfaces was started at the beginning of the 20th century by H. MINKOWSKI and W. BLASCHKE in relation to the theory of convexity. We will prove the J. HADAMARD theorem (1857), which states that ovaloids are topologically spheres, and the STOKER theorem (1936), which shows that closed non-compact surfaces with positive Gauss curvature are diffeomorphic to $\mathbb{R}^2$. Furthermore, both are characterized as boundaries of convex domains of Euclidean space. Just this information about ovaloids will enable us to prove again the HILBERT-LIEBMANN theorem using as a tool some integral expressions known as the MINKOWSKI formulas. We will obtain as well a new version of the J. H. JELLETT result (1853) which, in this context, can be given by saying that a compact surface star-shaped with respect to a given point with constant mean curvature must be a sphere centred at this point.

These last results about ovaloids introduce the second question that we posed: what are the compact surfaces with constant mean curvature? We know, as mentioned in the introduction of Chapter 3, that the Gauss and mean curvatures were born in the 18th century and, since then, they have fought to prevail over each other. In spite of the fact that the initial battle was won, in some sense, by the Gauss curvature, once GAUSS proved its intrinsic character, the global problems on the mean curvature have proved to be more complicated and, so, more interesting. In fact, it was only between 1956 and 1962 that A. D. ALEXANDROV divulged the proof of his celebrated theorem that, in our situation, asserts that the only compact surfaces with constant mean curvature are, again, spheres. This theorem was really a surprising result, not because of its contents, but because of the method that ALEXANDROV invented to show it, completely different from those of LIEBMANN and JELLETT, which is based on an ingenious use of properties of some partial differential equations. Including this proof is not feasible for the level of this book, and that is also why this theorem, though it is basic to the global theory of surfaces, does not appear in even the most prestigious texts. In 1978, R. REILLY found a new proof that, although less geometrical, was easier to understand and, especially, to explain. Unfortunately, it is also difficult to include in a first course in differential geometry. In the last ten years, the authors have dealt with some problems related to the mean curvature which have allowed them to outline the proof that we will present here, based on a paper by H. HEINTZE and H. KARCHER.

Finally, the determination of compact surfaces with constant mean curvature will lead us to answer one of the oldest questions in geometry, already posed by the Greeks: the isoperimetric problem, which can be given as follows. Among all the compact surfaces in $\mathbb{R}^3$ whose inner domains have a given volume, which has the least area? The ancient Greeks themselves knew that the answer was spheres, as was first shown by H. A. SCHWARZ according to earlier ideas of STEINER and MINKOWSKI, who had looked for the solution among convex domains. Our proof will consist, following [5] and [6], of obtaining the BRUNN-MINKOWSKI inequality, which is nothing more than another property of Lebesgue measure, and of seeing that surfaces reaching equality must have constant mean curvature.

Though we think that this chapter may be a sufficient introduction to global geometry of surfaces in Euclidean space, the compacity hypothesis demanded by the theory of integration and our desire to not complicate this text excessively will keep us from studying negatively curved surfaces or surfaces with vanishing Gauss curvature. Thus, a suitable continuation for the reader would be the search for readable proofs of the HILBERT and EFIMOV theorems on the inexistence of closed negatively curved surfaces in $\mathbb{R}^3$ or for the HARTMAN-NIRENBERG and MASSEY theorems asserting that the only ones with null curvature are cylinders having any plane curve as a directrix. (See Appendix 7.7.)

6.2. Positively curved surfaces

Before determining compact surfaces with constant Gauss or mean curvature, we will stop to consider surfaces whose Gauss curvature does not change sign. Recall—Exercise 3.42—that, on any compact surface, there is always an elliptic point. Therefore, if the sign of the Gauss curvature

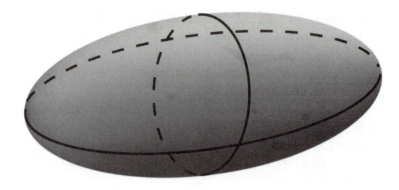

Figure 6.1. *Ovaloid*

does not change, then it must be positive everywhere. We will use the term *ovaloid* for any compact connected surface with positive Gauss curvature everywhere. This is equivalent, according to definition (3.5), to the fact that the second fundamental form is definite at each point and, taking Exercise (4) of Chapter 4 into account, to the statement that the second fundamental form relative to the inner normal is positive definite at each point. For example, according to Exercise 3.24, all ellipsoids are ovaloids.

More generally, in this section, we will give a complete description of the shape that *positively curved closed surfaces* adopt. In the following, S will denote a connected surface which is closed as a subset of the Euclidean space $\mathbb{R}^3$ and has positive Gauss curvature everywhere. By Exercise (6) of Chapter 4, which generalizes the Jordan-Brower and Brower-Samelson theorems to closed surfaces, we know that the surface S separates $\mathbb{R}^3$ into two connected components and that it is orientable. Since the second fundamental form of S, relative to any choice of normal field, is definite, we may choose as the orientation of S the one corresponding to the Gauss map $N : S \to \mathbb{S}^2$ for which the second fundamental form is *positive* definite at each point. Then, all the half-lines starting at a point $p \in S$ with direction $N(p)$ lie locally on the same side of the surface, since any two points of S belong to a relatively compact open subset of S, which possesses a tubular neighbourhood. This privileged connected component Ω of $\mathbb{R}^3 - S$ towards all the half-lines in the direction of the so-chosen Gauss map point will be called the *inner domain* of the surface S. It is clear that, if S is compact, this notion of inner domain coincides with what we have already defined in Remark 4.20. Examples of closed connected surfaces with positive Gauss curvature are, according to Example 3.45, the elliptic paraboloids.

A concept related, in a natural way, to the Gauss curvature is that of the convexity of surfaces. Usually, a surface is said to be *convex* when it lies on one of the closed half-spaces determined by the affine tangent plane at

Figure 6.2. *Inner domain*

each of its points. We will say that it is *strictly convex* when it is convex and touches each tangent plane only at one point. It is worth pointing out that, at first, this notion of convexity does not have much to do with the notion of affine convexity. However, we have the following result relating them. That which refers to compact surfaces was first shown by Hadamard in 1857 and the analogous statements for closed surfaces are consequences of a result due to Stoker in 1936.

Theorem 6.1 (First part of Hadamard-Stoker's theorem). *Let S be a connected surface closed as a subset of $\mathbb{R}^3$ with positive Gauss curvature everywhere, and let Ω be its inner domain. Then the following hold.*

> **A:** *For each $x, y \in \overline{\Omega}$ we have $]x, y[\subset \Omega$. In particular Ω is a convex subset of $\mathbb{R}^3$.*

> **B:** *The intersection of S with the affine tangent plane of S at any of its points is only the point itself. Moreover $\overline{\Omega}$ is contained in all the closed half-spaces of $\mathbb{R}^3$ determined by the affine tangent planes and the inner normals of the surface.*

Proof. Part **A**. First, let us see that Ω is convex. Consider the subset $A = \{(x, y) \in \Omega \times \Omega \,|\, [x, y] \subset \Omega\}$ of the product $\Omega \times \Omega$. Since Ω is connected, if it were not convex, then the boundary of A in $\Omega \times \Omega$ would be non-empty. Take a pair (x, y) in Bdry A. By definition of the topological boundary, there are sequences $\{x_n\}_{n \in \mathbb{N}}$, $\{x'_n\}_{n \in \mathbb{N}}$, $\{y_n\}_{n \in \mathbb{N}}$, and $\{y'_n\}_{n \in \mathbb{N}}$ of points of Ω such that $\lim_{n \to \infty} x_n = \lim_{n \to \infty} x'_n = x$, $\lim_{n \to \infty} y_n = \lim_{n \to \infty} y'_n = y$, and $[x_n, y_n] \subset \Omega$, $[x'_n, y'_n] \not\subset \Omega$, for each $n \in \mathbb{N}$. This implies that one could find a point $z \in]x, y[\cap S$ of first contact between the segment and the surface. From Lemma 4.15, we deduce that $]x, y[\subset \overline{\Omega}$ would lie in the affine tangent plane of S at z and that the height function relative to this plane, when restricted to the normal section at z in the direction of the tangent segment $]x, y[$, would attain a maximum at z. But, on the other hand, we said that

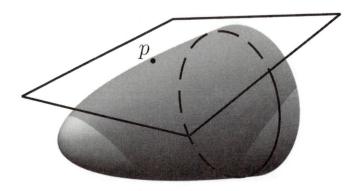

Figure 6.3. *Hadamard-Stoker's theorem*

the positivity of the Gauss curvature forces the second fundamental form relative to the orientation chosen on S to be positive definite at each point. Then, Remark 3.28—see also Proposition 3.37—says that this same height function reaches a local strict maximum at z. This is a contradiction and, so, Ω is a convex subset of Euclidean space. Now, an elementary fact on convex sets is that the adherence of a convex set is convex as well (the reader should try to prove this as an exercise). Therefore, if part **A** were not true, then there would exist two points $x, y \in S$ such that $]x, y[\cap S \neq \emptyset$ and this is impossible from the above. In fact, arguing in the same way, we would have a first contact point at each $z \in]x, y[\cap S$ and this is again a contradiction.

Part **B**. Let $p \in S$ be a point of the surface. Represent by Π_p the affine tangent plane of S at p and by Π_p^+ the closed half-space of $\mathbb{R}^3$ determined by this plane Π_p and the inner normal vector $N(p)$. In other words,

$$\Pi_p^+ = h_p^{-1}[0, +\infty) \quad \text{where} \quad h_p(q) = \langle q - p, N(p) \rangle,$$

for each $q \in \mathbb{R}^3$, is the height function relative to Π_p.

Suppose now that $q \in \Pi_p \cap S$ for some $p \in S$. Then p and q belong to $\overline{\Omega}$ and, from part **A**, we would have $]p, q[\subset \Omega \cap \Pi_p$. But, since p is an elliptic point, Proposition 3.37 and Remark 3.28 tell us again that all the points in a neighbourhood of p in Π_p, except p itself, lie in the outer domain of S, and this leads us to a contradiction, unless $p = q$ and so $]p, q[= \emptyset$. Therefore, $\Pi_p \cap S = \{p\}$, for each $p \in S$, which is the first assertion of part **B**.

To see that the second assertion is also true, observe that, for $p \in S$, the set $S - \{p\}$ is connected and does not intersect Π_p. But, since there are points of S near p which lie in Π_p^+, because p is an elliptic point, we have $S - \{p\} \subset \Pi_p^+$ and, since $p \in \Pi_p$, $S \subset \Pi_p^+$. Hence $\mathbb{R}^3 - \Pi_p^+ \subset \mathbb{R}^3 - S$. Thus, the connected open set $\mathbb{R}^3 - \Pi_p^+$ belongs to one of the two connected components of the complement of S and has points in the outer domain of

S, again because p is elliptic. Consequently, $\mathbb{R}^3 - \Pi_p^+ \subset \mathbb{R}^3 - \overline{\Omega}$. Taking complements, we conclude that $\overline{\Omega} \subset \Pi_p^+$, and this occurs for each $p \in S$. $\quad\square$

Remark 6.2. If $p \in S$, assertion **B** of Theorem 6.1, $\overline{\Omega} \subset \Pi_p^+$, is equivalent to
$$h_p(q) = \langle q - p, N(p) \rangle \geq 0, \qquad \forall q \in \overline{\Omega},$$
and the equality implies that $q = p$. Hence
$$\langle p - q, N(p) \rangle < 0, \qquad \forall p \in S \quad \forall q \in \Omega,$$
that is, the support function of S corresponding to the inner normal based on any point of Ω is negative everywhere. In particular—see Exercise (4) of Chapter 3—we have that S is star-shaped with respect to any point of its inner domain.

Remark 6.3 (Intersection of a straight line and an ovaloid). Assume that the surface S is an ovaloid and that Ω is the inner domain determined by it. Let R be a straight line of $\mathbb{R}^3$ cutting the inner domain, that is, with $R \cap \Omega \neq \emptyset$. As a consequence of the Hadamard-Stoker Theorem 6.1, $R \cap \Omega$ is a non-empty bounded convex set, which is open as a subset of the line R. So, it has to be an open interval. This, along with the fact that R cannot be tangent at any point of S, by part **B** of Theorem 6.1, implies that $R \cap S$ consists of exactly two points, that is, *a straight line touching the inner domain of an ovaloid cuts it at exactly two point*; see Exercise (6) at the end of this chapter.

Theorem 6.1 above asserts that the inner domain of a closed connected surface with positive Gauss curvature is convex and that, furthermore, the surface itself is strictly convex. To proceed with this type of result, we will need a lemma, in which we will use the following notation. If $x \in \mathbb{R}^3$ is a point of Euclidean space and $v \in \mathbb{R}^3 - \{0\}$ is a non-null vector, we will denote by $R(x, v)$ the straight line passing through x in the direction of v and by $R^+(x, v)$ the set
$$R^+(x, v) = \{x + tv \,|\, t \geq 0\},$$
that is, the half-line with origin at x and direction given by v.

Lemma 6.4. *Let S be a closed connected surface with positive Gauss curvature, and let Ω be its inner domain. Then the following hold.*

   a: *If $R^+(x, v) \subset \overline{\Omega}$ for some $x \in \overline{\Omega}$ and $v \in \mathbb{R}^3 - \{0\}$, then $R^+(y, v) \subset \overline{\Omega}$ and, moreover, $R^+(y, v) - \{y\} \subset \Omega$, for each $y \in \overline{\Omega}$.*

   b: *There are no straight lines contained in $\overline{\Omega}$.*

Proof. Part **a**. If we suppose that, for some $x \in \overline{\Omega}$ and a certain non-null vector v, $R^+(x, v) \subset \overline{\Omega}$, then $x + tv \in \overline{\Omega}$ for each $t \geq 0$. If $y \in \overline{\Omega}$, by part

A of the Hadamard-Stoker Theorem 6.1 above, we have $]y, x + tv[\subset \Omega$, for $t \geq 0$. Taking the limit as $t \to \infty$, one obtains that $R^+(y, v) \subset \overline{\Omega}$. If this half-line had some point z of S different from its origin, we would have a contradiction to part **A** of Theorem 6.1.

Part **b**. Suppose that there is a straight line R in $\overline{\Omega}$. We pick a point $x \in R$ and let $v \in \mathbb{S}^2$ be a direction vector of R. Then the two half-lines $R^+(x, v)$ and $R^+(x, -v)$ are in $\overline{\Omega}$. Thus, if $p \in S \subset \overline{\Omega}$, by part **a**, the half-lines $R^+(p, v)$ and $R^+(p, -v)$ are contained in $\overline{\Omega}$. Therefore

$$p + v \in R^+(p, v) \subset \overline{\Omega} \quad \text{and} \quad p - v \in R^+(p, -v) \subset \overline{\Omega}.$$

Using again part **A** of the Hadamard-Stoker theorem, we would get

$$p \in]p - v, p + v[\subset \Omega,$$

which is impossible.   □

Theorem 6.5 (Second part of Hadamard-Stoker's theorem). *Let S be a connected surface, closed as a subset of $\mathbb{R}^3$ and with positive Gauss curvature everywhere.*

> **C:** *If S is an ovaloid, that is, if S is compact, then the Gauss map $N : S \to \mathbb{S}^2$ is a diffeomorphism. Consequently, S is diffeomorphic to a sphere.*
>
> **D:** *If S is not compact, the Gauss map $N : S \to \mathbb{S}^2$ is injective and S is a graph on an open convex subset of a plane.*

Proof. We are assuming that the surface S has positive Gauss curvature everywhere. Then, equation (3.4) implies that the Gauss map $N : S \to \mathbb{S}^2$ is a local diffeomorphism, by the inverse function Theorem 2.75. Let us see now that, under our hypotheses, N must be injective. Indeed, if $p, q \in S$ are two points of the surface with $N(p) = N(q)$, the two corresponding affine tangent planes Π_p and Π_q are parallel. By part **B** of Theorem 6.1, we have $S \subset \overline{\Omega} \subset \Pi_p^+ \cap \Pi_q^+$. This intersection coincides with one of the two half-spaces, because the boundary planes are parallel, say, with Π_p^+. Hence $q \in \Pi_q^+ \subset \Pi_p^-$ and so $q \in \Pi_p^+ \cap \Pi_p^- = \Pi_p$ and $\Pi_p = \Pi_q$. Assertion **B** also gives us that

$$\{p\} = S \cap \Pi_p = S \cap \Pi_q = \{q\}.$$

Then $p = q$ and N is injective, whether or not S is compact.

Part **C**. If we suppose now that S is compact, then $N(S) \subset \mathbb{S}^2$ is compact, and so, it is closed as a subset of the unit sphere. Since N is a local diffeomorphism, this image is also an open subset of $\mathbb{S}^2$ and, by connectedness, $N(S) = \mathbb{S}^2$, that is, N is surjective. In conclusion, the Gauss map N is a diffeomorphism between S and the sphere $\mathbb{S}^2$.

Part **D**. Consider now the case where S is not compact and, hence, it is not bounded. Then, we may select a sequence $\{q_n\}_{n\in\mathbb{N}}$ of points $q_n \in S$ such that $\lim_{n\to\infty} |q_n| = +\infty$. Thus, if we fix every $q \in S$, since $q, q_n \in S \subset \overline{\Omega}$, by part **A** of Theorem 6.1, we have $[q, q_n] \subset \overline{\Omega}$ for each $n \in \mathbb{N}$. Extracting a subsequence, if necessary, we may find $a \in \mathbb{S}^2$ such that $R^+(q, a) \subset \overline{\Omega}$. We apply part **a** of Lemma 6.4 above to deduce that

$$R^+(p, a) - \{p\} \subset \Omega, \qquad \forall p \in S.$$

From this, if $p \in S$ and if there were another point $p' \in S$ in $R(p, a)$, either p or p' would be on the open half-line starting from the other point in the direction of a and, so, it would belong to Ω. Therefore

$$R(p, a) \cap S = \{p\}, \qquad \forall p \in S.$$

Furthermore, none of these lines may be tangent to S. The reason for this is that, if $R(p, a)$ were tangent to S, of course at p, there would be a neighbourhood of p in the line which would be contained, except for p itself, in the outer domain of S—we have already argued in this way several times above—and this would contradict the fact that $R^+(p, a) - \{p\} \subset \Omega$. Consequently, the orthogonal projection P of $\mathbb{R}^3$ to the plane

$$\Pi = \{x \in \mathbb{R}^3 \mid \langle x, a \rangle = 0\}$$

passing through the origin and perpendicular to a is injective when restricted to S. Moreover, if $p \in S$ and $v \in T_pS$ is in the kernel of $(dP)_p$, we have

$$0 = (dP)_p(v) = P(v) = v - \langle v, a \rangle a.$$

Since $a \notin T_pS$, we infer that $v = 0$. That is, the projection P is also a local diffeomorphism, by the inverse function Theorem 2.75. The final conclusion is that P is a diffeomorphism between S and its image in Π, which is an open convex subset of this plane. In other words, S is a graph over this open set—of a function which is necessarily convex, because Ω is convex. $\qquad\square$

Remark 6.6. If S is an ovaloid, assertion **C** of this second part of the Hadamard-Stoker Theorem 6.5 says that $N : S \to \mathbb{S}^2$ is a diffeomorphism. We can apply the change of variables formula Theorem 5.14 to the constant function 1 on the unit sphere and obtain

$$4\pi = A(\mathbb{S}^2) = \int_{\mathbb{S}^2} 1 = \int_S |\text{Jac } N|.$$

But—see Exercise (2) at the end of Chapter 5—if $p \in S$,

$$|(\text{Jac } N)(p)| = |(dN)_p(e_1) \wedge (dN)_p(e_2)|,$$

where $\{e_1, e_2\}$ is any orthonormal basis of T_pS. For example, taking e_1 and e_2 as principal directions of S at the point p, we have

$$|(\text{Jac } N)(p)| = |k_1(p)k_2(p)||e_1 \wedge e_2| = K(p),$$

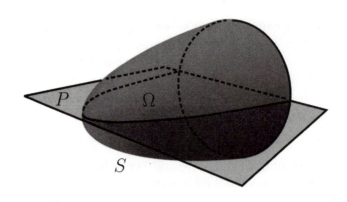

Figure 6.4. *Lemma 6.7*

because the Gauss curvature is positive. It follows that

$$\int_S K(p)\,dp = 4\pi$$

on any ovaloid S.

These two Hadamard-Stoker theorems are the strongest results that one can expect for positively curved closed surfaces. However, we can get more information about this type of surface if we assume that some of the curvatures of the surface, Gauss or mean curvature, do not come near zero. To arrive at this kind of result we will need to show that some pieces of the surfaces that we are dealing with can be expressed as graphs.

Lemma 6.7. *Let P be a plane cutting the inner domain Ω of a closed connected surface S with positive Gauss curvature. Then there exists an open subset V of S which is a graph over $P \cap \Omega$.*

Proof. Denote by U the non-empty open subset $P \cap \Omega$ of the plane P and suppose that, given $v \in \mathbb{S}^2$, normal to P, there is a point $y \in U$ such that the half-line $R^+(y, v)$ does not intersect the surface S. Then $R^+(y, v)$, which is connected, must lie in one of the two connected components of $\mathbb{R}^3 - S$ and, since $y \in R^+(y, v) \cap \Omega$, we will have $R^+(y, v) \subset \Omega$. Thus, by part **a** of Lemma 6.4, we will obtain that $R^+(x, v) \subset \Omega$ for each $x \in U$. But part **b** of the same lemma asserts that Ω contains no straight lines. Consequently, any half-line $R^+(x, -v)$ cannot be entirely included in Ω, whatever $x \in U$ is. However, each of them has the point $x \in R^+(x, -v)$ in Ω. Hence, we have as a consequence that $R^+(x, -v) \cap S \neq \emptyset$ for each $x \in U$.

We may assume from the above that there exists a vector $v \in \mathbb{S}^2$, normal to P, such that all the half-lines $R^+(x, v)$ starting of the points x of U in the direction of v meet the surface S. Thus, we have, by Theorem 6.1, that each intersection $R^+(x, v) \cap \Omega$ is a non-void bounded convex open subset of the half-line $R^+(x, v)$, which moreover contains its origin x, for every $x \in U$. Therefore, each $R^+(x, v) \cap S$ has a unique point for each $x \in U$ and also the straight line $R(x, v)$ is not tangent to S at this point, because otherwise, by part **B** of Theorem 6.1, $R(x, v)$ could not have any points of the inner domain Ω. This means that, if $f : S \to P$ is the restriction to S of the orthogonal projection onto the plane P, we have that f is a diffeomorphism from the open subset $V = f^{-1}(U)$ of S to U, that is, V is a graph over the open set U.   □

We will show now that, on a closed connected positively curved surface which is not compact, the Gauss curvature must approach zero at infinity and the fact that the mean curvature does not come near zero on this class of surfaces supplies us with some bounding properties.

Theorem 6.8 (Bonnet's theorem). *Let S be a connected surface which is closed as a subset of $\mathbb{R}^3$. The following assertions are true:*

A: *If $\inf_{p \in S} K(p) > 0$, then S is compact, that is, S is an ovaloid.*

B: *If S is non-compact, has positive Gauss curvature, and its mean curvature satisfies $\inf_{p \in S} H(p) = h > 0$, then S is a graph over a convex open subset of a plane, whose diameter is less than or equal to $2/h$.*

Proof. Part **A**. Imagine that S is not compact and denote by $k > 0$ the infimum of the Gauss curvature on S. By part **D** of the Hadamard-Stoker Theorem 6.5, we know that S is a graph over a convex open subset of a plane P and that the Gauss map is a diffeomorphism of S over an open subset of $\mathbb{S}^2$. Thus, if $\mathcal{R}$ is any region of the surface S, one has

$$k\, A(\mathcal{R}) = \int_{\mathcal{R}} k \leq \int_{\mathcal{R}} K(p)\, dp = \int_{\mathcal{R}} |(\operatorname{Jac} N)(p)|\, dp.$$

Using change of variables Theorem 5.14, we see that

$$k\, A(\mathcal{R}) \leq \int_{N(\mathcal{R})} 1 = A(N(\mathcal{R})) \leq 4\pi,$$

and this bound serves for each region $\mathcal{R}$ of the surface S. On the other hand, if $]a, b[$ is an open segment of the inner domain Ω of S, we will have $R^+(x, v) \subset \Omega$ for some $v \in \mathbb{S}^2$ normal to the plane P and each $x \in]a, b[$. But, if $n \in \mathbb{N}$, the set

$$U_n = \{x + tv \mid x \in]a, b[,\ 0 < t < n\}$$

is an open subset of the plane Q determined by $]a, b[$ and the vector v. Applying Lemma 6.7 to this plane Q, we deduce that, for each $n \in \mathbb{N}$, there exists an open subset $\mathcal{R}_n$ of the surface which is a graph over the open subset

U_n of the plane Q. Therefore, if we recall Exercise (13) of Chapter 5, we obtain that

$$\frac{4\pi}{k} \geq A(\mathcal{R}_n) \geq A(U_n) = n|a - b|$$

for all natural number n. This is absurd and, hence, S must be a compact surface.

Part **B**. As in the previous case **A**, our hypotheses and part **D** of the Hadamard-Stoker Theorem 6.5 imply that S is a graph over a convex open subset U of some plane P. Let q_1 and q_2 be any two points of U and p_1 and p_2 the two points of S projecting onto them. One of the two unit vectors normal to the plane P, which we will represent by v, satisfies $R^+(p_1, v) \subset \overline{\Omega}$ and, from Lemma 6.4, $R^+(x, v) \subset \Omega$ for each $x \in]p_1, p_2[$. Then, the set

$$V = \{x + tv \mid x \in \,]p_1, p_2[, \ t > 0\}$$

is an open subset of the plane Q determined by the segment $]p_1, p_2[$ and the vector v. Applying Lemma 6.7 to this plane Q, we infer the existence of an open subset S' of the surface S which is a graph over this open subset V of Q. Note now that this V belongs to a plane strip whose width is $|q_1 - q_2|$. We can finish using Exercise (18) at the end of Chapter 3. $\qquad\square$

Remark 6.9. The result above implies that, for positively curved closed surfaces, there are two or three independent directions, depending on whether we do or do not allow the mean curvature or the Gauss curvature to come near zero, with respect to which the surface is bounded. Assertion **A** concerning the Gauss curvature is a weak version of a result by Bonnet which has a very well-known generalization to Riemannian geometry: the Bonnet-Myers theorem.

6.3. Minkowski formulas and ovaloids

Let S be a compact surface. We will work with differentiable fields V on this surface, as introduced in Definition 3.1. That is, $V : S \rightarrow \mathbb{R}^3$ is a differentiable map. If $N_\varepsilon(S) \subset \mathbb{R}^3$, with $\varepsilon > 0$, is a tubular neighbourhood of the surface, we define a vector field X on it—see Definition 5.30—by extending V as a constant along the normal segments of $N_\varepsilon(S)$. That is,

$$(6.1) \quad (X \circ F)(p, t) = X(p + tN(p)) = V(p), \qquad \forall (p, t) \in S \times (-\varepsilon, \varepsilon),$$

where the map F was introduced in (4.1). As a consequence from this definition, X is differentiable as well. Now, we apply to this field X the divergence Theorem 5.31 on the regular domain $V_a(S)$, $a \in (0, \varepsilon)$—see Remark 5.33—and obtain

$$\int_{V_a(S)} \operatorname{div} X = -\int_S \langle X, N \rangle - \int_{S_a} \langle X, N_a \rangle,$$

where N and N_a are, respectively, the inner and the outer normal fields of the surface S and of the parallel S_a at distance $a \in (0, \varepsilon)$. We use the change of variables Theorem 5.14 in the second integral on the right-hand side for the diffeomorphism $F_a : S \to S_a$ that we already studied in Remark 4.29. Moreover, if we take into account that $N_a \circ F_a = -N$ and that $X \circ F_a = V$, we have—see also Example 5.17—

$$\int_{S_a} \langle X, N_a \rangle = - \int_S \langle V, N \rangle (1 - 2aH + a^2 K).$$

Therefore

$$(6.2) \qquad \int_{V_a(S)} \mathrm{div}\, X = -2a \int_S \langle V, N \rangle H + a^2 \int_S \langle V, N \rangle K,$$

for all a with $0 < a < \varepsilon$. On the other hand, we will now compute the divergence of the field X that we have built on the tubular neighbourhood $N_\varepsilon(S)$ from the field V given on the surface S. For each $p \in S$ and $t \in (-\varepsilon, \varepsilon)$, one has, from definition (6.1),

$$(dX)_{p+tN(p)}(N(p)) = 0 \quad \text{and} \quad (dX)_{p+tN(p)}(e_i) = \frac{(dV)_p(e_i)}{1 - tk_i(p)}, \quad i = 1, 2,$$

where $\{e_1, e_2, N(p)\}$ is the basis of $\mathbb{R}^3$ formed by the unit normal of S at p and the principal directions of the surface at this point. Thus

$$(\mathrm{div}\, X)(p + tN(p)) = \mathrm{trace}\, (dX)_{p+tN(p)} = \frac{\langle (dV)_p(e_1), e_1 \rangle}{1 - tk_1(p)} + \frac{\langle (dV)_p(e_2), e_2 \rangle}{1 - tk_2(p)},$$

for all $p \in S$ and $t \in (-\varepsilon, \varepsilon)$. The left-hand side of (6.2), after using (5.1) and the Fubini Theorem 5.18, becomes

$$\int_{V_a(S)} \mathrm{div}\, X = \int_0^a \int_S [(\mathrm{div}\, X) \circ F](p, t) |\mathrm{Jac}\, F|(p, t)\, dp\, dt.$$

But, if we substitute the expression $|\mathrm{Jac}\, F|(p, t) = (1 - tk_1(p))(1 - tk_2(p))$ and the value of the divergence of X that we have just computed, we get

$$[(\mathrm{div}\, X) \circ F](p, t) |\mathrm{Jac}\, F|(p, t) = \langle (dV)_p(e_1), e_1 \rangle + \langle (dV)_p(e_2), e_2 \rangle$$

$$-t[k_2(p)\langle (dV)_p(e_1), e_1 \rangle + k_1(p)\langle (dV)_p(e_2), e_2 \rangle].$$

By the way, this calculation shows that the functions

$$p \in S \longmapsto \langle (dV)_p(e_1), e_1 \rangle + \langle (dV)_p(e_2), e_2 \rangle,$$

$$p \in S \longmapsto k_2(p)\langle (dV)_p(e_1), e_1 \rangle + k_1(p)\langle (dV)_p(e_2), e_2 \rangle,$$

defined on the surface S from the vector field V are differentiable, where $\{e_1, e_2\}$ is an orthonormal basis of the principal directions of T_pS. The first function will be denoted by $\mathrm{div}\, V$ and will be called the divergence of the field V. The reason is obvious if we recall Definition 5.30 of the divergence of a vector field defined on Euclidean space. Putting all this into (6.2), it follows that

$$a \int_S \operatorname{div} V - \frac{1}{2} a^2 \int_S [k_2(p)\langle (dV)_p(e_1), e_1 \rangle + k_1(p)\langle (dV)_p(e_2), e_2 \rangle]\, dp$$

$$= -2a \int_S \langle V, N \rangle H + a^2 \int_S \langle V, N \rangle K,$$

and this is valid for each $a \in (0, \varepsilon)$. Comparing coefficients in this polynomical equality, we may establish the following theorem.

Theorem 6.10 (Divergence theorem for surfaces). *Let S be a compact surface and $V : S \to \mathbb{R}^3$ a differentiable vector field defined on S. Then the following hold.*

> **A:** $\displaystyle \int_S \operatorname{div} V = -2 \int_S \langle V, N \rangle H.$

> **B:** $\displaystyle \int_S [k_2(p)\langle (dV)_p(e_1), e_1 \rangle + k_1(p)\langle (dV)_p(e_2), e_2 \rangle]\, dp = -2 \int_S \langle V, N \rangle K,$
> *where $\{e_1, e_2\}$ is an orthonormal basis of principal directions at the point $p \in S$.*

As usual, H and k_1, k_2 stand for the mean and the principal curvature functions of the surface.

As a direct consequence we can obtain the Minkowski integral formulas; the reader should try to deduce them also from Exercise (9) of Chapter 5 and Example 5.19.

Theorem 6.11 (Minkowski formulas). *Let S be any compact surface, N its inner Gauss map, and H and K its mean and Gauss curvatures. Then, we have the following two integral equalities:*

> **A:** $\displaystyle \int_S (1 + \langle p, N(p) \rangle H(p))\, dp = 0,$

> **B:** $\displaystyle \int_S (H(p) + \langle p, N(p) \rangle K(p))\, dp = 0.$

Proof. It suffices to apply Theorem 6.10 to the vector field $V : S \to \mathbb{R}^3$ given by $V(p) = p$ for all $p \in S$. $\qquad\square$

From these two Minkowski formulas we can deduce, in an alternative way, the two global theorems that we showed in Section 3.6: the Jellett-Liebmann and Hilbert-Liebmann results in Corollaries 3.48 and 3.49.

Corollary 6.12 (Hilbert-Liebmann's theorem). *A compact connected surface with constant Gauss curvature must be a sphere.*

Proof. Let S be a surface under our hypotheses. We know, by Exercise 3.42, that K is a positive constant. Then S is an ovaloid that we can apply to the Hadamard-Stoker Theorem 6.5. After doing this, we may be assured that the support function based on any inner point is negative. Now, we may also suppose that the origin of $\mathbb{R}^3$ is in Ω. If not, a suitable translation would take it to this situation without changing either the hypotheses or the conclusions of the theorem. The Minkowski formulas—the second one divided by $\sqrt{K}$—give

$$\int_S (1 + \langle p, N(p)\rangle H(p))\, dp = 0 \quad \text{and} \quad \int_S \left(\frac{H(p)}{\sqrt{K}} + \langle p, N(p)\rangle \sqrt{K} \right) dp = 0.$$

Subtracting these two equalities, we have

$$\int_S \left((1 - \frac{H(p)}{\sqrt{K}}) + \langle p, N(p)\rangle (H(p) - \sqrt{K}) \right) dp = 0.$$

But, since $H^2 \geq K$, we obtain that either $H(p) \geq \sqrt{K}$ or $H(p) \leq -\sqrt{K}$, for all $p \in S$. The correct alternative is the first one, as Exercise (4) of Chapter 4 shows. Therefore

$$1 - \frac{H(p)}{\sqrt{K}} \leq 0 \quad \text{and} \quad (H(p) - \sqrt{K})\langle p, N(p)\rangle \leq 0$$

at each point p of S. Then, the integrand above is non-positive and has vanishing integral. Using Proposition 5.6, we infer that the integrand is zero. This implies that $H^2 = K$ and then S is totally umbilical. Thus, one finishes by applying Corollary 3.31. □

Corollary 6.13 (Jellett's theorem). *The only compact connected surfaces which are star-shaped with respect to a given point and have constant mean curvature are the spheres centred at this point.*

Proof. If S is a compact connected surface and it is star-shaped with respect to, say, the origin of $\mathbb{R}^3$, then $0 \in \Omega$, where Ω is the inner domain of S. Otherwise, from Exercise (9) of Chapter 6, there would exist some straight line tangent to S passing through the origin. Since the mean curvature H of S is constant, we write the Minkowski formulas—the first one multiplied by the constant H—and we have

$$\int_S (H + H^2\langle p, N(p)\rangle)\, dp = 0 \quad \text{and} \quad \int_S (H + K(p)\langle p, N(p)\rangle)\, dp = 0.$$

Subtracting the two equalities, one gets

$$\int_S (H^2 - K(p))\langle p, N(p)\rangle\, dp = 0.$$

But the integrand is non-positive, from inequality (3.6) and Exercise (4) of Chapter 3. Now we may conclude as in the proof of the previous result. □

Exercise 6.14. ↑ Use the Minkowski formulas to show that, if on an ovaloid the Gauss and mean curvatures coincide everywhere, then it is necessarily a unit sphere.

Exercise 6.15. ↑ Manipulate the Minkowski formulas to prove that an ovaloid such that $HK = 1$ has to be a unit sphere.

6.4. The Alexandrov theorem

The study and characterization of compact surfaces with constant mean curvature are perhaps more involved than the case of constant Gauss curvature. The Alexandrov theorem solves the problem completely. As we did in the case of the Gauss curvature, before starting with the problem of constant mean curvature, we will examine the case in which this curvature does not change its sign. It is worth noting that the fact that this sign is either positive or negative is not significant, but it depends on the orientation that we have chosen on the surface. In any case, when we take the inner normal field on a compact surface, if the mean curvature does not vanish anywhere, it must be positive, as Exercise (4) of Chapter 4 shows.

Theorem 6.16 (Heintze-Karcher's inequality). *Let S a compact surface whose (inner) mean curvature H is positive everywhere. Then*

$$\operatorname{vol} \Omega \leq \frac{1}{3} \int_S \frac{1}{H(p)} \, dp,$$

where Ω is the inner domain determined by S. Moreover, equality holds if and only if S is a sphere.

Proof. Define a subset A of $S \times \mathbb{R}$ by

$$A = \{(p, t) \in S \times \mathbb{R} \mid 0 \leq t \leq \frac{1}{k_2(p)}\},$$

where k_2 is the biggest principal curvature of S corresponding to the inner normal which, since $k_1 \leq H \leq k_2$, is positive everywhere. If we denote by $a > 0$ a real number bigger than the maximum on S of the continuous function $1/k_2$, then A is a compact set contained in $S \times [0, a)$. Exercise (8) of Chapter 4 says that the inner domain Ω determined by S satisfies

$$\Omega \subset F(A),$$

where $F : S \times \mathbb{R} \to \mathbb{R}^3$ is the map defined in (4.1) by $F(p, t) = p + tN(p)$ for all $p \in S$ and $t \in \mathbb{R}$. Now, apply the area formula of Theorem 5.27, on $S \times (0, a)$, to this map F and the characteristic function χ_A of the set A and get the integral equality

$$\int_{\mathbb{R}^3} n(F, \chi_A) = \int_{S \times (0, a)} \chi_A \, |\operatorname{Jac} F|.$$

Clearly $n(F, \chi_A) \geq 1$ on $F(A) \supset \Omega$ and, then, also using the Fubini Theorem 5.18,

$$\operatorname{vol}\Omega \leq \int_S \left(\int_0^a \chi_A(p,t) \, |\operatorname{Jac} F|(p,t) \, dt \right) dp.$$

Now, we substitute the absolute value of the Jacobian of F and take into account the definition of the set A. Thus we have

$$\operatorname{vol}\Omega \leq \int_S \left(\int_0^{1/k_2(p)} |(1 - tk_1(p))(1 - tk_2(p))| \, dt \right) dp.$$

But when $0 \leq t \leq 1/k_2(p)$, the previous integrand is non-negative. Moreover, using inequality (3.6),

$$|(1 - tk_1(p))(1 - tk_2(p))| = 1 - 2tH(p) + t^2 K(p) \leq (1 - tH(p))^2$$

and equality is attained only at the umbilical points. Hence, since $1/k_2(p) \leq 1/H(p)$ for each $p \in S$ and since the function $(1 - tH(p))^2$ is non-negative,

$$\operatorname{vol}\Omega \leq \int_S \left(\int_0^{1/H(p)} (1 - tH(p))^2 \, dt \right) dp,$$

with equality only if the surface S is totally umbilical. The proof is completed by computing the integral on the right-hand side and recalling Corollary 3.31 which classified closed totally umbilical surfaces. $\qquad\square$

This Heintze-Karcher inequality that we have just proved, in the case of surfaces of constant mean curvature, and the first Minkowski formula supply complementary information. They can be combined to yield one of the most beautiful theorems of classical differential geometry.

Theorem 6.17 (Alexandrov's theorem). *If a compact connected surface has constant mean curvature, then it is a sphere.*

Proof. If S is a surface satisfying the conditions of the theorem, Theorem 6.16 implies that $\operatorname{vol}\Omega \leq A(S)/3H$, where Ω is the inner domain of S and $H > 0$ is the value taken by the mean curvature relative to the inner normal. Moreover, if equality occurs, S is a sphere. But, indeed, equality does occur, since in our case the first Minkowski formula of Theorem 6.11 gives

$$A(S) + H \int_S \langle p, N(p) \rangle \, dp = 0.$$

It is enough to mention that

$$\int_S \langle p, N(p) \rangle \, dp = -3\operatorname{vol}\Omega,$$

according to Remark 5.32. $\qquad\square$

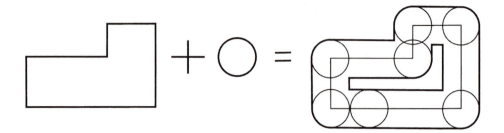

Figure 6.5. *Sum of sets*

6.5. The isoperimetric inequality

Given a compact connected surface S in Euclidean space, it makes sense to talk, after Chapter 4 where we proved the Jordan-Brower theorem, about the volume enclosed by S, and, after Chapter 5 where we introduced integration, about the area of the surface S. Thus, it is natural in this context to pose the isoperimetric question, which can also be thought of in the case of plane curves (see Chapter 9): Among all the compact surfaces enclosing a given volume, which has the least area? To give an answer, we will first give a useful geometrical property of the Lebesgue integral on $\mathbb{R}^3$: the Brunn-Minkowski inequality.

If A and B are two arbitrary subsets of Euclidean space $\mathbb{R}^n$, we define its *sum* $A + B$ as the set

$$A + B = \{a + b \mid a \in A, \ b \in B\}.$$

For example, if A is the open interval (a_1, a_2) of $\mathbb{R}$ and B is the open interval (b_1, b_2) of $\mathbb{R}$, the sum is

$$A + B = (a_1, a_2) + (b_1, b_2) = (a_1 + a_2, b_1 + b_2).$$

The following exercises, which point out some of the properties of this sum of sets, will be used in the text that follows.

Exercise 6.18. ↑ Let $A, B \subset \mathbb{R}^n$. If some of these subsets are open, prove that the sum $A + B$ is open as well.

Exercise 6.19. Suppose that $A, B \subset \mathbb{R}^n$ are two bounded subsets of Euclidean space. Show that $A + B$ is bounded.

Exercise 6.20. ↑ Let I_1, I_2, I_3 and J_1, J_2, J_3 be six bounded open intervals of the real line. The two subsets of $\mathbb{R}^3$ given by

$$A = I_1 \times I_2 \times I_3 \quad \text{and} \quad B = J_1 \times J_2 \times J_3$$

are called *parallelepipeds with edges parallel to the coordinate axes*. Prove that

$$A + B = (I_1 + J_1) \times (I_2 + J_2) \times (I_3 + J_3)$$

and that, if A and B are disjoint, there exists a plane parallel to one of the coordinate planes separating A and B.

Exercise 6.21. If A and B are two arc-connected subsets of $\mathbb{R}^3$, prove that the sum $A + B$ is also an arc-connected subset.

Theorem 6.22 (Brunn-Minkowski's inequality). *Let A and B be two bounded open subsets of Euclidean space $\mathbb{R}^3$. Then*

$$(\operatorname{vol} A)^{1/3} + (\operatorname{vol} B)^{1/3} \le (\operatorname{vol}(A + B))^{1/3}.$$

Proof. Suppose first that A and B are the parallelepipeds $I_1 \times I_2 \times I_3$ and $J_1 \times J_2 \times J_3$, respectively, where the I_i and the J_i, $i = 1, 2, 3$, are bounded open intervals of $\mathbb{R}$ with lengths a_i and b_i. Then

$$\frac{(\operatorname{vol} A)^{1/3} + (\operatorname{vol} B)^{1/3}}{(\operatorname{vol}(A + B))^{1/3}} = \frac{\left(\prod_{i=1}^3 a_i\right)^{1/3} + \left(\prod_{i=1}^3 b_i\right)^{1/3}}{\left(\prod_{i=1}^3 (a_i + b_i)\right)^{1/3}},$$

and thus

$$\frac{(\operatorname{vol} A)^{1/3} + (\operatorname{vol} B)^{1/3}}{(\operatorname{vol}(A + B))^{1/3}} = \left(\prod_{i=1}^3 \frac{a_i}{a_i + b_i}\right)^{1/3} + \left(\prod_{i=1}^3 \frac{b_i}{a_i + b_i}\right)^{1/3}.$$

But the geometrical mean of three numbers is always less than or equal to their arithmetical mean. Hence

$$\frac{(\operatorname{vol} A)^{1/3} + (\operatorname{vol} B)^{1/3}}{(\operatorname{vol}(A + B))^{1/3}} \le \frac{1}{3} \sum_{i=1}^3 \frac{a_i}{a_i + b_i} + \frac{1}{3} \sum_{i=1}^3 \frac{b_i}{a_i + b_i} = 1.$$

So, in this particular case, the inequality claimed in the theorem is verified.

As a second case, suppose now that the bounded open sets A and B are disjoint finite unions of bounded open parallelepipeds whose edges are parallel to the coordinate axes. That is,

$$A = \bigcup_{i=1}^n A_i \quad \text{and} \quad B = \bigcup_{j=1}^m B_j,$$

where A_i and B_j are as in the first case above. We are going to prove the Brunn-Minkowski inequality for this kind of set using induction on the total number $n + m$ of involved parallelepipeds. It is clear that the result is true if $n + m = 2$, because this is the particular case above. Assume that $n + m \ge 3$ and that the desired inequality is true for disjoint finite unions of bounded open parallelepipeds with edges parallel to the coordinate axes and total number less than $n + m$. Since $n + m > 2$, we have either $n > 1$ or $m > 1$. Suppose the first case. From Exercise 6.20, there exists a plane P parallel to the coordinate axes separating A_1 and A_2. Let P^+ and P^- be the two open half-spaces in which P divides $\mathbb{R}^3$ and let $A^+ = A \cap P^+$ and $A^- = A \cap P^-$.

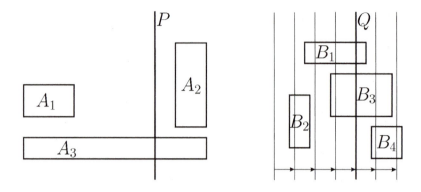

Figure 6.6. *Choice of P and Q*

Then A^+ and A^- are also finite unions of parallelepipeds whose faces are parallel to the coordinate planes, namely

$$A^+ = \bigcup_{i=1}^{n^+} A_i^+ \quad \text{and} \quad A^- = \bigcup_{i=1}^{n^-} A_i^-,$$

where $n^+ < n$ and $n^- < n$, since P separates at least A_1 and A_2. On the other hand, we can find another plane Q parallel to P such that

(6.3)
$$\frac{\text{vol}\,A^+}{\text{vol}\,A} = \frac{\text{vol}\,B^+}{\text{vol}\,B},$$

where we have used the same notation as before. The reason is that, as soon as P is chosen, the fraction on the left-hand side is a number between 0 and 1 and, if one takes Q on the left of all the blocks of B and displaces it to the right until surpassing all of them, the fraction on the right side is a continuous function of the position of Q taking values from 0 to 1; see Figure 6.6.

In this way, since $\text{vol}\,A^+ + \text{vol}\,A^- = \text{vol}\,A$ and $\text{vol}\,B^+ + \text{vol}\,B^- = \text{vol}\,B$, we obtain

(6.4)
$$\frac{\text{vol}\,A^-}{\text{vol}\,A} = \frac{\text{vol}\,B^-}{\text{vol}\,B}.$$

Moreover, B^+ and B^- are also disjoint finite unions of open parallelepipeds with faces parallel to the coordinate planes

$$B^+ = \bigcup_{j=1}^{m^+} B_j^+ \quad \text{and} \quad B^- = \bigcup_{j=1}^{m^-} B_j^-,$$

where $m^+ \leq m$ and $m^- \leq m$, because we may not suppose that Q separates two blocks of B. Hence we can apply the induction hypothesis to the pairs

A^+, B^+ and A^-, B^- because the total numbers of blocks satisfy $n^+ + m^+ < n + m$ and $n^- + m^- < n + m$, respectively. As a consequence

$$\text{vol}\,(A^+ + B^+) \geq [(\text{vol}\,A^+)^{1/3} + (\text{vol}\,B^+)^{1/3}]^3,$$

(6.5) $$\text{vol}\,(A^- + B^-) \geq [(\text{vol}\,A^-)^{1/3} + (\text{vol}\,B^-)^{1/3}]^3.$$

On the other hand, since $A^+ \subset P^+$ and $B^+ \subset Q^+$, then $A^+ + B^+ \subset P^+ + Q^+ = (P + Q)^+$ and, analogously, $A^- + B^- \subset (P + Q)^-$. Bearing in mind that $P + Q$ is another plane of $\mathbb{R}^3$, we conclude that $A^+ + B^+$ and $A^- + B^-$ are disjoint. Thus, taking (6.5) into account,

$$\text{vol}\,(A + B) \geq \text{vol}\,(A^+ + B^+) + \text{vol}\,(A^- + B^-)$$
$$\geq [(\text{vol}\,A^+)^{1/3} + (\text{vol}\,B^+)^{1/3}]^3 + [(\text{vol}\,A^-)^{1/3} + (\text{vol}\,B^-)^{1/3}]^3.$$

But, from (6.3) and (6.4),

$$\text{vol}\,(A + B) \geq \text{vol}\,A^+ \left[1 + \left(\frac{\text{vol}\,B}{\text{vol}\,A}\right)^{1/3}\right]^3 + \text{vol}\,A^- \left[1 + \left(\frac{\text{vol}\,B}{\text{vol}\,A}\right)^{1/3}\right]^3$$

$$\geq \text{vol}\,A \left[1 + \left(\frac{\text{vol}\,B}{\text{vol}\,A}\right)^{1/3}\right]^3 = [(\text{vol}\,A)^{1/3} + (\text{vol}\,B)^{1/3}]^3,$$

and so the induction is complete and the result is also true for this second particular case.

Now consider the general case. For this, denote by A and B any two bounded open sets in $\mathbb{R}^3$. Using the elementary theory of Lebesgue integration, there are two sequences A_n and B_n, $n \in \mathbb{N}$, of open sets in $\mathbb{R}^3$ of the type that we have considered in the second particular case, such that $A_n \subset A$ and $B_n \subset B$ and

$$\lim_{n \to \infty} \text{vol}\,A_n = \text{vol}\,A \quad \text{and} \quad \lim_{n \to \infty} \text{vol}\,B_n = \text{vol}\,B.$$

Hence $A_n + B_n \subset A + B$, for every $n \in \mathbb{N}$, and so

$$\text{vol}\,(A + B)^{1/3} \geq \text{vol}\,(A_n + B_n)^{1/3} \geq (\text{vol}\,A_n)^{1/3} + (\text{vol}\,B_n)^{1/3},$$

for every $n \in \mathbb{N}$. Taking limits as n tends to infinity, one gets the Brunn-Minkowski inequality that we wanted to prove. $\square$

Now, given a compact connected surface, let us see that there is a natural situation where we can apply the Brunn-Minkowski inequality. In fact, take $\varepsilon > 0$ small enough for $N_\varepsilon(S)$ to be a tubular neighbourhood. For every $t \in (0, \varepsilon)$, the parallel surface S_t at oriented distance t is in the inner domain Ω determined by S. Hence, also $\Omega_t \subset \Omega$, where Ω_t is the inner domain determined by S_t; see Remark 5.33. Denote by B_t^3 the open ball of radius t centred at the origin of $\mathbb{R}^3$. Then, if $p \in (\Omega_t + B_t^3) \cap S$, we have $p = q + v$ with $q \in \Omega_t$ and $v \in B_t^3$. Thus $|p - q| = |v| < t$, and so $\text{dist}\,(q, S) < t$. But this is impossible because the minimum of the square of the distance function from the points of S to q has to be reached at some point $p_0 \in S$ such that q is on the normal line of S at p_0 and, since $q \in \Omega_t$, we have $|p_0 - q| \geq t$. Thus, the sum set $\Omega_t + B_t^3$, which is connected by Exercise 6.21 above, is included in one of the two connected components of $\mathbb{R}^3 - S$. Since $\Omega_t \subset \Omega$, we finally have that

$$\Omega_t + B_t^3 \subset \Omega, \qquad \forall t \in (0, \varepsilon).$$

Now applying the Brunn-Minkowski inequality that we have just proved, it can be seen that

$$\text{vol} \, \Omega \geq \text{vol} \, (\Omega_t + B_t^3) \geq [(\text{vol} \, \Omega_t)^{1/3} + (\text{vol} \, B_t^3)^{1/3}]^3,$$

for each $t \in (0, \varepsilon)$. Developing the cube on the right-hand side and removing the terms with exponent greater than $1/3$ for the volume of the ball B_t^3, we conclude that

$$\text{vol} \, \Omega \geq \text{vol} \, \Omega_t + 3(\text{vol} \, B_t^3)^{1/3}(\text{vol} \, \Omega_t)^{2/3}, \qquad \forall t \in (0, \varepsilon).$$

Taking into account that, as shown in Remark 5.32, $\text{vol} \, B_t^3 = (4\pi/3)t^3$, we get

$$\frac{\text{vol} \, \Omega - \text{vol} \, \Omega_t}{t} \geq 3 \left(\frac{4\pi}{3} \right)^{1/3} (\text{vol} \, \Omega_t)^{2/3}, \qquad \forall t \in (0, \varepsilon).$$

In Example 5.19 we computed the volume $\text{vol} \, \Omega_t$ enclosed by the parallel surface S_t. Then, taking the limit in the previous inequality as t goes to zero, we have

$$A(S) \geq 3 \left(\frac{4\pi}{3} \right)^{1/3} (\text{vol} \, \Omega)^{2/3}.$$

This inequality involving the area of a compact connected surface and the volume of its inner domain is the *isoperimetric inequality*, which can also be written as

$$(6.6) \qquad \frac{A(S)^3}{(\text{vol} \, \Omega)^2} \geq 36\pi = \frac{A(\mathbb{S}^2)^3}{(\text{vol} \, B^3)^2}.$$

That is, the function $S \mapsto A(S)^3/(\text{vol} \, \Omega)^2$, defined for each compact connected surface S, attains its minimum at any sphere. But the proof that we have just given does not permit us to show that these are indeed the only minima. However, we can do this in a small roundabout way.

For each differentiable function $f : S \to \mathbb{R}$ we may find a number $\delta > 0$ such that, if $|t| < \delta$, then $tf(S) \subset (-\varepsilon, \varepsilon)$, where $\varepsilon > 0$ is chosen so that $N_\varepsilon(S)$ is a tubular neighbourhood. In this way, it makes sense to consider the set

$$S_t(f) = \{x \in N_\varepsilon(S) \, | \, x = p + tf(p)N(p), \ p \in S, \ |t| < \delta\},$$

which is a compact surface, according to Exercise (3) of Chapter 4, such that $S_0(f) = S$, for any f. The family of surfaces $S_t(f)$ with $|t| < \delta$ will be called the *variation* of S corresponding to the function f. Note that each $S_t(f)$ is diffeomorphic to S by means of the map $\phi_t : S \to S_t(f)$ given by

$$\phi_t(p) = p + tf(p)N(p) = F(p, tf(p)), \qquad \forall p \in S,$$

where $F : S \times \mathbb{R} \to \mathbb{R}^3$ is, as usual, the map defined in (4.1), which is injective and which has as an inverse the projection π of the tubular neighbourhood $N_\varepsilon(S)$ restricted to $S_t(f)$; see Exercise (1) at the end of Chapter 4. The following two examples, which are studied by using the change of variables formula, the divergence theorem, and the differentiable dependence of parameters, will finish to help us see what happens with the case of equality in the isoperimetric inequality.

Example 6.23 (First variation of area). Let $S_t(f)$, $|t| < \delta$, for some $\delta > 0$, be the variation of the surface S corresponding to the function $f : S \to \mathbb{R}$. Let us see that the function

$$t \in (-\delta, \delta) \longmapsto A(t) = A(S_t(f))$$

is differentiable and that

$$\frac{d}{dt}\bigg|_{t=0} A(S_t(f)) = -2 \int_S f(p)H(p)\, dp.$$

In fact, if e_1, e_2 are the principal directions of S at the point $p \in S$, we have

$$(d\phi_t)_p(e_i) = (1 - tf(p)k_i(p))e_i + t(df)_p(e_i)N(p), \qquad i = 1, 2.$$

Then, we can see that

$$(6.7) \quad (d\phi_t)_p(e_1) \wedge (d\phi_t)_p(e_2) = [1 - 2tf(p)H(p)]N(p) - t(\nabla f)_p + t^2 G(p, t),$$

where $p \in S$, $|t| < \delta$, ∇f is the gradient field of the function f—see Exercise (17) at the end of this chapter—and G is a certain differentiable function defined on $S \times (-\delta, \delta)$ taking values in $\mathbb{R}^3$. Thus, since from the change of variables Theorem 5.14

$$A(t) = A(S_t(f)) = \int_{S_t(f)} 1 = \int_S |\text{Jac } \phi_t|,$$

equation (6.7) says, bearing in mind the differentiable dependence of parameters of Remark 5.22, that $A(t)$ is differentiable at $t = 0$ and

$$A'(0) = \int_S \frac{d}{dt}\bigg|_{t=0} |\text{Jac } \phi_t| = \int_S \frac{d}{dt}\bigg|_{t=0} |(d\phi_t)_p(e_1) \wedge (d\phi_t)_p(e_2)|\, dp.$$

Therefore, using (6.7),

$$\frac{d}{dt}\bigg|_{t=0} |(d\phi_t)_p(e_1) \wedge (d\phi_t)_p(e_2)| = -2f(p)H(p).$$

From this we can easily get the required formula.

Example 6.24 (First variation of volume). Represent by $\Omega_t(f)$ the inner domain determined by the compact surface $S_t(f)$. We will prove that the function

$$t \in (-\delta, \delta) \longmapsto V(t) = \text{vol}\,\Omega_t(f)$$

is differentiable and that

$$\frac{d}{dt}\bigg|_{t=0} \text{vol}\,\Omega_t(f) = -\int_S f(p)\,dp.$$

For this, recall that

$$V(t) = -\frac{1}{3}\int_{S_t} \langle N_t(p), p\rangle\,dp,$$

by Remark 5.32, where N_t is the inner normal of $S_t(f)$. Again using the change of variables formula for the diffeomorphism ϕ_t, we obtain

$$V(t) = -\frac{1}{3}\int_S \langle N_t \circ \phi_t, \phi_t\rangle |\text{Jac}\,\phi_t|.$$

Since the diffeomorphism ϕ_t between S and $S_t(f)$ is the identity map when $t = 0$, we have

$$(N_t \circ \phi_t)(p) = \frac{(d\phi_t)_p(e_1) \wedge (d\phi_t)_p(e_2)}{|(d\phi_t)_p(e_1) \wedge (d\phi_t)_p(e_2)|} = \frac{(d\phi_t)_p(e_1) \wedge (d\phi_t)_p(e_2)}{|\text{Jac}\,\phi_t|(p)}$$

and so, from (6.7),

$$V(t) = -\frac{1}{3}\int_S \langle (d\phi_t)_p(e_1) \wedge (d\phi_t)_p(e_2), \phi_t(p)\rangle\,dp$$

$$= -\frac{1}{3}\int_S tf(p)\,dp - \frac{1}{3}\int_S [(1 - 2tf(p)H(p))\langle N(p), p\rangle - t\langle(\nabla f)_p, p\rangle]\,dp$$

$$-\frac{t^2}{3}\int_S D(p, t)\,dp,$$

where D is a differentiable function defined on $S \times (-\delta, \delta)$ and ∇f is again the gradient of f; see Exercise (17) at the end of this chapter. Hence V is differentiable at $t = 0$ and

$$V'(0) = \frac{1}{3}\int_S [-f(p) + 2f(p)H(p)\langle N(p), p\rangle + \langle(\nabla f)_p, p\rangle]\,dp.$$

Now, if we solve Exercise (20) at the end of this chapter, one has that

$$2\int_S f(p)\langle p, N(p)\rangle H(p)\,dp = -\int_S [2f(p) + \langle(\nabla f)_p, p\rangle]\,dp.$$

Substituting into the previous equality, we get the integral formula that we called the first variation of volume formula.

Theorem 6.25 (Space isoperimetric inequality). *Let S be a compact connected surface. Then $A(S)^3 \geq 36\pi\,(\mathrm{vol}\,\Omega)^2$, where Ω is its inner domain. Moreover, equality occurs if and only if S is a sphere.*

Proof. Inequality was already proved in (6.6). Furthermore, we know that equality is satisfied for spheres of arbitrary radius. If S is a surface which attains equality and $f : S \to \mathbb{R}$ is any differentiable function, we consider the variation $S_t(f)$ of S relative to f, which is defined for $|t| < \delta$ with $\delta > 0$. Then, the function $h : (-\delta, \delta) \to \mathbb{R}$ given by

$$h(t) = A(S_t(f))^3 - 36\pi\,(\mathrm{vol}\,\Omega_t(f))^2$$

is differentiable from the above examples and has a minimum at $t = 0$ by our hypothesis. Consequently

$$0 = h'(0) = 3A(S)^2 \left.\frac{d}{dt}\right|_{t=0} A(S_t(f)) - 72\pi\,\mathrm{vol}\,\Omega \left.\frac{d}{dt}\right|_{t=0} \mathrm{vol}\,\Omega_t(f).$$

That is, if S attains equality in the isoperimetric inequality, one has, using the first two variation formulas computed earlier,

$$\int_S f(p)[12\pi\,\mathrm{vol}\,\Omega - A(S)^2 H(p)]\,dp = 0$$

for each differentiable function f defined on S. For instance, choosing f as

$$f(p) = 12\pi\,\mathrm{vol}\,\Omega - A(S)^2 H(p), \qquad \forall p \in S,$$

we deduce that f itself must be identically zero. In conclusion, the surface S has constant mean curvature and we conclude by applying the Alexandrov Theorem 6.17. $\qquad\square$

Exercises

(1) Let S be an ovaloid in $\mathbb{R}^3$. Show that

$$\int_S H^2 \geq 4\pi$$

and that equality occurs if and only if S is a sphere.

(2) $\uparrow$ Suppose that there exists some $c \in \mathbb{R}$ such that

$$cA(S) \leq \left(\int_S H(p)\,dp\right)^2$$

for all ovaloids S and that there exists an ovaloid attaining equality. Prove, using the parallel surfaces of S, that $c = 4\pi$.

(3) ↑ Let S be an ovaloid and $N : S \to \mathbb{S}^2$ its Gauss map. Show that there is a unique diffeomorphism $\phi : S \to S$ such that $N \circ \phi = -N$. If $|\phi(p) - p|$ is the constant function $c \in \mathbb{R}$ on S, we say that the ovaloid has constant *width* c. Prove that, in this case, $k_i \circ \phi = k_i/(1 - ck_i)$, $i = 1, 2$.

(4) ↑ Let S be a compact connected surface and let N be a Gauss map for S. Let s and k_1, k_2 be the corresponding support function based on the origin and the principal curvatures. If there is a real number $c \in \mathbb{R} - \{0\}$ such that neither s takes the value c nor k_1, k_2 the value $-1/c$, prove—applying the *monodromy theorem*, which asserts that a local diffeomorphism from a compact surface to a sphere has to be a diffeomorphism, to the map $p \in S \mapsto (p - cN(p))/|p - cN(p)| \in \mathbb{S}^2$— that S is diffeomorphic to the sphere.

(5) ↑ Let S be a compact connected surface contained in an open ball of radius $r > 0$. If its principal curvatures satisfy $k_1, k_2 > -1/r$, then S is diffeomorphic to the sphere. In particular, prove that a compact connected surface whose principal curvatures are non-negative everywhere must be diffeomorphic to a sphere.

(6) ↑ Let S be an ovaloid and R a straight line of $\mathbb{R}^3$. Demonstrate that $S \cap R = \emptyset$, or $S \cap R = \{p\}$ and R is tangent to S at p, or $S \cap R$ has exactly two points.

(7) Show that the following integral formulas are always valid:

$$\int_S \langle N, a \rangle H = 0 \quad \text{and} \quad \int_S \langle N, a \rangle K = 0,$$

where S is a compact surface, N its Gauss map, H and K its mean and Gauss curvatures, and $a \in \mathbb{R}^3$ an arbitrary vector.

(8) ↑ Let S be a compact surface contained in a closed ball of radius $r > 0$ and such that its mean curvature satisfies $|H| \leq 1/r$. Prove that S is a sphere of radius r.

(9) ↑ Let S be an ovaloid and $F_t : S \to \mathbb{R}^3$ the map given by $F_t(p) = p + tN(p)$, where N is the outer Gauss map of S. Show that, for each $t > 0$, the image $S_t = F_t(S)$ is an ovaloid and moreover $F_t : S \to S_t$ is a diffeomorphism. (Use the monodromy theorem—see Exercise (4)—for the map $\phi = F_t/|F_t|$.)

(10) ↑ Use the isoperimetric inequality to prove that, on each ovaloid S, we have

$$\left(\int_S H(p) \, dp \right)^2 \geq 4\pi A(S).$$

(11) ↑ Let S be a compact connected surface with $K \geq 1$. If one can find an open unit ball inside of the inner domain of S, then S is a unit sphere.

(12) ↑ An ovaloid whose (inner) mean and Gauss curvatures satisfy $H+K = 2$ must be a unit sphere.

(13) An ovaloid whose mean and Gauss curvatures satisfy $H^2 + K^2 = 2$ must also be a unit sphere.

(14) ↑ Let S be an ovaloid with $H \geq 1/r$ and $K \leq H/r$, where $r > 0$. Show that S is a sphere of radius r.

(15) Let S be a compact surface. Show that S has constant mean curvature if and only if, for each differentiable function f defined on S, such that $\mathrm{vol}\,\Omega_t(f) = \mathrm{vol}\,\Omega$, $|t| < \delta$, one has that

$$\frac{d}{dt}\bigg|_{t=0} A(S_t(f)) = 0.$$

(16) Let S be a compact surface and $V : S \to \mathbb{R}^3$ a tangent vector field. Prove that

$$\int_S (\mathrm{div}\, V)(p)\, dp = 0,$$

$$\int_S \{k_1(p)\langle (dV)_p(e_1), e_1\rangle + k_2(p)\langle (dV)_p(e_2), e_2\rangle\}\, dp = 0,$$

where $\{e_1, e_2\}$ is a basis of principal directions at $T_p S$ for each $p \in S$.

(17) Let $f : S \to \mathbb{R}$ be a differentiable function defined on a surface S. We use the term *gradient* of f for the field of tangent vectors denoted by $\nabla f : S \to \mathbb{R}^3$ and given by

$$\begin{cases} \langle (\nabla f)(p), v\rangle = (df)_p(v), & v \in T_p S, \\ \langle (\nabla f)(p), N(p)\rangle = 0, \end{cases}$$

where $N(p)$ is a unit normal to S at p. Prove that ∇f is a differentiable field and that, if it is identically zero, f is constant on each connected component of S.

(18) If $f : S \to \mathbb{R}$ is a differentiable function on a surface S, we define the *Laplacian* of f by the expression $\Delta f = \mathrm{div}\, \nabla f$. Prove that

$$\Delta(fg) = f\,\Delta g + g\,\Delta f + 2\,\langle \nabla f, \nabla g\rangle,$$

for each two functions f and g defined on S.

(19) ↑ Let S be a compact surface and $f : S \to \mathbb{R}$ a differentiable function. Prove that, if $\Delta f \geq 0$, then f is constant on each connected component of S.

(20) If $f : S \to \mathbb{R}$ is a differentiable function on a compact surface, apply the divergence theorem to the field V defined on S by $V(p) = f(p)p$ for

each $p \in S$ and show that

$$\int_S \langle (\nabla f)_p, p \rangle \, dp + 2 \int_S f(p) \, dp = -2 \int_S f(p) H(p) \langle p, N(p) \rangle \, dp.$$

Hints for solving
the exercises

Exercise 6.14: Subtracting the two Minkowski formulas, if $H = K$, one has

$$\int_S 1 = \int_S H.$$

On the other hand, inequality (3.6) gives $H = K \le H^2$ and, so, $H \ge 1$, because $H = K > 0$. We finish by using Proposition 5.6 about properties of integration.

Exercise 6.15: Subtract the two Minkowski formulas again. If $HK = 1$, we see that

$$\int_S \left(1 - H(p) + \langle N(p), p \rangle \left(H(p) - \frac{1}{H(p)} \right) \right) dp = 0.$$

Again inequality (3.6) and the hypothesis $HK = 1$ imply that $H \ge 1$. To finish, take into account that the support function based on the origin—assumed to be in Ω—is strictly negative.

Exercise 6.18: Suppose that $A \subset \mathbb{R}^n$ is open. By definition

$$A + B = \bigcup_{b \in B} (A + b).$$

But each $A + b$, with $b \in B$, is open because translations of $\mathbb{R}^n$ are homeomorphisms. Then $A + B$ is open.

Exercise 6.20: The required equality follows directly from the definition of the sum of sets. Suppose that $A \cap B = \emptyset$. Then, at least one of the intersections $I_1 \cap J_1$, $I_2 \cap J_2$, $I_3 \cap J_3$ has to be empty. Suppose that it is the first one. Thus, there exists $a \in \mathbb{R}$ such that the intervals I_1 and J_1 are on different sides of a. Hence, the plane of the equation $x = a$ separates the two sets A and B.

Exercise (2): Let S be an ovaloid attaining equality and let S_t be its parallel surface at distance t, with t small. The function

$$t \longmapsto \left(\int_{S_t} H_t(p)\, dp \right)^2 - cA(S_t)$$

has a minimum at $t = 0$. Using Exercise (16) of Chapter 3, Theorem 5.14 of the change of variables, and Example 5.17, we can explicitly compute this function, namely

$$t \longmapsto \left(\int_S H - t \int_S K \right)^2 - c \left(A(S) - 2t \int_S H + t^2 \int_S K \right).$$

This function is differentiable. Thus, its derivative at $t = 0$ vanishes. Hence

$$\left(\int_S H \right) \left(c - \int_S K \right) = 0.$$

Since S is an ovaloid, $H^2 \geq K > 0$. Therefore

$$c = \int_S K = 4\pi,$$

where the last equality follows from the Hadamard-Stoker Theorem 6.5.

Exercise (3): It is clear that the required involution $\phi : S \to S$ exists because the Gauss map $N : S \to \mathbb{S}^2$ is a diffeomorphism and, so, it is enough to put $\phi = N^{-1} \circ A \circ N$, where A is the antipodal map of the sphere. Applying the chain rule to the equality $N \circ \phi = -N$, one has

$$(*) \qquad\qquad (dN)_{\phi(p)} \circ (d\phi)_p = -(dN)_p, \qquad \forall p \in S.$$

On the other hand, if $|\phi(p) - p| = c$ for each $p \in S$ and some $c \in \mathbb{R}$, one has that $c > 0$ and

$$(**) \qquad \langle \phi(p) - p, (d\phi)_p(v) - v \rangle = 0, \qquad \forall p \in S \quad \forall v \in T_pS.$$

Let us see that $(d\phi)_p - I_{T_pS}$ is trivial. In fact, if $(d\phi)_p(v) = v$ for some $v \in T_pS$, from $(*)$ we have $(dN)_{\phi(p)}(v) = -(dN)_p(v)$. After scalar multiplication by v, we can deduce that

$$\sigma_{\phi(p)}(v, v) = -\sigma_p(v, v).$$

Since S is an ovaloid, σ is definite and, so, $v = 0$. Then $(d\phi)_p - I_{T_pS}$ is injective and, as a consequence, its image is the whole of the plane T_pS. Equality $(**)$ may then be rewritten in another way:

$$\phi(p) - p \perp T_pS, \qquad \forall p \in S.$$

Therefore $\phi(p) - p = cN(p)$, where N is the inner normal (for the choice of sign, consult Remark 6.2). Therefore, if $e_1, e_2 \in T_pS$ are principal directions, we obtain

$$(d\phi)_p(e_i) - e_i = -ck_i(p)e_i, \qquad i = 1, 2.$$

Hence $e_1, e_2 \in T_p S = T_{\phi(p)} S$ are eigenvectors for $(dN)_{\phi(p)}$ and we are finished.

Exercise (4): The map $\phi : S \to \mathbb{S}^2$ given by

$$\phi(p) = \frac{p - cN(p)}{|p - cN(p)|}, \qquad \forall p \in S$$

is well defined and is differentiable, because $p - cN(p)$ cannot vanish anywhere since the support function does not take the value c. From the definition we obtain

$$(d\phi)_p(v) = \frac{v - c(dN)_p(v)}{|p - cN(p)|} - \frac{\langle p - cN(p), v - c(dN)_p(v) \rangle}{|p - cN(p)|^3}(p - cN(p)),$$

for each $p \in S$ and $v \in T_p S$. Hence, if $e_1, e_2 \in T_p S$ are principal directions at $p \in S$,

$$|\mathrm{Jac}\, \phi|(p) = \frac{1}{|p - cN(p)|^3}|1 + ck_1(p)||1 + ck_2(p)||\langle N(p), p \rangle - c|.$$

By our hypotheses, this Jacobian never vanishes. Then ϕ is a local diffeomorphism, by the inverse function theorem. One concludes the exercise using the monodromy theorem cited above.

Exercise (5): If S is contained in an open ball of radius $r > 0$, let us put the origin as the centre of this ball. Thus, if $p \in S$,

$$|\langle p, N(p) \rangle| \leq |p| < r.$$

Then, the support function based on the origin never takes the value r. Now apply Exercise (4) above.

Exercise (6): By Remark 6.3, if $R \cap \Omega \neq \emptyset$, then $R \cap S$ consists of exactly two points. Suppose, then, that $R \cap \Omega = \emptyset$ and that $R \cap S \neq \emptyset$. Hence $R \cap S = R \cap \overline{\Omega}$ must be, by Theorem 6.1, a non-empty closed convex subset of R, that is, a non-empty closed interval. If it had more than one point, by Exercise 3.33, S would have non-positive Gauss curvature at the interior points of this interval. Since S is an ovaloid, $R \cap S$ has only a point. Finally, apply Exercise (7) of Chapter 4 to see that R is tangent to S at this point.

Exercise (8): If the centre of the ball where S is included is the origin, we use the first Minkowski formula and

$$A(S) = -\int_S \langle p, N(p) \rangle H(p)\, dp \leq \int_S |p||H(p)|\, dp \leq A(S).$$

Then we may conclude, for example, using the Alexandrov Theorem 6.17 or Exercise (4) of Chapter 2.

Exercise (9): The same calculation as in Exercise (4) gives

$$\left|\text{Jac}\,\frac{F_t}{|F_t|}\right|(p) = \frac{1}{|F_t(p)|^3}|1 - tk_1(p)||1 - tk_2(p)||\langle p, N(p)\rangle + t|,$$

which, by hypothesis, does not vanish, since $k_1, k_2 \leq 0$ and $\langle p, N(p)\rangle > 0$, $p \in S$, because N is the outer normal. The monodromy theorem implies that $\phi = F_t/|F_t|$ is a diffeomorphism for each $t \geq 0$. As a consequence, ϕ and, so, each F_t is injective. From Exercise (9) of Chapter 2, $S_t = F_t(S)$ is a compact surface and $F_t : S \rightarrow S_t$ is a diffeomorphism. Exercise (16) of Chapter 3 asserts that its Gauss curvature is always positive.

Exercise (10): We know, after solving Exercise (9), that it makes sense to consider outer parallel surfaces S_t at arbitrary distance $t \geq 0$. As each S_t is compact and connected, we can apply to each of them the isoperimetric inequality (6.6) and obtain

$$A(S_t)^3 \geq 36\pi(\text{vol}\,\Omega_t)^2, \qquad \forall t \geq 0.$$

Taking into account Example 5.17 and the analogue, for outer parallel surfaces, of Example 5.19, we get the polynomical inequality

$$\left(A(S) + 2t\int_S H + 4\pi t^2\right)^3 \geq 36\pi\left(\text{vol}\,\Omega + tA(S) + t^2\int_S H + \frac{4\pi}{3}t^3\right)^2,$$

which is valid for each $t > 0$. Dividing by t^6 and denoting by s the inverse $1/t$, we have that the function

$$f(s) = \left(4\pi + 2s\int_S H + s^2 A(S)\right)^3 - 36\pi\left(\frac{4\pi}{3} + s\int_S H + s^2 A(S) + s^3\text{vol}\,\Omega\right)^2$$

is non-negative for $s > 0$. One can easily see that $f(0) = 0$ and $f'(0) = 0$. Therefore, we have

$$0 \leq f''(0) = 24\pi\left(\int_S H\right)^2 - 96\pi^2 A(S).$$

Exercise (11): Integrating the inequality $K \geq 1$, we obtain

$$A(S) \leq 4\pi = A(\mathbb{S}^2),$$

because S is an ovaloid. On the other hand, if an open unit ball B is in the inner domain Ω of S,

$$\text{vol}\,\Omega \geq \text{vol}\,B.$$

Consequently

$$A(S)^3 \leq A(\mathbb{S}^2)^3 = 36\pi(\text{vol}\,B)^2 \leq 36\pi(\text{vol}\,\Omega)^2,$$

which, along with the isoperimetric inequality, allows us to finish the exercise.

Exercise (12): Substituting $K = 2 - H$ into inequality (3.6), one has that $H^2 + H - 2 \geq 0$ and, so, that $H \geq 1$ (with respect to the inner normal), with equality only at the umbilical points. Going back to the original equality, one sees that $K \leq 1$. Then $K \leq 1 \leq H$. Recalling that the support function is strictly negative, the two Minkowski formulas give

$$A(S) = -\int_S H(p)\langle p, N(p)\rangle \, dp \geq -\int_S K(p)\langle p, N(p)\rangle \, dp = \int_S H \geq A(S).$$

Therefore S is totally umbilical.

Exercise (14): Since the support function is negative, bearing in mind the hypotheses, we have

$$\frac{1}{r} \leq H \qquad \text{and} \qquad -K(p)\langle p, N(p)\rangle \leq -\frac{1}{r}H(p)\langle p, N(p)\rangle,$$

for each $p \in S$. Integrating and using the Minkowski formulas,

$$\frac{1}{r}A(S) \leq \int_S H = -\int_S K(p)\langle p, N(p)\rangle \, dp \leq -\frac{1}{r}\int_S H(p)\langle p, N(p)\rangle \, dp = \frac{1}{r}A(S).$$

Thus S has constant mean curvature $1/r$ and constant Gauss curvature $1/r^2$. We conclude from Corollary 3.49 by Hilbert and Liebmann or from the Alexandrov Theorem 6.17

Exercise (19): The definition of Laplacian and Exercise (16) above imply directly that the integral on S of Δf vanishes. Therefore, since Δf is continuous, we have $\Delta f = 0$. From Exercise (18),

$$\Delta f^2 = 2f\Delta f + 2|\nabla f|^2 = 2|\nabla f|^2.$$

Integrating this equality and taking into account the fact that the integral of Δf^2 is zero, we arrive at

$$\int_S |\nabla f|^2 = 0.$$

Then $\nabla f = 0$ and f is constant on each connected component of S.

Chapter 7

Intrinsic Geometry
of Surfaces

7.1. Introduction

We have been studying differential geometry of curves and surfaces of Euclidean space $\mathbb{R}^3$ and, until now, the only distinction that we have made about the results obtained is on their *local* or *global* nature. It is true that we entitled Chapter 6 with the adjective *extrinsic*—referring to the geometry of surfaces—and that we start this chapter using the term *intrinsic*. This label of intrinsic, applied to the geometry of surfaces, might not appear to be appropriate, since the very essence of surfaces, as they were defined, is *to be embedded in* $\mathbb{R}^3$. That is, the geometry of surfaces consists of studying how they bend and extend in Euclidean space and, in this sense, it is extrinsic. This assertion is clearly confirmed if we consider that both the separation properties and the existence of tubular neighbourhoods, established in Chapter 4, which could be extrinsic properties par excellence, have been the key in the proof of several of the theorems that we have considered. Then, why use the expression *intrinsic geometry*, and what does it refer to? The answer, as with so many matters in this field, goes back to GAUSS and requires further explanation.

Since we are studying the differential geometry of curves and surfaces in Euclidean space $\mathbb{R}^3$, it seems to be logical, and indeed we have done it until now, to check that the notions that we introduce and the results that we state are invariant under rigid motions of $\mathbb{R}^3$, which are the isomorphisms of Euclidean geometry. We will begin in Section 7.2 by seeing that, for a map between surfaces to be extended to a rigid motion of $\mathbb{R}^3$, there are two necessary and sufficient conditions: to preserve the arc length of curves and the second fundamental forms. This is a version of the so-called fundamental theorem of the theory of surfaces, first proved by O. BONNET (1819–1892) in 1867. In other words, it is clear that, if a surface were made from a flexible but inelastic material, any deformation that we produced on it would preserve the length of curves. Let us assume that, in fact, surfaces are made from such a material, and let us try to perform a deformation of the surface which preserves the curvatures of all its normal sections. The only way to accomplish this is to move the surface rigidly in $\mathbb{R}^3$. That is, the deformation would be impossible. But, what happens if we remove the condition imposed on the second fundamental form? It seems intuitively clear that, if we restrict ourselves to a small enough piece of a surface, we could deform it, preserving the length of curves. But, it is difficult to assume that, for instance, an entire compact surface might be deformed in this way without translating or rotating in $\mathbb{R}^3$.

If a surface can be mapped to another one preserving lengths of curves, we say, using current terminology, that they are locally isometric. In GAUSS's time, they said that the first surface could be *developed onto* the second one. Topographers, and GAUSS was one of them, trying to draw maps of the Earth's surface, thought of surfaces, unlike our usual view, not as a two-dimensional film in three-space, but as the boundary of a three-dimensional rigid body. This idea leads in a natural way to considering deformations of surfaces preserving lengths of curves. They knew, for example, that a piece of sphere—the Earth's surface—cannot be developed onto a plane. On the other hand, many people have had the experience of making a cylinder by bending a piece of paper—a flexible but inelastic material—and also of having made conical hats from paper. A good question relative to these situations is: is there some quantity or geometrical object remaining invariant under local isometries? If so, we would find some obstructions to the existence of this class of maps between two given surfaces. On page twenty, Art. 12, of the *Disquisitiones* of 1827 GAUSS says textually:

> *Formula [(4.6)] itaque art. praec. sponte perducit ad egregium theorema. Si superficies curva in quamcunque aliam superficiem explicatur, mensura curvaturae in singulis punctis invariata manet.*

This statement, in spite of the rigidity of the Latin—that GAUSS often complained about—to express some mathematical nuances, can be translated as follows:

> *Then the formula in the previous article leads by itself to an excellent theorem. If a curved surface is developed onto*

> *any other, the measure of the curvature at each point re-*
> *mains unchanged.*

This is the Gauss Theorema Egregium and asserts that the Gauss curvature is invariant, not only through the group of rigid motions of $\mathbb{R}^3$, but through a very much wider class of transformations: local isometries between surfaces. We will prove this famous result in Section 7.3 from an expression for the Gauss curvature, in terms of an arbitrary parametrization, which does not make use of the normal field of the surface, that is, an expression which can be calculated *inside* of the surface, that is, an *intrinsic* expression. One wonders if such an expression—see Lemma 7.11—leading to an invariance greater than Euclidean invariance, might be taken as a more appropriate definition than the original one as the product of the two principal curvatures. But, on one hand, the original definition is easy; it is also clear and esthetically more agreeable than the alternative offered by Lemma 7.11. On the other hand, it is quite an awkward exercise to see that the expression for the Gauss curvature contained in the proof of the Theorema Egregium does not depend on the chosen parametrization. The problem is not that the original definition was bad, but that, as Spivak points out in [**17**], the defined notion—the Gauss curvature—was too good.

It turns out in this way, from the Theorema Egregium, that there exists an *intrinsic geometry* in the realm of surfaces in Euclidean space. From this perspective we are interested in all properties which are preserved through local isometries, or, in other words, through differentiable maps leaving lengths of curves unchanged, for example, the Gauss curvature. This situation does not have an analogue in the case of curves and that is why we did not talk about their intrinsic geometry. Indeed, all regular curves are locally isometric to each other: the arc length function permits us to establish the corresponding isometries. For this reason, there are no obstructions to deforming inelastically plane or space curves.

As a consequence of the Theorema Egregium and of global results in previous chapters, we will be able to show another property of spheres relative to the deformation problem that we have considered: its rigidity. It is impossible to deform a sphere preserving lengths of curves. This was suggested by F. MINDING (1806–1885) as early as 1838 and was proved by H. LIEBMANN, as a corollary to his well-known result obtained in 1899. This rigidity of spheres can be viewed in quite a wider context, since it is really a property shared by all positively curved compact surfaces, that is, all ovaloids are rigid. This statement is the so-called rigidity theorem by S. COHN-VOSSEN, proved in 1927, that we will deduce from an integral equality due to HERGLOTZ and which had been likely announced, according to a statement by CAUCHY, by LAGRANGE in 1812.

Finally, in this chapter we will deal with the other intrinsic invariants par excellence: *geodesics*. It seems to be logical, and indeed so it is, that the least length curves joining two given points of a surface, if they exist, should be preserved through local isometries. It was probably JOHN BERNOUILLI, in 1697, who was the first mathematician to study their existence and to see that they are paths of vanishing tangential acceleration—in the particular case of ovaloids. L. EULER was however the first geometer to write their equations, in 1732, in the case of surfaces given in implicit form. Its current name and definition, apart from their minimizing properties, are due to J. LIOUVILLE (1844) and the beginning of the study of these minimizing features is due to C. G. J. JACOBI (1804–1851) in 1843. We will finish this chapter by proving an easy version of the existence theorem for geodesics which minimize the length between any two points of a surface *closed* as a subset of $\mathbb{R}^3$. In a more general form this is nothing more than the famous result proved in 1931 by H. HOPF and W. RINOW.

7.2. Rigid motions and isometries

Let S be a surface and $\phi : \mathbb{R}^3 \to \mathbb{R}^3$ a rigid motion of Euclidean space—see Definition 1.7—given by $\phi(x) = Ax + b$, for each $x \in \mathbb{R}^3$, where A is an orthogonal matrix of order three and $b \in \mathbb{R}^3$ is a fix vector. We know, from Example 2.12, that the image $S' = \phi(S)$ of S through ϕ is another surface and that

$$f = \phi_{|S} : S \longrightarrow S'$$

is a diffeomorphism; see Exercise 2.45. Now then, diffeomorphisms between surfaces obtained by restricting a rigid motion of Euclidean space, which we will also call *congruences*, keep invariant, not only the topological and differentiable structures, but their geometry. We set out now to clarify this assertion. First, we have

$$(df)_p(v) = (d\phi_{|S})_p(v) = (d\phi)_p(v) = Av,$$

for each $p \in S$ and each vector $v \in T_pS$. Since A is an orthogonal matrix, it preserves the scalar product in $\mathbb{R}^3$. Then

$$\langle (df)_p(u), (df)_p(v) \rangle = \langle u, v \rangle, \qquad \forall p \in S \quad \forall u, v \in T_pS.$$

That is, the differential $(df)_p$ at each point of S is a linear isometry between the corresponding tangent planes. On the other hand, if $N : S \to \mathbb{S}^2$ is a Gauss map defined on S, the new map $N' : S' \to \mathbb{S}^2$ defined by the equality

$$N' \circ f = A \circ N$$

is a Gauss map for S'. Thus, the relation between the differentials is given by

$$(dN')_{f(p)} \circ A = A \circ (dN)_p$$

for each $p \in S$. One arrives then at the following relation between the two second fundamental forms σ and σ' of S and S':

$$\sigma'_{f(p)}((df)_p(u), (df)_p(v)) = \sigma_p(u, v), \qquad \forall p \in S \quad \forall u, v \in T_p S,$$

as established in Exercise 3.25, called the invariance of the second fundamental form under rigid motions. As a consequence, the mean and Gauss curvatures of S and of S' are transposed into each other through f, that is, $H' \circ f = H$ and $K' \circ f = K$. We can summarize, then, the effect of the rigid motions of Euclidean space on surfaces in the following result.

Proposition 7.1. *Let S be a surface and $\phi : \mathbb{R}^3 \to \mathbb{R}^3$ a rigid motion. Let $S' = \phi(S)$ and $f = \phi_{|S} : S \to \mathbb{R}^3$. Then the following hold.*

A: *S' is a surface and $f : S \to S'$ is a diffeomorphism.*

B: *For each $p \in S$ and $u, v \in T_p S$ one has*

$$\langle (df)_p(u), (df)_p(v) \rangle = \langle u, v \rangle.$$

C: *For each point $p \in S$ and for tangent vectors $u, v \in T_p S$ the equality*

$$\sigma'_{f(p)}((df)_p(u), (df)_p(v)) = \sigma_p(u, v)$$

holds, where σ and σ' are the second fundamental forms of S and of S', respectively.

In general, a diffeomorphism $f : S \to S'$ between two surfaces is said to *preserve the first fundamental form* when it satisfies condition **B** of the proposition and is said to *preserve the second fundamental form* if it obeys **C**. Then, we may affirm that congruences between surfaces preserve the first and second fundamental forms or, equivalently, that congruent surfaces have the same first and second fundamental forms.

On the other hand, but in relation to the above, if $f : S \to S'$ is a differentiable map between two surfaces S and S' satisfying requirement **B** of Proposition 7.1, we will say that f is a *local isometry*. The reason for giving these names is the following proposition.

Proposition 7.2. *Let $f : S \to S'$ be a differentiable map between two surfaces. Then $(df)_p : T_p S \to T_{f(p)} S'$ is a linear isometry, for each $p \in S$, that is, f is a local isometry, if and only if f preserves the length of curves.*

Proof. Let $\alpha : [a, b] \to S$ be a differentiable curve. Then the length of the image curve on S' satisfies

$$L_a^b(f \circ \alpha) = \int_a^b |(f \circ \alpha)'(t)| \, dt.$$

Since $(f \circ \alpha)'(t) = (df)_{\alpha(t)}(\alpha'(t))$ for each $t \in [a, b]$, if we suppose that the differential of f at each point of S is a linear isometry,

$$|(f \circ \alpha)'(t)| = |(df)_{\alpha(t)}(\alpha'(t))| = |\alpha'(t)|,$$

and so

$$L_a^b(f \circ \alpha) = \int_a^b |\alpha'(t)| \, dt = L_a^b(\alpha),$$

as required. Conversely, suppose that the map f between S and S' keeps the length of curves invariant. Let $p \in S$ be an arbitrary point, and let $v \in T_pS$. There exists a differentiable curve $\alpha : (-\varepsilon, \varepsilon) \to S$ for some $\varepsilon > 0$ such that $\alpha(0) = p$ and $\alpha'(0) = v$. By the hypothesis on f, we have, for each $t \in (0, \varepsilon)$,

$$\int_0^t |(f \circ \alpha)'(u)| \, du = L_0^t(f \circ \alpha) = L_0^t(\alpha) = \int_0^t |\alpha'(u)| \, du.$$

Taking derivatives with respect to t at $t = 0$—the functions are differentiable by the fundamental theorem of calculus—we obtain

$$|(df)_p(v)| = |(f \circ \alpha)'(0)| = |\alpha'(0)| = |v|.$$

Therefore $(df)_p$ preserves the scalar product.   □

It is convenient to realize that, since a linear map between Euclidean vector spaces with the same dimension which preserves the scalar product must be an isomorphism, any local isometry between surfaces is necessarily a local diffeomorphism, by the inverse function Theorem 2.75. A local isometry which is moreover a diffeomorphism will be called an *isometry*. If there exists an isometry between two surfaces, we will say that they are *isometric*.

Exercise 7.3. Prove that, if $f : S \to S'$ is an isometry between two surfaces, the absolute value of the Jacobian of f is identically 1. As a consequence, if S and S' are compact, their areas coincide.

Exercise 7.4. Let P be the plane of the equation $z = 0$ in $\mathbb{R}^3$ and C the right unit cylinder given by the implicit equation $x^2 + y^2 = 1$. Show that the map $f : P \to C$ given by $f(x, y, 0) = (\cos x, \sin x, y)$, for each $(x, y, 0) \in P$, is a local isometry.

Exercise 7.5. Prove that the composition of local isometries between surfaces is also a local isometry and that the inverse map of an isometry is an isometry as well. Conclude that the set of all the isometries from a surface onto itself is a group relative to the composition law, which is called the *group of isometries* of the surface.

Exercise 7.6. ↑ Let $F : O \to O'$ be a diffeomorphism between two connected open subsets of $\mathbb{R}^3$ such that its differential at each point of O is a linear isometry of $\mathbb{R}^3$. Demonstrate that F is the restriction to O of a rigid motion of $\mathbb{R}^3$.

It can be deduced, from the second of these exercises, that there are open subsets of the plane and of the cylinder which are isometric. Such an isometry cannot come from a rigid motion of $\mathbb{R}^3$, that is, it is not a congruence, because congruences take planes into planes. However, we have the following converse of Proposition 7.1.

Theorem 7.7 (Fundamental theorem of the local theory of surfaces). *Let S and S' be two orientable surfaces, with S connected, and let N and N' be Gauss maps for each of them with associated second fundamental forms σ and σ', respectively. If $f : S \to S'$ is a local isometry preserving the second fundamental forms, then there exists a rigid motion $\psi : \mathbb{R}^3 \to \mathbb{R}^3$ such that $\psi_{|S} = f$, that is, f is a congruence.*

Proof. Let $p_0 \in S$ be an arbitrary point on the first surface and $f(p_0) \in S'$ its image. Lemma 4.25 along with the fact that f is a local diffeomorphism imply that there exist two connected open neighbourhoods V of p_0 in S and V' of $f(p_0)$ in S' such that $f : V \to V'$ is an isometry and there is a number $\varepsilon > 0$ such that $N_\varepsilon(V)$ and $N_\varepsilon(V')$ are tubular neighbourhoods. We define a map $\phi : N_\varepsilon(V) \to N_\varepsilon(V')$ by

$$\phi(p + tN(p)) = f(p) + t(N' \circ f)(p), \qquad \forall p \in V \quad \forall |t| < \varepsilon.$$

This ϕ is clearly differentiable and bijective because it is nothing more than the composite map $F' \circ (f \times \mathrm{Id}_{(-\varepsilon,\varepsilon)}) \circ F^{-1}$, where $F : V \times (-\varepsilon, \varepsilon) \to N_\varepsilon(V)$ and $F' : V' \times (-\varepsilon, \varepsilon) \to N_\varepsilon(V')$ were defined in (4.1). Using the definition itself or the commutativity relation $\phi \circ F = F' \circ (f \times \mathrm{Id}_{(-\varepsilon,\varepsilon)})$, one can compute the differential of ϕ and see that

(7.1) $$(d\phi)_{p+tN(p)}(N(p)) = N'(f(p))$$

$$(d\phi)_{p+tN(p)} \circ [I_{T_pS} + t(dN)_p] = [I_{T_{f(p)}S'} + t(dN')_{f(p)}] \circ (df)_p,$$

for each $p \in V$ and $|t| < \varepsilon$. But we are assuming that f preserves the second fundamental form, so then

$$\langle (dN')_{f(p)}(df)_p(u), (df)_p(v) \rangle = -\sigma'_{f(p)}((df)_p(u), (df)_p(v))$$
$$= -\sigma_p(u, v) = \langle (dN)_p u, v \rangle,$$

for each $p \in S$ and $u, v \in T_pS$. Therefore, since $(df)_p$ is a linear isometry,

$$(dN')_{f(p)} \circ (df)_p = (df)_p \circ (dN)_p, \qquad \forall p \in S.$$

Applying this commutativity rule to the second equality in (7.1), we get

$$(d\phi)_{p+tN(p)} \circ [I_{T_pS} + t(dN)_p] = (df)_p \circ [I_{T_pS} + t(dN)_p].$$

This, together with the fact that $I_{T_pS} + t(dN)_p$ is an isomorphism of the plane T_pS for all $p + tN(p)$ in the tubular neighbourhood, leads us to the equalities

$$(d\phi)_{p+tN(p)}(N(p)) = N'(f(p)) \quad \text{and} \quad (d\phi)_{p+tN(p)|T_pS} = (df)_p,$$

valid for each $p \in V$ and $|t| < \varepsilon$. Consequently, we have that ϕ is a diffeomorphism between two open subsets of $\mathbb{R}^3$ whose differential is a linear isometry at each point of its domain $N_\varepsilon(V)$. Thus, from Exercise 7.6, ϕ must be the restriction to $N_\varepsilon(V)$ of a rigid motion ψ of $\mathbb{R}^3$. So, since the initial point $p_0 \in S$ was arbitrarily chosen, we have shown that, for each $p \in S$, there exist an open neighbourhood V of p in S and a rigid motion $\psi(x) = Ax + b$, $x \in \mathbb{R}^3$, such that $\psi_{|V} = f_{|V}$ and $N' \circ f_{|V} = A \circ N_{|V}$. Bearing in mind that S is connected, we deduce that all these rigid motions corresponding to the diverse neighbourhoods V coincide. $\qquad\square$

Example 7.8. Let P and P' be two planes of $\mathbb{R}^3$ and $f : U \to P'$ a local isometry defined on a connected open subset U of P. Since the second fundamental forms of P and of P' are identically zero—for any orientation—f always preserves the second fundamental forms. Therefore, f is the restriction of a rigid motion.

Example 7.9. If S and S' are two spheres with the same radius $r > 0$, $U \subset S$ is a connected open subset of the first one, and $f : U \to S'$ is a local isometry, then f necessarily preserves the second fundamental forms σ and σ', associated, for instance, to the two inner normal fields, because, according to Example 3.20,

$$\sigma'_{f(p)}((df)_p(u), (df)_p(v)) = \frac{1}{r}\langle (df)_p(u), (df)_p(v)\rangle = \frac{1}{r}\langle u, v\rangle = \sigma_p(u, v),$$

for each $p \in S$ and $u, v \in T_pS$. Then f is the restriction of a rigid motion.

7.3. Gauss's Theorema Egregium

As we already pointed out in the introduction to this chapter, intrinsic geometry of surfaces is the study of those properties remaining invariant through isometries, or, more generally, through maps preserving the first fundamental form. It is quite remarkable that the Gauss curvature, as defined in (3.4) and (3.5) from the second fundamental form, turns out to be the main invariant of this intrinsic geometry. Gauss wrote in the *Disquisitiones* that the following theorem, due to himself, is *certainly excellent*.

Theorem 7.10 (Theorema Egregium). *Let $f : S \to S'$ be a local isometry between two surfaces S and S', and let K and K' be their respective Gauss curvatures. Then $K = K' \circ f$. Equivalently, local isometries preserve the Gauss curvature.*

Proof. Let us take a parametrization $X : U \to S$ of our surface S, and let us work on the open set $X(U)$ covered by it. We will use the same notation as in the proof of Proposition 3.43. We know that on the open set $X(U)$ we dispose of a Gauss map $N = N^X \circ X^{-1}$, where $N^X = X_u \wedge X_v / |X_u \wedge X_v|$, and that, at each of its points the partial derivatives $\{X_u, X_v\}$ supply us a basis of the corresponding tangent plane $T_X S$—we will work all the time with functions defined on the open subset U of $\mathbb{R}^2$, although without explicitly indicating the point. With respect to this basis, the two fundamental forms of the surface are represented by symmetric matrices whose coefficients were denoted by E, F, G and e, f, g, respectively. Now, we may prove the following preliminary equality.

Lemma 7.11. *Let U be an open subset of $\mathbb{R}^2$ and $F : U \to \mathbb{R}^3$ a differentiable map taking values on a surface S, for example, a parametrization of S. If K is the Gauss curvature of S, one has the following expression, valid at the points of U:*

$$\langle (F_{vv}^T)_u - (F_{uv}^T)_v, F_u \rangle = (K \circ F)(|F_u|^2 |F_v|^2 - \langle F_u, F_v \rangle^2),$$

where the superscript T means the part of the corresponding vector belonging to the tangent plane of the surface at the given point.

Proof. In fact, we have, if N is a local unit normal field and $N^F = N \circ F$,

$$F_{vv}^T = F_{vv} - \langle F_{vv}, N^F \rangle N^F = F_{vv} + \langle F_v, N^F{}_v \rangle N^F,$$

$$F_{uv}^T = F_{uv} - \langle F_{uv}, N^F \rangle N^F = F_{uv} + \langle F_u, N^F{}_v \rangle N^F,$$

because $\langle F_u, N^F \rangle = \langle F_v, N^F \rangle \equiv 0$. Taking partial derivatives of these two expressions relative to u and to v, we obtain

$$(F_{vv}^T)_u = F_{vvu} + \langle F_v, N^F{}_v \rangle_u N^F + \langle F_v, N^F{}_v \rangle N^F{}_u,$$

$$(F_{uv}^T)_v = F_{uvv} + \langle F_u, N^F{}_v \rangle_v N^F + \langle F_u, N^F{}_v \rangle N^F{}_v.$$

After scalar multiplication by F_u and subtracting, we have, by the Schwarz theorem of elementary calculus, that

$$\langle (F_{vv}^T)_u - (F_{uv}^T)_v, F_u \rangle = \langle N^F{}_u \wedge N^F{}_v, F_u \wedge F_v \rangle.$$

Now then,

$$N^F{}_u \wedge N^F{}_v = (dN)_F(F_u) \wedge (dN)_F(F_v) = \det(dN)_F(F_u \wedge F_v),$$

and, using (3.4),

$$N^F{}_u \wedge N^F{}_v = (K \circ F)(F_u \wedge F_v),$$

and we conclude.   □

If we apply Lemma 7.11 to the parametrization X, we obtain

$$(7.2) \qquad \langle (X_{vv}^T)_u - (X_{uv}^T)_v, X_u \rangle = (K \circ X)(EG - F^2).$$

Now, the composite map $X' = f \circ X$ is a parametrization of the second surface S'. Then, applying Lemma 7.11 to the second surface S' and to the parametrization X', we can write

$$(7.3) \qquad \langle (X'^T_{vv})_u - (X'^T_{uv})_v, X'_u \rangle = (K' \circ X')(E'G' - F'^2).$$

On the other hand, taking partial derivatives in the definition of X' and using the chain rule, one has

$$X'_u = (df)_X(X_u) \quad \text{and} \quad X'_v = (df)_X(X_v),$$

and so, computing scalar products and taking into account that f is a local isometry,

$$(7.4) \qquad E' = |X'_u|^2 = |X_u|^2 = E, \quad G' = |X'_v|^2 = |X_v|^2 = G,$$
$$F' = \langle X'_u, X'_v \rangle = \langle X_u, X_v \rangle = F.$$

Differentiating again the three equalities in (7.4) with respect to the two variables, one obtains that the six scalar products between the second partial derivatives and the first partial derivatives of X' and of X coincide. Hence, since $(df)_X$ is a linear isometry at each point,

$$(7.5) \qquad (X'_{uu})^T = (df)_X(X_{uu}^T), \qquad (X'_{vv})^T = (df)_X(X_{vv}^T),$$
$$(X'_{uv})^T = (df)_X(X_{uv}^T).$$

Consequently, since

$$\langle (X'^T_{vv})_u, X'_u \rangle = \langle X'^T_{vv}, X'_u \rangle_u - \langle X'^T_{vv}, X'^T_{uu} \rangle$$

$$= \langle (df)_X(X_{vv}^T), (df)_X(X_u) \rangle_u - \langle (df)_X(X_{vv}^T), (df)_X(X_{uu}^T) \rangle$$

$$= \langle X_{vv}^T, X_u \rangle_u - \langle X_{vv}^T, X_{uu}^T \rangle = \langle (X_{vv}^T)_u, X_u \rangle,$$

and, analogously,

$$\langle (X'^T_{uv})_v, X'_u \rangle = \langle (X_{uv}^T)_v, X_u \rangle,$$

these two equalities, together with (7.3), (7.4), and (7.5), give us

$$K \circ X = K' \circ X' = K' \circ f \circ X$$

on U. Therefore, $K' \circ f = K$ on the open subset $X(U)$ of S. Since the surface may be covered by open sets of this type, the theorem is proved. $\square$

As an immediate consequence of Gauss's Theorema Egregium, one sees, for instance, that an open subset of a sphere cannot be isometric to an open piece of a plane. Furthermore, we deduce—again—a new property of spheres that, in some texts but not in all of them, is called *rigidity*. This implies that it is impossible to deform a sphere made from a material which can

Figure 7.1. *Non-congruent isometric surfaces*

be bent—flexible—but which, by doing it, keeps the length of each curve invariant—inelastic. For example, if we try to bend a sphere made from paper, we will break it.

Theorem 7.12 (Rigidity of spheres). *Let $\mathbb{S}^2(r)$ be a sphere of radius $r > 0$ and S a connected surface. Each local isometry $f : \mathbb{S}^2(r) \to S$ is the restriction of a rigid motion of $\mathbb{R}^3$. In particular, S is another sphere of radius r.*

Proof. Since f is a local isometry, in particular, it is a local diffeomorphism and so it is an open map. Thus, its image $f(\mathbb{S}^2(r))$ is an open subset of S. It is a closed subset as well, because $\mathbb{S}^2(r)$ is compact. Since S is connected, f is surjective. On the other hand, we know that the sphere $\mathbb{S}^2(r)$ has constant Gauss curvature equal to $1/r^2$. The Gauss Theorema Egregium that we have just proved implies that S also has constant Gauss curvature $1/r^2$. Hence, applying the Hilbert and Liebmann theorem stated either in Corollary 3.49 or in Corollary 6.12, S is also a sphere of radius r. Then $f : \mathbb{S}^2(r) \to S$ is a local isometry between two spheres of the same radius. From the fundamental theorem of the local theory of surfaces—Example 7.9—we deduce that f is the restriction of a rigid motion. $\square$

This rigidity of the sphere means that its geometry is completely determined only by its first fundamental form. In this sense, it is a property which is not exclusive of spheres, although there are compact surfaces which do not possess it. For example, the surfaces in Figure 7.1—which appear in almost all books dealing with this topic—are two compact surfaces of revolution which are isometric but not congruent. The isometry insinuated in Figure 7.1 fixes all the points under the tangent plane of the surface along the circle of vanishing Gauss curvature and coincides, when restricted to the points over this plane, with the reflection about that tangent plane. Obviously, this isometry does not come from a rigid motion of $\mathbb{R}^3$ since it has too many fixed points. Another question is whether we can pass continuously from one to the other of these surfaces through isometric surfaces or whether we can pass from the identity map to that isometry through a continuous family of isometries—and, maybe, this should be the fact leading us to define the lack of rigidity. In fact, this is impossible, as shown in Exercise (6) at the end of this chapter. In the following section, we will see that rigidity of spheres is a property shared by all positively curved compact surfaces, but the proof of this is considerably more involved than in the case of spheres.

7.4. Rigidity of ovaloids

Suppose that S and S' are two orientable surfaces and that $f : S \to S'$ is a diffeomorphism. If $V : S \to \mathbb{R}^3$ is a differentiable field of tangent vectors on S, we may define another field $W : S \to \mathbb{R}^3$ on the same surface by

$$W(p) = A(f)_p V(p), \qquad \forall p \in S,$$

where $A(f)_p : T_p S \to T_p S$ is the linear endomorphism given by

$$A(f)_p = -(df)_p^{-1} \circ (dN')_{f(p)} \circ (df)_p, \qquad p \in S,$$

and N' is a Gauss map for S'. This is a well-defined field because $(dN')_q$ can be thought of as an endomorphism of $T_q S'$ for each $q \in S'$. Furthermore, it is also clear that

$$\det A(f)_p = K'(f(p)) \quad \text{and} \quad \text{trace}\, A(f)_p = 2\, H'(f(p)),$$

since similar matrices have the same trace and determinant, and that

$$W(p) \in T_p S, \qquad \forall p \in S.$$

That is, W is another field of vectors tangent to S depending on the original field V and on the diffeomorphism f. Let us see now that this field W is differentiable. For this, let $X : U \to S$ be any parametrization defined on an open subset U of $\mathbb{R}^2$. We know that $X' = f \circ X$ is a parametrization of S' with the same domain as X and that the partial derivatives $\{X_u, X_v\}$ and $\{X'_u, X'_v\}$ provide, at each point of U, bases of the tangent planes $T_X S$ and $T_{X'} S'$, respectively. If we use the usual notation, we may set

$$W^X = W \circ X, \quad V^X = V \circ X, \quad \text{and} \quad A(f)^X = -(df)_X^{-1} \circ (dN')_{X'} \circ (df)_X.$$

Then, since the matrix of the linear map $(df)_X : T_X S \to T_{X'} S'$, relative to these bases, is the identity matrix because

$$X'_u = (f \circ X)_u = (df)_X(X_u) \quad \text{and} \quad X'_v = (f \circ X)_v = (df)_X(X_v),$$

the matrix $a(f)$ of $A(f)^X$ is the same one as the matrix of the endomorphism $(dN')_{X'}$, that is, according to (3.11),

$$a(f) = -\begin{pmatrix} E' & F' \\ F' & G' \end{pmatrix}^{-1} \begin{pmatrix} e' & f' \\ f' & g' \end{pmatrix},$$

whose coefficients are differentiable functions defined on the open subset U of $\mathbb{R}^2$. In conclusion, if

$$V^X = bX_u + cX_v,$$

for some differentiable functions b, c defined on U, one has that

(7.6) $$W^X = mX_u + nX_v,$$

where the new coefficients m, n are functions defined on U satisfying

(7.7) $$\begin{pmatrix} m \\ n \end{pmatrix} = a(f) \begin{pmatrix} b \\ c \end{pmatrix} = - \begin{pmatrix} E' & F' \\ F' & G' \end{pmatrix}^{-1} \begin{pmatrix} e' & f' \\ f' & g' \end{pmatrix} \begin{pmatrix} b \\ c \end{pmatrix}.$$

This proves that $W^X = W \circ X$ is differentiable on $U \subset \mathbb{R}^2$ and, hence, that the field W is differentiable.

Once we have checked that this new tangent field W, which we have defined on the first surface S from the field V and the diffeomorphism $f : S \to S'$, is differentiable, we intend to compute its divergence, in the particular case where f is a local isometry. After that, we will apply, when S is compact, the divergence Theorem 6.10 on surfaces. To do all of this, it is worth recalling the following facts.

Lemma 7.13. *Let $p \in S$ be an arbitrary point of a surface. There exists a parametrization $X : U \to S$ of S defined on an open subset U of $\mathbb{R}^2$ such that $X(q) = p$ for some $q \in U$ and such that the following hold.*

a: *The vectors $\{X_u(q), X_v(q)\}$ form an orthonormal basis of T_pS.*

b: *The vector $X_{uv}(q)$ vanishes and the vectors $X_{uu}(q)$ and $X_{vv}(q)$ are perpendicular to T_pS.*

c: *The coefficients E, F, G of the first fundamental form associated to X have vanishing first partial derivatives at q.*

d: *The first partial derivatives of the coefficients e, f, g of the second fundamental form associated to X satisfy*

$$e_v(q) = f_u(q) \quad and \quad g_u(q) = f_v(q).$$

Proof. In fact, it is enough to take a suitable neighbourhood of the point p in the surface S as a graph on the tangent plane at this point and to take the corresponding parametrization, as we did in the proof of the Hilbert Theorem 3.47. □

Now, we can proceed to calculate the divergence of the vector field W. Pick $p \in S$. Then, we apply Lemma 7.13 to the second surface S' at the point $p' = f(p)$ and we get a parametrization X' of S' with the required properties **a**, **b**, **c**, and **d**. We choose as a parametrization X of the first surface S the map $X = f^{-1} \circ X'$. Hence, we have $X' = f \circ X$. From property **a** of X' and the fact that $(df)_p$ is a linear isometry, the vectors $X_u(q)$ and

$X_v(q)$ are an orthonormal basis of T_pS. Consequently, by the definition of divergence of a field of tangent vectors on a surface, given in Section 6.3, we have

$$(\operatorname{div} W)(p) \;=\; \langle (dW)_p X_u(q), X_u(q) \rangle + \langle (dW)_p X_v(q), X_v(q) \rangle,$$

$$=\; \langle W_u^X, X_u \rangle(q) + \langle W_v^X, X_v \rangle(q).$$

But, from (7.6), we have

$$W_u^X = m_u X_u + n_u X_v + m X_{uu} + n X_{vu}, \quad W_v^X = m_v X_u + n_v X_v + m X_{uv} + n X_{vv}.$$

Taking Lemma 7.13 and (7.5) into account, the two equalities above imply

$$(\operatorname{div} W)(p) = (m_u + n_v)(q).$$

On the other hand, recalling the relation (7.7), this expression becomes

$$(\operatorname{div} W)(p) = \left[(e'_u + f'_v)b + (f'_u + g'_v)c + e'b_u + f'(c_u + b_v) + g'c_v \right](q).$$

From the particular way in which we have chosen the parametrization X', according to Lemma 7.13, we have

$$f'(q) = 0, \qquad f'_v(q) = g'_u(q), \quad \text{and} \quad f'_u(q) = e'_v(q).$$

Substituting, it follows that

$$(\operatorname{div} W)(p) = \left[(e'_u + g'_u)b + (e'_v + g'_v)c + e'b_u + g'c_v \right](q).$$

Finally, the properties mentioned in the lemma relative to the coefficients e', f', g' at the point q also imply that

$$e'(q) = (k'_1 \circ f)(p), \quad g'(q) = (k'_2 \circ f)(p),$$

$$2\,(H' \circ X')_u(q) = (e'_u + g'_u)(q), \quad 2\,(H' \circ X')_v(q) = (e'_v + g'_v)(q).$$

Then, we may conclude that

$$
\begin{aligned}
(\operatorname{div} W)(p) \;=\;& 2\langle \nabla(H' \circ f)(p), V(p) \rangle \\
(7.8) \qquad +\;& (k'_1 \circ f)(p)\langle (dV)_p(e_1), e_1 \rangle + (k'_2 \circ f)(p)\langle (dV)_p(e_2), e_2 \rangle,
\end{aligned}
$$

for each $p \in S$ and where $\{e_1, e_2\}$ is a basis taken by $(df)_p$ into the principal directions of S' at the point $p' = f(p)$. We can, then, announce the following proposition as an immediate consequence of the first part of the divergence Theorem 6.10 for surfaces.

Proposition 7.14. *Let $f : S \to S'$ be a local isometry between two compact surfaces and V a differentiable field of tangent vectors of S. Then*

$$2 \int_S \langle \nabla(H' \circ f), V \rangle = -\int_S \sum_{i=1}^{2} (k'_i \circ f)(p)\langle (dV)_p(e_i), e_i \rangle \, dp,$$

where $\{e_1, e_2\}$ is an orthonormal basis of T_pS at the point p of S which is mapped through $(df)_p$ to the principal directions of S' at $f(p)$ and k'_1 and k'_2 are the corresponding principal curvatures of S'.

Now consider the particular case where the field $V : S \to \mathbb{R}^3$ is given by

$$V(p) = p^T = p - \langle p, N(p) \rangle N(p), \qquad \forall p \in S,$$

where N is the Gauss map of S. Then

$$(dV)_p(v)^T = v - \langle p, N(p) \rangle (dN)_p(v), \qquad \forall v \in T_pS.$$

Hence

$$\sum_{i=1}^{2} (k'_i \circ f)(p) \langle (dV)_p(e_i), e_i \rangle = 2 \left(H' \circ f \right)(p) - \langle p, N(p) \rangle \operatorname{trace}(dN)_p A(f)_p.$$

From this, Proposition 7.14 above leads us, in this particular case, to the integral formula

$$2 \int_S \langle \nabla (H' \circ f)_p, p \rangle \, dp = -2 \int_S (H' \circ f)(p) \, dp + \int_S \langle p, N(p) \rangle \operatorname{trace}(dN)_p A(f)_p \, dp.$$

But, on the other hand,

$$\int_S \langle \nabla (H' \circ f)_p, p \rangle \, dp = -2 \int_S (H' \circ f)(p) \, dp - 2 \int_S \langle p, N(p) \rangle H(p)(H' \circ f)(p) \, dp,$$

as seen in Exercise (20) of Chapter 6. Eliminating the gradient term with the two last equalities, we obtain

$$2 \int_S (H' \circ f)(p) \, dp$$
$$= - \int_S \langle p, N(p) \rangle \operatorname{trace}(dN)_p A(f)_p \, dp - 4 \int_S \langle p, N(p) \rangle H(p)(H' \circ f)(p) \, dp.$$

From this, taking the definition of mean curvature into account, we deduce the following result.

Proposition 7.15 (Herglotz's integral formula). *If $f : S \to S'$ is a local isometry between two compact surfaces in Euclidean space, the integral formula*

$$2 \int_S H' \circ f = \int_S \langle p, N(p) \rangle \left[\operatorname{trace}(dN)_p \operatorname{trace} A(f)_p - \operatorname{trace}(dN)_p A(f)_p \right] dp$$

is valid, where H' is the mean curvature of S' and $A(f)_p$ is the endomorphism

$$A(f)_p = -(df)_p^{-1} \circ (dN')_{f(p)} \circ (df)_p, \qquad \forall p \in S,$$

and N' is the Gauss map of S'.

As a consequence of this integral formula due to Herglotz and of the following easy, but non-trivial, exercises about endomorphisms of a Euclidean vector plane, we will able to deduce the rigidity result for positively curved surfaces that we talked about after the rigidity of spheres Theorem 7.12.

Exercise 7.16. ↑ Let ϕ and ψ be two self-adjoint endomorphisms of a Euclidean vector plane. Show that $\det(\phi+\psi) = \det\phi + \det\psi + \text{trace }\phi\,\text{trace }\psi - \text{trace }\phi\circ\psi$.

Exercise 7.17. ↑ If ϕ and ψ are two definite self-adjoint endomorphisms of a Euclidean vector plane and $\det\phi = \det\psi$, prove that $\det(\phi+\psi) \le 0$ and that equality occurs if and only if $\phi = -\psi$.

Theorem 7.18 (Cohn-Vossen's rigidity theorem). *Any isometry between two ovaloids is necessarily the restriction of a rigid motion of Euclidean space.*

Proof. Let $f : S \to S'$ be an isometry between two ovaloids S and S'. Since the map f is a diffeomorphism, one can construct from it the symmetric endomorphism $A(f)_p$ given by

$$A(f)_p = -(df)_p^{-1} \circ (dN')_{f(p)} \circ (df)_p, \qquad \forall p \in S,$$

which, as seen at the beginning of this section, has trace $2(H' \circ f)$ and determinant $K' \circ f$. Now then, using the Gauss Theorema Egregium 7.10, we have

$$\det A(f) = K' \circ f = K,$$

because f is an isometry. Then, the second Minkowski formula of Theorem 6.11 can be written in this case as

$$2\int_S H = -2\int_S \langle p, N(p)\rangle K(p)\, dp$$

$$= -\int_S \langle p, N(p)\rangle \det(dN)_p\, dp - \int_S \langle p, N(p)\rangle \det A(f)_p\, dp.$$

On the other hand, the Herglotz integral formula gives

$$2\int_S H' \circ f = \int_S \langle p, N(p)\rangle[\,\text{trace}\,(dN)_p\text{trace}\,A(f)_p - \text{trace}\,(dN)_p A(f)_p]\, dp.$$

Subtracting these two equalities and applying Exercise 7.16—notice that both $(dN)_p$ and $A(f)_p$ are self-adjoint endomorphisms of the plane T_pS—we obtain

$$2\int_S (H' \circ f - H)(p)\, dp = \int_S \langle p, N(p)\rangle \det[A(f)_p + (dN)_p]\, dp.$$

The integrand on the right-hand side in this integral equality is non-negative everywhere. The reason is that, on one hand, the support function based on the origin—which we may consider in the inner domain of S—is negative, according to Remark 6.2, and, on the other hand, the function taking each point $p \in S$ into the determinant of the endomorphism $A(f)_p + (dN)_p$ is non-positive by Exercise 7.17 above. Consequently, we have

$$(7.9) \qquad \int_S (H' \circ f - H)(p)\, dp \geq 0,$$

and, if equality is attained, then $A(f)_p = -(dN)_p$ at each point $p \in S$, that is, $(dN')_{f(p)} \circ (df)_p = (df)_p \circ (dN)_p$ for all $p \in S$. This is equivalent to saying that f preserves the second fundamental form. Therefore, if equality holds in the integral inequality above, then f is the restriction of a rigid motion, by the Fundamental Theorem 7.7. Let us see now that, indeed, equality does occur in this inequality. The reason is that, if we interchange the roles of S and S' and substitute the isometry f by its inverse f^{-1}, we get another integral inequality on S', namely

$$\int_{S'} (H \circ f^{-1} - H')(p)\, dp \geq 0.$$

Applying to this integral the change of variables Theorem 5.14 for the diffeomorphism $f : S \to S'$ and taking into account that the absolute value of its Jacobian is identically 1, we have

$$\int_S (H - H' \circ f)(p)\, dp \geq 0,$$

which, together with (7.9), gives the required equality. □

7.5. Geodesics

We have already mentioned throughout this chapter that the subject of the intrinsic geometry of surfaces is the study of those properties and objects that isometries keep invariant, in other words, according to Proposition 7.2, all the geometrical objects preserved by diffeomorphisms which keep the length of curves invariant. We have known, until now, only one invariant of this class: the Gauss curvature, and we have profited from its invariance in helping us to show the rigidity of positively curved compact surfaces, that is, the fact that their intrinsic geometry determines their manner of being embedded in three-space.

Other objects which should be, in a first approach, invariant through maps preserving lengths of curves are least length curves joining two given points, if they exist. We will start now the study of certain curves of a special nature on the surfaces which are fundamental in intrinsic geometry: *geodesics*. We pose a question: in the case when they exist, how are the curves which minimize the length between two points of a surface? First,

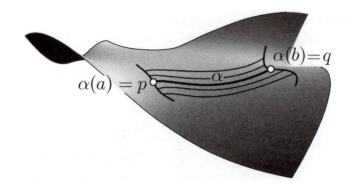

Figure 7.2. *Variation of a curve*

we will only think about curves joining two points which have shorter length than the curves close to them.

Definition 7.19. Let $\alpha : [a, b] \to S$ be a differentiable curve on a surface $S \subset \mathbb{R}^3$. A *variation* of the curve α is a differentiable map $F : [a, b] \times (-\varepsilon, \varepsilon) \to S$ such that $F(t, 0) = \alpha(t)$ for each $t \in [a, b]$ where ε is a positive number. For each $s \in (-\varepsilon, \varepsilon)$, we have a curve $F_s : [a, b] \to S$ given by $F_s(t) = F(t, s)$. We will call it a longitudinal curve of the variation. When all these curves have the same end points as $F_0 = \alpha$, that is, when $F(a, s) = \alpha(a)$ and $F(b, s) = \alpha(b)$ for each $s \in (-\varepsilon, \varepsilon)$, we will say that the variation is *proper*. Associated to the variation F, we define $V : [a, b] \to \mathbb{R}^3$ by

$$V(t) = \frac{\partial F}{\partial s}(t, 0) = \left.\frac{d}{ds}\right|_{s=0} F(t, s).$$

V is clearly differentiable and, moreover,

$$V(t) \in T_{\alpha(t)}S, \qquad \forall t \in [a, b].$$

The map V is called the *variational field* of the variation F. Of course, if F is proper, we have $V(a) = V(b) = 0$.

Let us see that variations of curves on a surface and variational fields associated to them have a deeper relation.

Proposition 7.20. *Let $\alpha : [a, b] \to S$ be a differentiable curve on the surface S and $V : [a, b] \to \mathbb{R}^3$ a differentiable map such that*

$$V(t) \in T_{\alpha(t)}S, \qquad \forall t \in [a, b].$$

Then there exist $\varepsilon > 0$ and a variation

$$F : [a, b] \times (-\varepsilon, \varepsilon) \longrightarrow S$$

of α whose variational field is V. Furthermore, if $V(a) = V(b) = 0$, F can be chosen to be proper.

Proof. Since the trace $K = \alpha([a, b])$ of the curve is a compact subset of S, by Lemma 4.25, we may find a neighbourhood W of K and a number $\delta > 0$ such that we have defined on $N_\delta(W)$ a differentiable projection $\pi : N_\delta(W) \to W \subset S$ of the tubular neighbourhood, with the properties pointed out in Exercise (1) of Chapter 4. Since $N_\delta(W)$ is an open subset of $\mathbb{R}^3$ containing the compact set K, there exists a number $\varepsilon' > 0$ such that, if dist $(p, K) < \varepsilon'$, then $p \in N_\delta(W)$. Put $\varepsilon = \varepsilon'/(1 + M)$ with $M = \max_{t \in [a,b]} |V(t)|$. Thus, if $s \in (-\varepsilon, \varepsilon)$, we have

$$\text{dist}\,(\alpha(t) + sV(t), K) \leq |sV(t)| \leq \varepsilon\,M < \varepsilon',$$

for each $t \in [a, b]$. Hence

$$\alpha(t) + sV(t) \in N_\delta(W), \qquad \forall (t, s) \in [a, b] \times (-\varepsilon, \varepsilon).$$

So, we define the required variation $F : [a, b] \times (-\varepsilon, \varepsilon) \to S$ by the expression

$$F(t, s) = \pi(\alpha(t) + sV(t)), \qquad \forall (t, s) \in [a, b] \times (-\varepsilon, \varepsilon).$$

It is clear that F is differentiable. When $s = 0$, one has

$$F(t, 0) = \pi(\alpha(t)) = \alpha(t),$$

since $\alpha(t) \in S$ and, moreover, its variational field is

$$\frac{\partial F}{\partial s}(t, 0) = \left.\frac{d}{ds}\right|_{s=0} \pi(\alpha(t) + sV(t))$$

$$= (d\pi)_{\alpha(t)}(V(t)) = V(t),$$

because $\alpha(t) \in S$ and $V(t) \in T_{\alpha(t)}S$ and, according to the exercise cited before, the differential of π at the points of the surface, restricted to the corresponding tangent plane, is the identity map. Finally, if $V(a) = V(b) = 0$,

$$F(a, s) = \pi(\alpha(a) + sV(a)) = \pi(\alpha(a)) = \alpha(a),$$

for each $s \in (-\varepsilon, \varepsilon)$. Analogously, $F(b, s) = \alpha(b)$. $\qquad\square$

We can associate to each variation $F : [a, b] \times (-\varepsilon, \varepsilon) \to S$ of the curve $\alpha = F_0$ a function

$$L_F : (-\varepsilon, \varepsilon) \longrightarrow \mathbb{R}$$

by the expression

$$L_F(s) = L_a^b(F_s) = \int_a^b |F_s'(t)|\, dt = \int_a^b \left|\frac{\partial F}{\partial t}(t, s)\right|\, dt.$$

This is the *length function* of the variation and clearly $L_F(0) = L_a^b(\alpha)$ for each variation F. Now suppose that the curve α is p.b.a.l. If we define

$G : [a, b] \times (-\varepsilon, \varepsilon) \to \mathbb{R}$ by

$$G(t, s) = \left| \frac{\partial F}{\partial t}(t, s) \right|,$$

we have that G is continuous and that

$$G(t, 0) = |\alpha'(t)| = 1 > 0, \qquad \forall t \in [a, b].$$

Hence, there exists a number δ such that $0 < \delta < \varepsilon$ and such that

$$|s| < \delta \implies G(t, s) > 0, \qquad \forall t \in [a, b].$$

Thus $G_{|[a,b] \times (-\delta, \delta)}$ is positive and, so, differentiable. As a consequence, the length function of the variation F, suitably restricted, $L_F : [a, b] \times (-\delta, \delta) \to \mathbb{R}$, is differentiable and, moreover,

$$L_F'(s) = \int_a^b \frac{\partial G}{\partial s}(t, s)\, dt$$

for all $s \in (-\delta, \delta)$. Our aim now is to compute $L_F'(0)$. For this, we have

$$(7.10) \qquad \frac{\partial G}{\partial s}(t, 0) = \left\langle \frac{\partial F}{\partial t}, \frac{\partial^2 F}{\partial s \partial t} \right\rangle (t, 0), \qquad \forall t \in [a, b].$$

With these preliminaries, one can already learn something about curves minimizing the length between its end points, even among only close curves.

Theorem 7.21 (First variation of the length). *Let $\alpha : [a, b] \to S$ be a curve p.b.a.l. on a surface S and $F : [a, b] \times (-\delta, \delta) \to S$ a variation of α. Then, if L_F is the associated length function,*

$$L_F'(0) = \langle V(b), \alpha'(b) \rangle - \langle V(a), \alpha'(a) \rangle - \int_a^b \langle V(t), \alpha''(t) \rangle\, dt,$$

where V is the variational field of F.

Proof. From (7.10) it follows that

$$L_F'(0) = \int_a^b \left\langle \frac{\partial F}{\partial t}, \frac{\partial^2 F}{\partial s \partial t} \right\rangle (t, 0)\, dt.$$

Integrating by parts, we have

$$L_F'(0) = \left\langle \frac{\partial F}{\partial t}, \frac{\partial F}{\partial s} \right\rangle \Bigg|_{(a,0)}^{(b,0)} - \int_a^b \left\langle \frac{\partial^2 F}{\partial t^2}, \frac{\partial F}{\partial s} \right\rangle (t, 0)\, dt.$$

One finishes by bearing in mind that

$$\frac{\partial F}{\partial t}(t, 0) = \alpha'(t), \quad \frac{\partial^2 F}{\partial t^2}(t, 0) = \alpha''(0), \quad \text{and} \quad \frac{\partial F}{\partial s}(t, 0) = V(t).$$

$\square$

Corollary 7.22. *A regular curve* $\alpha : [a, b] \to S$ *on a surface* S *has critical length, that is,* $L'_F(0) = 0$, *for every proper variation* F *if and only if* $[\alpha''(t)]^T \wedge \alpha'(t) = 0$ *for each* $t \in [a, b]$, *in other words, if and only if its tangential acceleration* $[\alpha''(t)]^T$ *at each point of the surface is proportional to the velocity* $\alpha'(t)$. *If we assume that* α *is parametrized proportionally to the arc length, then this is equivalent to the fact that* $[\alpha''(t)]^T = 0$.

Proof. First, it is an elementary exercise to observe that the condition $[\alpha''(t)]^T \wedge \alpha'(t) = 0$ is invariant under reparametrizations of α and the same applies to the fact of having critical length for all proper variations. Consequently, we suppose from now on that the curve α is p.b.a.l. In this case, since $|\alpha'(t)|^2 = 1$, the acceleration $\alpha''(t)$ is perpendicular to the velocity $\alpha'(t)$ at each time t. This is why the two conditions $[\alpha''(t)]^T \wedge \alpha'(t) = 0$ and $[\alpha''(t)]^T = 0$ are equivalent. By the formula of the first variation of the length in Theorem 7.21 above, if $L'_F(0) = 0$ for every proper variation F of α, one has

$$\int_a^b \langle V(t), \alpha''(t) \rangle \, dt = 0,$$

where V is the corresponding variational field. But, using Proposition 7.20, there exists a proper variation whose variational field is $V(t) = h(t)[\alpha''(t)]^T$, where h is a function differentiable on $[a, b]$, positive on (a, b), and with $h(a) = h(b) = 0$. Hence

$$\int_a^b h(t) |[\alpha''(t)]^T|^2 \, dt = 0.$$

Thus the acceleration $\alpha''(t)$ does not have a component tangent to the surface. Conversely, if the curve p.b.a.l. α satisfies $[\alpha''(t)]^T = 0$ for each $t \in [a, b]$, the formula of the first variation of the length in Theorem 7.21 implies that $L'_F(0) = 0$ for each F proper. $\square$

In particular, the condition supplied by Corollary 7.22 is a necessary condition for a regular curve to minimize the length between its two end points: it must be the path of a particle subjected exclusively to the force needed to remain at each time on the surface. That is, from the point of view of a two-dimensional physicist living on the surface, it is the path of a particle which is not subjected to any exterior perturbation—an inertial trajectory, if our physicist accepts Newton's laws.

Definition 7.23 (Geodesic). In general, any differentiable curve $\alpha : [a, b] \to S$ on a surface S will be called a *geodesic* when

$$\alpha''(t) \perp T_{\alpha(t)} S, \qquad \forall t \in [a, b],$$

independently of having or not having the property of minimizing the length with respect to close curves or distant curves with the same end points.

Remark 7.24. If $\alpha : [a, b] \to S$ is a geodesic of S, then

$$\frac{d}{dt}|\alpha'(t)|^2 = 2\langle \alpha''(t), \alpha'(t) \rangle = 0$$

and, so, *any geodesic of a surface has constant length velocity vector at each time t.* That is, any geodesic is either a constant curve or it is parametrized proportionally to the arc length. Notice that being geodesic is not a geometrical property because it is a property of curves which does depend on the parametrization, and so it is a physical property. In fact, if we reparametrize a geodesic curve, the new curve that we obtain is also geodesic if and only if the reparametrization is homothetic. Instead, it is true that any regular curve having the geometrical property of minimizing—in general, making critical—the length between its ends must be the trace of a geodesic, that is, there are ways of moving along it—inertial or with constant speed—which make it a geodesic.

The first property arising from the relation between the definition of a geodesic and the length of curves is its invariance under isometries, which converts them to an object of intrinsic geometry.

Theorem 7.25. *Let $f : S \to S'$ be a local isometry between two surfaces S and S', and let $\alpha : [a, b] \to S$ be a differentiable curve on S. Then α is a geodesic of S if and only if $f \circ \alpha$ is a geodesic of S'.*

Proof. Since the fact that a curve is a geodesic has a local nature, we may restrict ourselves to pieces of the curve and of the surface small enough for f to be a diffeomorphism. Thus, we may assume that f is a global isometry. Then, suppose that α is a geodesic of S and let $G : [a, b] \times (-\delta, \delta) \to S'$ be a variation of $f \circ \alpha$ on S' with $\delta > 0$. Then $F = f^{-1} \circ G$ is a variation of α. Moreover, since f^{-1} preserves the length of curves, one has

$$L_F(s) = L_a^b(F_s) = L_a^b(f^{-1} \circ G_s) = L_a^b(G_s) = L_G(s)$$

for each $s \in (-\delta, \delta)$. Hence $L'_G(0) = L'_F(0) = 0$ and $f \circ \alpha$ is a geodesic because it has velocity vector of constant length and so we may apply Corollary 7.22. The converse can be shown in the same way by changing the roles played by S and S' and using f^{-1} instead of f. $\qquad \square$

Example 7.26 (Geodesics of a plane). Let $P = \{x \in \mathbb{R}^3 \, | \, \langle x, a \rangle = b\}$ be a plane perpendicular to the unit vector $a \in \mathbb{R}^3$. If $\alpha : [a, b] \to P$ is an arbitrary differentiable curve on P, we have $\langle \alpha(t), a \rangle = b$ for each $t \in [a, b]$. Thus $\langle \alpha''(t), a \rangle = 0$, that is,

$$\alpha''(t) \in T_{\alpha(t)}P, \qquad \forall t \in [a, b].$$

Hence α is geodesic if and only if $\alpha'' = 0$, that is, $\alpha(t) = tc + d$ where $c, d \in \mathbb{R}^3$. Therefore, we can say that *the geodesics of a plane are just its straight lines parametrized proportionally to the arc length.*

Example 7.27 (Geodesics of a sphere). Let α be a differentiable curve p.b.a.l. on a sphere $\mathbb{S}^2(r)$ centred at a point $a \in \mathbb{R}^3$ with radius $r > 0$. We have $|\alpha(t) - a|^2 = r^2$ for all t. Taking derivatives of this expression twice, we obtain $\langle \alpha(t) - a, \alpha''(t) \rangle = -1$ at each instant. Since the plane $T_{\alpha(t)}\mathbb{S}^2(r)$ is the orthogonal complement of the vector $\alpha(t) - a$, one has that

$$[\alpha''(t)]^T = \alpha''(t) + \frac{1}{r^2}(\alpha(t) - a).$$

Thus, α is a geodesic if and only if it satisfies the equations

$$r^2\alpha''(t) + \alpha(t) - a = 0, \quad |\alpha(t) - a|^2 = r^2, \quad \text{and} \quad |\alpha'(t)|^2 = 1.$$

It is easy to see that the solutions are

$$\alpha(t) = a + p\cos\frac{t}{r} + rv\sin\frac{t}{r}, \quad |p|^2 = r^2, \quad |v|^2 = 1, \quad \langle p, v \rangle = 0,$$

that is, *the great circle determined by the plane spanned by p and v run with constant speed.*

Example 7.28 (Geodesics of a cylinder). Let C be the right cylinder of unit radius whose axis is the z-axis of $\mathbb{R}^3$, and let $\alpha : [a, b] \to C$ be a differentiable curve on C given by $\alpha(t) = (x(t), y(t), z(t))$. Then α is a geodesic if and only if

$$(x''(t), y''(t), z''(t)) \perp T_{(x(t),y(t),z(t))}C$$

for each $t \in [a, b]$. Using Exercise 2.55, this is equivalent to

$$(x''(t), y''(t), z''(t)) \| (x(t), y(t), 0).$$

Since $x(t)^2 + y(t)^2 = 1$, differentiating twice and assuming that α is p.b.a.l., one has that $x(t)x''(t) + y(t)y''(t) = -1 + (z'(t))^2$. Hence the curve is a geodesic of the cylinder if and only if

$$x''(t) + (1 - z'(t)^2)x(t) = 0,$$
$$y''(t) + (1 - z'(t)^2)y(t) = 0,$$
$$z''(t) = 0.$$

Suppose now that $\alpha(0) = (1, 0, 0)$ and that $\alpha'(0) = (0, a, b)$ with $a^2 + b^2 = 1$. Then

$$z(t) = t\,b \quad \text{and} \quad x''(t) + a^2\,x(t) = y''(t) + b^2\,y(t) = 0,$$

and so x and y are of the form

$$\lambda \sin at + \mu \cos at, \qquad \lambda, \mu \in \mathbb{R},$$

and, therefore,

$$\alpha(t) = (\cos at, \sin at, bt) \qquad \text{with } a^2 + b^2 = 1.$$

Figure 7.3. *Geodesics of the cylinder*

That is, *the geodesics of the cylinder are circular helices including, as a limit case, its straight lines and circles.*

In every examples so far, two facts stand out. First, for each point of the surface being considered and for each tangent vector at the point, there is a geodesic passing through the point with this given velocity. Second, these surfaces mentioned above share the property that any of their geodesics are defined on the whole of $\mathbb{R}$. Which of these two facts could be generalized to an arbitrary surface? Of course, the answer is definitely not for the second one. To see this, consider the surface $S = P - \{p\}$ where P is a plane of $\mathbb{R}^3$ and $p \in P$ is any of its points. On this surface, the geodesics are straight lines parametrized with constant velocity. If $q \in S$ and if we take the vector $p - q \in T_q S$, the geodesic passing through q with velocity $p - q$ cannot be defined on the complete real line $\mathbb{R}$. However, we will see that the first property is not exclusive for the three examples studied above.

In fact, let S be an arbitrary surface. From Lemma 4.25, we know that, for each $p \in S$, there is an orientable open neighbourhood V_p of p in S and a number $\varepsilon_p > 0$ such that $N_{\varepsilon_p}(V_p)$ is a tubular neighbourhood. Let us represent by $N(V)$ any tubular neighbourhood of this type on an orientable open subset V of the surface. Such a subset of $\mathbb{R}^3$ is an open set containing the surface V and where the projection $\pi : N(V) \to V \subset S$ is defined, with the properties pointed out in Exercise (1) of Chapter 4, and also the oriented distance r. We define a map—really, a vector field on an open subset of $\mathbb{R}^6$—

$$\mathcal{G} : N(V) \times \mathbb{R}^3 \longrightarrow \mathbb{R}^3 \times \mathbb{R}^3$$

by means of the equality

$$\mathcal{G}(x, y) = (y, G(x, y)),$$

where $G : N(V) \times \mathbb{R}^3 \to \mathbb{R}^3$ is given by

$$G(x, y) = -\langle y, d(N \circ \pi)_x(y) \rangle \, (N \circ \pi)(x),$$

for each $x \in N(V)$ and $y \in \mathbb{R}^3$, where N is any local Gauss map for S. Furthermore, G is clearly differentiable and, thus, $\mathcal{G}$ is as well. Let us consider now the system of first order differential equations

(7.11) $$(x, y)' = \mathcal{G}(x, y).$$

Solving them is equivalent to solving the second order system of equations

(7.12) $$x'' = G(x, x').$$

The theorem of existence, uniqueness, and dependence of the initial conditions implies that, if $(a, b) \in N(V) \times \mathbb{R}^3$, there exist $\varepsilon(a, b) \in (0, +\infty]$ and a map $g_{a,b} : (-\varepsilon(a, b), \varepsilon(a, b)) \to N(V) \times \mathbb{R}^3$ such that

$$g_{a,b}(0) = (a, b), \quad g'_{a,b}(t) = \mathcal{G}(g_{a,b}(t)), \quad t \in (-\varepsilon(a, b), \varepsilon(a, b)),$$

that is, $g_{a,b}$ is a solution of (7.11) with initial condition (a, b), and moreover $g_{a,b}$ is maximal—with respect to the symmetric interval of definition—among all those having the same initial condition. On the other hand, the set

$$\mathcal{D} = \{(a, b, t) \in N(V) \times \mathbb{R}^3 \times \mathbb{R} \mid |t| < \varepsilon(a, b)\}$$

is an open subset of $N(V) \times \mathbb{R}^3 \times \mathbb{R} \subset \mathbb{R}^7$ and the map

$$\Gamma : \mathcal{D} \longrightarrow N(V) \times \mathbb{R}^3$$

given by

$$\Gamma(a, b, t) = g_{a,b}(t)$$

is differentiable.

Now let $g : (-\varepsilon, \varepsilon) \to N(V) \times \mathbb{R}^3$ be a solution of (7.11), with $\varepsilon > 0$ and initial conditions

$$g(0) = (p, v) \in TV = \{(p, v) \in V \times \mathbb{R}^3 \mid v \in T_pS\} \subset N(V) \times \mathbb{R}^3.$$

Denote by $x : (-\varepsilon, \varepsilon) \to N(V)$ the first component of g. Then x is a solution of the system (7.12) with $x(0) = p$ and $x'(0) = v$. To measure how much the trajectory x goes away from the surface S, consider the composition $f = r \circ x : (-\varepsilon, \varepsilon) \to \mathbb{R}$. We have $f(0) = r(x(0)) = r(p) = 0$ and

$$f'(t) = (dr)_{x(t)}(x'(t)), \quad t \in (-\varepsilon, \varepsilon),$$

and, using the properties given in Exercise (1) of Chapter 4,

$$f'(t) = \langle x'(t), (N \circ \pi)(x(t)) \rangle.$$

Hence, $f'(0) = \langle v, N(p) \rangle = 0$. Taking derivatives again

$$f''(t) = \langle x''(t), (N \circ \pi)(x(t)) \rangle + \langle x'(t), d(N \circ \pi)_{x(t)}(x'(t)) \rangle = 0,$$

because x is a solution of (7.12). As a consequence $f(t) = r(x(t)) = 0$ for each $t \in (-\varepsilon, \varepsilon)$, that is, $x(t) \in V \subset S$ at any time. But, moreover, in this case,

$$[x''(t)]^T = G(x(t), x'(t))^T = 0,$$

and so x is a geodesic of S with $x(0) = p$ and $x'(0) = v$. It is easy to see that the converse is also true. Indeed, if $x : (-\varepsilon, \varepsilon) \to V \subset S$ is a geodesic of S with $x(0) = p \in V$ and $x'(0) = v \in T_pS$, then $x''(t)$ is a vector at each t normal to the surface. Thus

$$x''(t) = \lambda(t)\,(N \circ x)(t)$$

for some function λ. After scalar multiplication by $(N \circ x)(t)$, one gets

$$\lambda(t) = \langle x''(t), (N \circ x)(t)\rangle = -\langle x'(t), (dN)_{x(t)}(x'(t))\rangle,$$

because $\langle x', N \circ x\rangle = 0$. Since $\pi(x(t)) = x(t)$ and $(d\pi)_{x(t)}$ is the identity map,

$$x''(t) = G(x(t), x'(t)).$$

Therefore, $g = (x, x')$ is a solution of (7.11) with initial condition $g(0) = (x(0), x'(0))$ and $x(0) = p$, $x'(0) = v$. Then, we can make the following statement.

Lemma 7.29. *The geodesics of the surface S are exactly the solutions of the system of second order differential equations (7.12) with initial conditions a point of the surface and a tangent vector at this point.*

From this, we may give the following fundamental result.

Theorem 7.30 (Existence and uniqueness of geodesics). *Let S be a surface of Euclidean space. For each point $p \in S$ and each tangent vector $v \in T_pS$ there exist a real number $0 < \varepsilon(p, v) \le +\infty$ and a unique geodesic $\gamma_{p,v} : (-\varepsilon(p, v), \varepsilon(p, v)) \to S$ such that*

$$\gamma_{p,v}(0) = p \quad and \quad \gamma'_{p,v}(0) = v,$$

which is maximal under these conditions. Furthermore, the set

$$\mathcal{E} = \{(p, v, t) \in TS \times \mathbb{R}\,|\,|t| < \varepsilon(p, v)\},$$

where $TS = \{(p, v) \in S \times \mathbb{R}^3\,|\,v \in T_pS\}$, is an open subset of $TS \times \mathbb{R} \subset S \times \mathbb{R}^3 \times \mathbb{R}$ and the map

$$\gamma : \mathcal{E} \subset S \times \mathbb{R}^3 \times \mathbb{R} \longrightarrow S$$

defined by $\gamma(p, v, t) = \gamma_{p,v}(t)$ is differentiable.

Proof. Given $p \in S$ and $v \in T_pS$, we have $(p, v) \in TV \subset N(V) \times \mathbb{R}^3$, for some orientable open set V containing the point p. Then, let $\delta > 0$ be the number corresponding to the solution of (7.11), $g : (-\delta, \delta) \to N(V) \times \mathbb{R}^3$ with initial condition (p, v). This g extends to a maximal solution on a symmetric interval $(-\varepsilon(p,v),\varepsilon(p,v))$. From Lemma 7.29, we know that $\gamma = p_1 \circ g$ is a geodesic of S with initial conditions $\gamma(0) = p$ and $\gamma'(0) = v$. This is the required $\gamma_{p,v}$. On the other hand, the subset $\mathcal{E}$ of $TS \times \mathbb{R}$ is, locally, nothing more than $(TS \times \mathbb{R}) \cap \mathcal{D}$, and so it is open in $TS \times \mathbb{R}$. Finally, the map $\gamma : \mathcal{E} \to S$ is only the restriction to $\mathcal{D}$ of the map Γ. Hence, it is differentiable. $\qquad\square$

Remark 7.31 (Homogeneity of geodesics). For a point $p \in S$ in the surface S and a vector $v \in T_p S$, set $\gamma = \gamma_{p,v}$ to be the geodesic defined on $(-\varepsilon(p,v),\varepsilon(p,v))$. Define

$$g : \left(-\frac{1}{s}\varepsilon(p,v), \frac{1}{s}\varepsilon(p,v) \right) \longrightarrow S,$$

where $s > 0$, by the relation

$$g(t) = \gamma(st).$$

Then g is another geodesic, because it is nothing more than a homothetic reparametrization of the first geodesic, and

$$g(0) = \gamma(0) = p \quad \text{and} \quad g'(0) = s\gamma'(0) = sv.$$

Consequently $\varepsilon(p,v)/s \le \varepsilon(p,sv)$ and g coincides with $\gamma_{p,sv}$ on this common domain. Since this is valid for any vector $v \in T_p S$ and any $s > 0$, we have

$$\frac{1}{s}\varepsilon(p,v) \le \varepsilon(p,sv) \le \frac{1}{s}\varepsilon\!\left(p,\frac{1}{s}sv\right).$$

That is, for each $p \in S$ and each $v \in T_p S$, we have

$$\varepsilon(p,sv) = \frac{1}{s}\varepsilon(p,v) \quad \gamma(p,sv,t) = \gamma(p,v,st)$$

for all $s > 0$ and all $|t| < \varepsilon(p,sv)$.

7.6. The exponential map

The fact that geodesics of a surface are solution curves of a second order differential equation, or components of solutions of a first order equation, as seen in the previous section, will permit us to handle at the same time sets of geodesics corresponding to sets of close initial conditions, through the map $\gamma : \mathcal{E} \to S$ which appeared in the existence and uniqueness Theorem 7.30, defined on the open subset $\mathcal{E}$ of $TS \times \mathbb{R}$.

Consider the following subset $\mathcal{E}_1$ of TS:

$$\mathcal{E}_1 = \{(p,v) \in TS \,|\, (p,v,1) \in \mathcal{E}\}.$$

Notice that this $\mathcal{E}_1$ is homeomorphic—through the first projection of $TS \times \mathbb{R}$ onto TS—to

$$\mathcal{E} \cap (TS \times \{1\}),$$

which is an open subset of $TS \times \{1\}$. Hence $\mathcal{E}_1$ is an open subset of $TS \subset S \times \mathbb{R}^3$. We will define on this subset the so-called *exponential map*

$$\exp : \mathcal{E}_1 \longrightarrow S$$

by the equality

$$\exp(p, v) = \gamma(p, v, 1),$$

which is, by construction, a differentiable map on $\mathcal{E}_1 \subset TS \subset S \times \mathbb{R}^3$.

Remark 7.32. For each point $p \in S$ we have that the geodesic $\gamma_{p,0_p}$, with initial conditions the point p and vanishing velocity at this point, is the constant curve p and, so, it is defined on the whole of $\mathbb{R}$. That is, $\varepsilon(p, 0_p) = +\infty$ for each $p \in S$. Thus, $1 < \varepsilon(p, 0_p)$ and then

$$(p, 0_p) \in \mathcal{E}_1 \quad \text{and} \quad \exp(p, 0_p) = p, \qquad \forall p \in S.$$

Remark 7.33. Suppose that $(p, v) \in \mathcal{E}_1$ is a point of TS in the domain of the exponential map. By definition $1 < \varepsilon(p, v)$. Hence, if $t \in (0, 1]$, it follows, by the homogeneity of geodesics, that

$$1 < \varepsilon(p, v) \le \frac{1}{t}\varepsilon(p, v) = \varepsilon(p, tv)$$

and, from this, that $(p, tv, 1) \in \mathcal{E}$ and that $(p, tv) \in \mathcal{E}_1$. On the other hand, if $t = 0$, we have $(p, tv) = (p, 0_p) \in \mathcal{E}_1$, as we saw in Remark 7.32. Summarizing,

$$(p, v) \in \mathcal{E}_1 \implies (p, tv) \in \mathcal{E}_1, \qquad t \in [0, 1].$$

Moreover

$$\exp(p, tv) = \gamma(p, tv, 1) = \gamma(p, v, t), \qquad t \in [0, 1].$$

Remark 7.34. As stated in Remark 7.32, $(p, 0_p) \in \mathcal{E}_1$ for each $p \in S$. But $\mathcal{E}_1$ is an open subset of TS, whose topology is induced from $S \times \mathbb{R}^3$. So, there exist an open subset V of S containing the point p and a number $\varepsilon > 0$ such that

$$T(V, \varepsilon) = \{(p, v) \in TV \,|\, |v| < \varepsilon\} \subset \mathcal{E}_1.$$

That is, given $p \in S$, the exponential map is defined, at least, for the points in a certain neighbourhood of p and the tangent vectors at these points with uniformly small length.

For any point $p \in S$ of a surface, one has that

$$\mathcal{E}(T_pS) = \{v \in T_pS \,|\, (p, v) \in \mathcal{E}_1\}$$

is clearly an open subset of the tangent plane T_pS including the origin 0_p and it is star-shaped with respect to it, from Remark 7.33 above. We will call the *exponential map at the point* $p \in S$ the map

$$\exp_p : \mathcal{E}(T_pS) \longrightarrow S$$

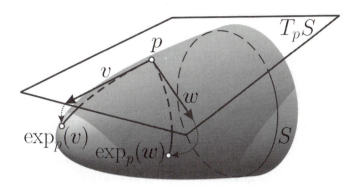

Figure 7.4. *Exponential map*

defined by

$$\exp_p(v) = \exp(p, v), \qquad \forall v \in \mathcal{E}(T_pS).$$

Clearly it is a differentiable map taking each vector tangent to the surface S at the point p, inside of its domain, into the point of the geodesic with initial conditions p and v which is attained at the time $t = 1$, that is, the point of the geodesic p.b.a.l. starting from p in the direction of v to where we arrive in a time $|v|$.

An immediate property of the exponential map $\exp_p : \mathcal{E}(T_pS) \to S$ of a surface S is

$$\exp_p(0_p) = \exp(p, 0_p) = \gamma(p, 0_p, 1) = p, \qquad \forall p \in S.$$

Moreover, if $v \in T_{0_p}T_pS = T_pS$, one has

$$(d\exp_p)_{0_p}(v) \;=\; \left.\frac{d}{dt}\right|_{t=0} \exp_p(tv)$$

$$=\; \left.\frac{d}{dt}\right|_{t=0} \gamma(p, tv, 1) = \left.\frac{d}{dt}\right|_{t=0} \gamma(p, v, t) = v,$$

where we have used the homogeneity of geodesics. Therefore

(7.13) $$\qquad\qquad (d\exp_p)_{0_p} = I_{T_pS}, \qquad \forall p \in S.$$

Then, in this situation we may apply the inverse function Theorem 2.75, assuring in this way that all points near $p \in S$ are attained by a unique geodesic starting from p.

Proposition 7.35. *For each point $p \in S$ of a surface, there is a number $\delta > 0$ such that the map*

$$\exp_p : B_\delta(0_p) \subset T_pS \longrightarrow S$$

is a diffeomorphism onto its image.

   This image, which we will denote by $B_\delta(p)$, will be called the *geodesic ball* with centre p and radius δ. One must realize that, if $B_\delta(p)$ is a geodesic ball, then the exponential map $\exp_p : B_\delta(0_p) \to S$ is a parametrization which exactly covers this open subset of the surface. Moreover, if $q \in B_\delta(p)$ is a point of the geodesic ball with centre p and radius δ, any geodesic joining p with q inside of $B_\delta(p)$ has to be a—homothetic—reparametrization of the curve $\exp_p(tv)$ where $v = (\exp_p)^{-1}(q)$, which is called the *radial geodesic* joining p with q. All these assertions follow from the fact that the differential of the exponential at $0_p \in T_pS$ is the identity map for each $p \in S$. We will now study the behaviour of this differential at other points.

Lemma 7.36 (Gauss's lemma). *Let $p \in S$ be a point of a surface and let $\varepsilon > 0$ be such that the exponential map $\exp_p$ is defined on $B_\varepsilon(0_p) \subset T_pS$. Then, if $v \in T_pS$ with $|v| < \varepsilon$, we have*

$$\begin{cases} |(d\exp_p)_v(u)| = |u| & \text{when } u\|v, \\[2mm] \langle (d\exp_p)_v(u), (d\exp_p)_v(w)\rangle = 0 & \text{when } u\|v \text{ and } w \perp v. \end{cases}$$

Proof. Let $\alpha : I \to T_pS$ be a differentiable curve defined on a neighbourhood of the origin in $\mathbb{R}$ such that $\alpha(0) = v$, $\alpha'(0) = w$, and $|\alpha(s)|^2 = |v|^2$ for all $s \in I$, that is, whose trace is the circle centred at the origin of T_pS passing through v. For $\delta > 0$ small enough one has that $t\alpha(s) \in B_\varepsilon(0_p)$ if $t \in (-1 - \delta, 1 + \delta)$. Then, we define a map

$$F : (-1 - \delta, 1 + \delta) \times I \longrightarrow S$$

by

$$F(t, s) = \exp_p(t\alpha(s)) = \gamma(p, t\alpha(s), 1) = \gamma(p, \alpha(s), t).$$

Differentiating with respect to t, one obtains

$$(d\exp_p)_{t\alpha(s)}(\alpha(s)) = \frac{\partial F}{\partial t}(t, s) = \gamma'(p, \alpha(s), t).$$

Thus

$$|(d\exp_p)_{t\alpha(s)}(\alpha(s))| = \left|\frac{\partial F}{\partial t}(t, s)\right| = |\gamma'(p, \alpha(s), t)| = |\alpha(s)| = |v|,$$

since the velocity vector of the geodesic $\gamma(p, \alpha(s), t)$ has length independent of t. Putting $t = 1$ and $s = 0$ in the above, one gets

$$|(d\exp_p)_v(v)| = |v|,$$

and, so, we have shown the first equality of the lemma. Another consequence of the same equality is, taking derivatives with respect to s, that

$$\left\langle \frac{\partial^2 F}{\partial s\partial t}, \frac{\partial F}{\partial t} \right\rangle = 0,$$

for $s \in I$ and $|t| < 1 + \delta$. Applying the Schwarz theorem of elementary calculus for mixed partial derivatives, we have

$$0 = \frac{\partial}{\partial t} \left\langle \frac{\partial F}{\partial s}, \frac{\partial F}{\partial t} \right\rangle - \left\langle \frac{\partial F}{\partial s}, \frac{\partial^2 F}{\partial t^2} \right\rangle.$$

Taking into account that

$$\frac{\partial^2 F}{\partial t^2}(t, s) = \gamma''(p, \alpha(s), t)$$

has no component tangent to the surface, because $\gamma(p, \alpha(s), t)$ is a geodesic, and that $(\partial F / \partial s)(t, s)$ is tangent to S, one obtains

$$\frac{\partial}{\partial t} \left\langle \frac{\partial F}{\partial s}, \frac{\partial F}{\partial t} \right\rangle = 0,$$

that is, that

$$\left\langle \frac{\partial F}{\partial s}, \frac{\partial F}{\partial t} \right\rangle$$

is constant in the variable t. But, since

$$\frac{\partial F}{\partial s}(t, s) = \frac{d}{ds} \exp_p(t\alpha(s)) = t(d \exp_p)_{t\alpha(s)}(\alpha'(s)),$$

one has that

$$t \langle (d \exp_p)_{t\alpha(s)}(\alpha(s)), (d \exp_p)_{t\alpha(s)}(\alpha'(s)) \rangle$$

does not depend on t. Computing this expression for $t = 1$ and $s = 0$ and, after that, for $t = 0$ and $s = 0$, it follows that

$$\langle (d \exp_p)_v(v), (d \exp_p)_v(w) \rangle = 0,$$

which is the second required equality. $\qquad \square$

Therefore, we have that the exponential map, at any point of the surface, preserves the length of radial curves starting from the origin of the tangent plane and the perpendicularity between these radii and the images of the circles centred at this origin. This property, discovered by Gauss in his *Disquisitiones*, implies the following result of local minimization of the length, which is a local converse of Corollary 7.22.

Proposition 7.37. *Let $B_\delta(p)$ be a geodesic ball of radius $\delta > 0$ with centre at a point p of a surface S. If $\alpha : [a, b] \to B_\delta(p) \subset S$ is a Lipschitz curve with $\alpha(a) = p$, then*

$$L_a^b(\alpha) \geq L_a^b(\gamma),$$

where γ is the only radial geodesic p.b.a.l. defined on $[a, b]$ joining p with $\alpha(b)$. If equality holds, then the traces of α and of γ coincide.

Proof. First, it is worth noting that a Lipschitz function is absolutely continuous, and thus it has a—bounded—derivative almost everywhere. This is why it makes sense to talk about the arc length of Lipschitz curves with the same definition as in the differentiable case. With this established, if the curve α passes again through the centre of the geodesic ball, for example, if $\alpha(c) = p$ for some $c \in (a, b]$, the radial geodesic with which we may compare the arc $\alpha_{|[a,c]}$ is the constant curve p. That is, the proof can be reduced to the case in which $\alpha(t) \neq p$ for each $t \in (a, b]$. Then we will have that $\alpha(b) \neq p$. Thus $\alpha(b) = \exp_p(v)$ with $0 < |v| < \delta$ and

$$\alpha(t) = \exp_p(\beta(t)), \qquad t \in [a, b],$$

where $\beta : [a, b] \to T_p S$ is a Lipschitz curve in the tangent plane of S at p with $\beta(a) = 0_p$, $\beta(t) \neq 0_p$ for each $t \in (a, b]$, and $\beta(b) = v$, and, hence,

$$
\begin{aligned}
L_a^b(\alpha) &= \int_a^b |\alpha'(t)|\, dt \\
&= \int_a^b |(\exp_p \circ \beta)'(t)|\, dt = \int_a^b |(d\exp_p)_{\beta(t)}(\beta'(t))|\, dt.
\end{aligned}
$$

If we choose $w(t) \in T_p S$ so that together with $\beta(t)/|\beta(t)|$ they form an orthonormal basis, one has, for $t \in (a, b]$, where the derivative of β exists,

$$\beta'(t) = \frac{1}{|\beta(t)|^2} \langle \beta'(t), \beta(t) \rangle\, \beta(t) + \langle \beta'(t), w(t) \rangle\, w(t).$$

Therefore, using Lemma 7.36,

$$|(d\exp_p)_{\beta(t)}(\beta'(t))| \geq \frac{1}{|\beta(t)|} |\langle \beta'(t), \beta(t) \rangle|,$$

and equality is attained for $t \in (a, b]$ if and only if $\beta'(t)$ is proportional to $\beta(t)$. Consequently,

$$
\begin{aligned}
L_a^b(\alpha) &\geq \int_a^b \frac{1}{|\beta(t)|} |\langle \beta'(t), \beta(t) \rangle|\, dt \\
&\geq \left| \int_a^b \frac{1}{|\beta(t)|} \langle \beta'(t), \beta(t) \rangle\, dt \right| = [|\beta(t)|]_{t=a}^{t=b} = |v| = L_a^b(\gamma).
\end{aligned}
$$

If equality occurs, $\beta'(t)$ is proportional to $\beta(t)$ for each $t \in (a, b]$ where this derivative exists. Thus, the trace of $\beta(t)$ is contained in the straight line starting from 0_p in $T_p S$ and passing through v. Then the curve $\alpha = \exp_p \circ \beta$ has the same trace as the geodesic γ. $\qquad \square$

Exercise 7.38. $\uparrow$ If $p \in \mathbb{S}^2$ is a point of the unit sphere, prove that $\mathbb{S}^2 - \{-p\}$ is a geodesic ball centred at p. What is its radius? Use Proposition 5.8 to obtain a formula for integrating functions on $\mathbb{S}^2$.

Exercise 7.39. Give an example of a geodesic on a surface having length greater than another differentiable curve with the same end points.

Let $p_0 \in S$ be a point of the surface, V an open neighbourhood of this point in S, and $\varepsilon > 0$ a number such that the exponential map exp is defined on $T(V, \varepsilon)$; see Remark 7.34. Consider now a parametrization $X : U \to S$ such that $p_0 = X(q_0) \in X(U) \subset V$ and a real number $\delta > 0$ such that, if $w \in \mathbb{R}^2$ and $|w| < \delta$, then $|(dX)_q(w)| < \varepsilon$, that is, $(dX)_q(w) \in T(V, \varepsilon)$ for each $q \in U$. From this, we define a map $E : U \times B_\delta \subset \mathbb{R}^2 \times \mathbb{R}^2 \to S \times S$ by

$$E(q, w) = (X(q), \exp_{X(q)}(dX)_q(w)), \qquad \forall (q, w) \in U \times B_\delta.$$

This map is differentiable because its first component is obviously differentiable and the second one can be written as

$$\exp_{X(q)}(dX)_q(w) = \exp\left(X(q), (dX)_q(w)\right) = \exp\left(X(q), w_1 X_u(q) + w_2 X_v(q)\right),$$

for each $q \in U$ and each $w = (w_1, w_2) \in \mathbb{R}^2$ with $|w| < \delta$. Clearly $(q_0, 0) \in U \times B_\delta$ and

$$E(q_0, 0) = (X(q_0), \exp_{X(q_0)} 0_p) = (p_0, p_0).$$

Let us compute now the differential of E at this point of its domain. This differential is a linear map

$$(dE)_{(q_0, 0)} : \mathbb{R}^2 \times \mathbb{R}^2 \longrightarrow T_{p_0} S \times T_{p_0} S.$$

Take $w \in \mathbb{R}^2$. Then

$$(dE)_{(q_0, 0)}(w, 0) = \left.\frac{d}{dt}\right|_{t=0} E(q_0 + tw, 0)$$

$$= \left.\frac{d}{dt}\right|_{t=0} (X(q_0 + tw), \exp_{X(q_0 + tw)} 0_{X(q_0 + tw)}).$$

Hence

$$(dE)_{(q_0, 0)}(w, 0) = ((dX)_{q_0}(w), (dX)_{q_0}(w)), \qquad \forall w \in \mathbb{R}^2.$$

On the other hand,

$$(dE)_{(q_0, 0)}(0, w) = \left.\frac{d}{dt}\right|_{t=0} E(q_0, tw) = \left.\frac{d}{dt}\right|_{t=0} (p_0, \exp_{p_0} t(dX)_{q_0}(w)),$$

and, consequently,

$$(dE)_{(q_0, 0)}(0, w) = (0, (dX)_{q_0}(w)), \qquad \forall w \in \mathbb{R}^2.$$

Since the differential $(dX)_q$ of the parametrization is injective at each $q \in U$, the differential $(dE)_{(q_0, 0)}$ is an linear isomorphism. Applying the inverse function theorem, we have the existence of an open neighbourhood of $(q_0, 0)$ in $U \times B_\delta$, which can be assumed to be of the form $U' \times B_\rho$, where U' is a relatively compact open neighbourhood of q_0 in U and ρ is a positive number less than δ, such that

$$E : U' \times B_\rho \longrightarrow S \times S$$

is a diffeomorphism onto its image, which is an open neighbourhood of (p_0, p_0) in $S \times S$. Since $|(dX)_q|$ and $|(dX)_q^{-1}|$ are bounded on U', we can choose two positive numbers $\eta < \rho$ and τ such that

$$(dX)_q(B_\eta) \subset B_\tau(0_{X(q)}) \subset (dX)_q(B_\rho),$$

for each $q \in U'$. Then, we may find a neighbourhood W of p_0 in S such that

$$W \times W \subset E(U' \times B_\eta).$$

A first conclusion is that, if $q \in U'$, the map

$$E : \{q\} \times B_\rho \longrightarrow \{X(q)\} \times S$$

is a diffeomorphism onto its image, that is,

$$\exp_{X(q)} : (dX)_q(B_\rho) \longrightarrow S$$

is also a diffeomorphism onto its image. Hence, the open ball $B_\tau(0_{X(q)})$ centred at the origin of $T_{X(q)}S$, which is contained in $(dX)_q(B_\rho)$, has as its image a geodesic ball $B_\tau(X(q))$ in S. A second conclusion is that, if p_1 and p_2 are points of W, one has

$$p_1 = X(q) \quad \text{and} \quad p_2 = \exp_{X(q)}(dX)_q(w),$$

for some $q \in U'$ and $w \in \mathbb{R}^2$ with $|w| < \eta$. Therefore

$$W \subset \exp_{p_1}[(dX)_q(B_\eta)] \subset B_\tau(p_1),$$

which is a valid inclusion for all $p_1 \in W$. We may, then, state the following improvement of Proposition 7.35.

Proposition 7.40 (Existence of normal neighbourhoods). *Each point of a surface S has an open neighbourhood which is contained in a geodesic ball centred at each of its points, with a common radius $\varepsilon > 0$. Such a neighbourhood will be called a* normal neighbourhood *of radius ε.*

The first consequence of this result will be an answer to the second question that we posed after Example 7.28: when can we assure that the geodesics of a surface extend indefinitely on it? Or, equivalently, what can we say about the positive number $\varepsilon(p, v) > 0$ that we associated to each $(p, v) \in TS$ in the existence and uniqueness Theorem 7.30?

Corollary 7.41. *If S is a surface which is closed as a subset of Euclidean space, then the geodesic $\gamma = \gamma_{p,v}$ with initial conditions $(p, v) \in TS$ is defined on the whole of $\mathbb{R}$. Thus, the exponential map of S is defined on the whole of TS.*

Proof. Let $(p, v) \in TS$, with $v \neq 0$, and suppose that the positive number $\varepsilon(p, v)$ given by Theorem 7.30 is finite. We choose a sequence $\{t_n\}_{n \in \mathbb{N}}$ of points of $(-\varepsilon(p, v), \varepsilon(p, v))$ tending to $\varepsilon(p, v)$. This is a Cauchy sequence and, since

$$|\gamma(t_n) - \gamma(t_m)| = \left| \int_{t_n}^{t_m} \gamma'(t)\, dt \right| \leq |v||t_n - t_m|,$$

the sequence $\{\gamma(t_n)\}_{n \in \mathbb{N}}$ of points of S is a Cauchy sequence as well. Hence, it converges to a point q of $\mathbb{R}^3$ that, since S is closed, must be a point of S. Consider a normal neighbourhood V of this point $q \in S$ with radius $\varepsilon > 0$, whose existence was given by Proposition 7.40. Since γ is continuous, we know that the image through γ of an interval of the form $(\delta, \varepsilon(p, v))$ is in V, for some $\delta > 0$. Choosing $t_0 \in (\delta, \varepsilon(p, v))$ so that $\varepsilon(p, v) - t_0 < \varepsilon$ and putting $r = \gamma(t_0) \in V$, one has that $\gamma(t) = \exp_r((t - t_0)u)$ for each $t \in (\delta, \varepsilon(p, v))$ and where $u \in T_r S$, $|u| = |v|$. This shows that γ extends on the right-hand side beyond $\varepsilon(p, v)$. Therefore $\varepsilon(p, v) = +\infty$ for all $(p, v) \in TS$ and so $\gamma_{p,v}$ is defined on the whole real line $\mathbb{R}$. As a consequence, $1 < \varepsilon(p, v)$ for all $(p, v) \in TS$ and so $(p, v) \in \mathcal{E}_1$. It follows that the domain of the exponential map is the whole of TS. $\qquad\square$

A second consequence of Proposition 7.40 of the existence of normal neighbourhoods will be to prove that all curves, not necessarily differentiable, minimizing the length between their end points have the same trace as regular geodesics. In other words, geodesics minimize the length in a much wider class than differentiable curves. This can be interpreted by considering that minimizing the length forces the curve to be smooth.

Theorem 7.42. *Let $\alpha : [a, b] \to S$ be a Lipschitz curve joining two points $p = \alpha(a)$ and $q = \alpha(b)$ of a surface S whose length is less than or equal to that of any piecewise differentiable curve joining them. Then there exists a geodesic—and so a differentiable curve—whose trace coincides with that of the curve α.*

Proof. For each $t \in [a, b]$, there exists, by Proposition 7.40, a normal neighbourhood in S of the point $\alpha(t)$, which will be denoted by V_t. Since α is continuous, each reciprocal image $\alpha^{-1}(V_t)$ is an open subset of the compact interval $[a, b]$. Let $\delta > 0$ be the Lebesgue number of this cover. Then, if $[c, d] \subset [a, b]$ is a subinterval with length less than δ, its image $\alpha([c, d])$ is contained in some of the normal neighbourhoods and so it is in a geodesic ball centred at $\alpha(c)$. Now then, the curve $\alpha_{|[c,d]}$ minimizes the length between its end points because, if not, there would exist a piecewise differentiable curve β defined on $[c, d]$ joining them with a shorter length and, then, replacing $\alpha_{|[c,d]}$ by β we would obtain another piecewise differentiable curve shorter than α joining $\alpha(a)$ and $\alpha(b)$. So, by Proposition 7.37, the restriction

$\alpha_{|[c,d]}$ has the same trace as a radial geodesic starting from $\alpha(c)$. Since this happens with any subinterval with length less than δ, we can conclude. $\square$

We can finally assure that, on any surface closed as a subset of $\mathbb{R}^3$, we can join any two points by means of a curve which minimizes the length and which must be a geodesic according to Corollary 7.22.

Theorem 7.43 (Hopf-Rinow's theorem). *Let S be a connected surface closed in $\mathbb{R}^3$, and let $p, q \in S$ be two points of S. Then there exists a geodesic $\gamma : [a, b] \to S$ with $\gamma(a) = p$ and $\gamma(b) = q$ and length less than or equal to the length of any piecewise differentiable curve joining the two points.*

Proof. The result is obvious when $p = q$. Suppose then that $p \neq q$ and take any two numbers $a \neq b$. Let $\Gamma_{p,q}$ be the set of all the piecewise differentiable curves γ defined on $[a, b]$ taking values on S and with $\gamma(a) = p$ and $\gamma(b) = q$. Represent by $d(p, q)$—see Exercise (5) of this chapter—the number

$$d(p, q) = \inf_{\gamma \in \Gamma_{p,q}} L_a^b(\gamma).$$

This infimum exists because all these lengths are non-negative numbers and moreover $\Gamma_{p,q}$ is non-empty—one can always join two points in a connected surface by means of differentiable curves. Let $\{\gamma_n\}_{n \in \mathbb{N}}$ be a sequence of $\Gamma_{p,q}$ such that

$$\lim_{n \to \infty} L_a^b(\gamma_n) = d(p, q).$$

We will consider all the curves in this sequence to be parametrized proportionally to the arc length on each regular piece, that is, $|\gamma_n'(t)| = C_n$ for each $t \in [a, b]$. Then

$$L_a^b(\gamma_n) = \int_a^b |\gamma_n'(t)| \, dt = (b - a)C_n.$$

In particular $\{C_n\}_{n \in \mathbb{N}}$ is a bounded sequence, because it converges. Thus, if $t_1, t_2 \in [a, b]$, we have

$$|\gamma_n(t_1) - \gamma_n(t_2)| = \left| \int_{t_1}^{t_2} \gamma_n'(t) \, dt \right| \leq C_n |t_1 - t_2|,$$

for each $n \in \mathbb{N}$. By the Ascoli-Arzelà theorem, there is a subsequence, which we will denote in the same way, which converges uniformly to a Lipschitz curve $\gamma : [a, b] \to \mathbb{R}^3$. Since S is closed in $\mathbb{R}^3$, the curve γ has its image on S and

$$L_a^b(\gamma) = \int_a^b |\gamma'(t)| \, dt \leq (b - a) \lim_{n \to \infty} C_n = \lim_{n \to \infty} L_a^b(\gamma_n) = d(p, q).$$

Therefore γ has length less than or equal to any other piecewise differentiable curve joining p and q. We finish the proof by applying Theorem 7.42. $\square$

Exercises

(1) ↑ Show that the group of isometries of a sphere coincides with the subgroup of the group of rigid motions of $\mathbb{R}^3$ fixing its centre and is isomorphic to the order three orthogonal group.

(2) ↑ Find a local isometry between the helicoid—defined in Example 3.46—and the catenoid, which was introduced—see Exercise (9) in Chapter 3—as the surface of revolution whose generating curve is a catenary.

(3) What can you say about a compact surface S whose Gauss map $N : S \to \mathbb{S}^2$ is an isometry?

(4) Consider the two following differentiable maps X and X':

$$X(u, v) = (u\cos v, u\sin v, \log u), \qquad \forall (u, v) \in \mathbb{R}^+ \times (0, 2\pi),$$
$$X'(u, v) = (u\cos v, u\sin v, v), \qquad \forall (u, v) \in \mathbb{R}^2.$$

Prove that $S = X(\mathbb{R}^+ \times (0, 2\pi))$ and $S' = X'(\mathbb{R}^2)$ are surfaces and that X and X' are parametrizations for each of them. Show that the map $f : S \to S'$ given by $f = X' \circ X^{-1}$ satisfies $K = K' \circ f$ and, however, that it is not a local isometry.

(5) ↑ Let S be a connected surface. For each $p, q \in S$, we define

$$d(p, q) = \inf_{\gamma \in \Gamma_{p,q}} L(\gamma),$$

where $\Gamma_{p,q}$ is the set of piecewise differentiable curves on S joining p with q. Demonstrate that $d : S \times S \to \mathbb{R}$ is a distance (the Riemannian or intrinsic distance). What relation does it have with the standard Euclidean distance? Prove also that, if S is closed in $\mathbb{R}^3$, then the metric space (S, d) is complete and that any isometry between surfaces is an isometry for this distance.

(6) ↑ Let S be a compact connected surface, and let $f : S \times [0, 1] \to \mathbb{R}^3$ be a differentiable map such that, for each $t \in [0, 1]$, $S_t = f(S \times \{t\})$ is a surface and the map $f_t : S \to S_t$ given by $f_t(p) = f(p, t)$ for all $p \in S$ is an isometry with $f_0 = I_S$. Suppose that σ^t is the second fundamental form of S_t relative to the inner normal of S_t. Prove that, if $K(p) > 0$ at a point $p \in S$ and if σ_p^0 is positive definite, then $\sigma_{f_t(p)}^t$ is positive definite for all $t \in [0, 1]$.

(7) Let S be a compact surface and $V : S \to \mathbb{R}^3$ a differentiable vector field. Calculate the divergence of the field JV defined by

$$JV = V \wedge N,$$

where N is the Gauss map of S and prove that the function
$$p \in S \longmapsto \langle (dV)_p(e_1), e_2 \rangle - \langle (dV)_p(e_2), e_1 \rangle,$$
where $\{e_1, e_2\}$ is any orthonormal basis of T_pS, is differentiable and has zero integral on S.

(8) ↑ Let $f, g : S \to \mathbb{R}^3$ be two differentiable functions on a surface S. Show that the subset of S determined by
$$\{p \in S \,|\, f(p) = g(p) \text{ and } (df)_p = (dg)_p\}$$
is closed.

(9) ↑ Use the previous exercise and the existence of geodesic balls on a surface to prove that, if two local isometries $f, g : S \to S'$ between two surfaces, from which S is connected, coincide along with their differentials at a point, then they coincide everywhere.

(10) *Darboux trihedron and Darboux equations.* Suppose that S is an orientable surface and $N : S \to \mathbb{S}^2$ is a Gauss map on S. If $\alpha : I \to S$ is a geodesic of S p.b.a.l. defined on an interval of $\mathbb{R}$, we define three maps $T, N, B : I \to \mathbb{R}^3$ by
$$T(s) = \alpha'(s), \quad N(s) = (N \circ \alpha)(s), \quad \text{and} \quad B(s) = T(s) \wedge N(s).$$
Then T, N, and B are differentiable and moreover $\{T(s), N(s), B(s)\}$ form an orthonormal basis of $\mathbb{R}^3$ for all $s \in I$ with $T(s), B(s) \in T_{\alpha(s)}S$ and $N(s) \perp T_{\alpha(s)}S$. Prove that the following equalities are true:
$$\begin{cases} T'(s) = \sigma_{\alpha(s)}(T(s), T(s)) \, N(s), \\ N'(s) = -\sigma_{\alpha(s)}(T(s), T(s)) \, T(s) - \sigma_{\alpha(s)}(T(s), B(s)) \, B(s), \\ B'(s) = \sigma_{\alpha(s)}(T(s), B(s)) \, N(s), \end{cases}$$
where σ is the second fundamental form of S corresponding to N.

(11) ↑ Find all the traces of the geodesics joining two given points of a plane, of a sphere, and of a right circular cylinder. What can you say about their number? Deduce that, in the case of the unit sphere, a geodesic of length less than π always minimizes the length between its ends.

(12) ↑ If all the geodesics of a connected surface are plane curves, show that the surface is contained either in a plane or in a sphere.

(13) ↑ Show that each meridian—generating curve—of a surface of revolution is a geodesic. On the other hand, a parallel is a geodesic if and only if it is at a critical distance from the axis of revolution.

(14) ↑ *Clairaut's relation.* Let $\alpha : I \to S$ be a geodesic p.b.a.l. on a surface of revolution whose axis is a straight line R. Represent by r and p the two functions taking each $t \in I$, respectively, to the distance from $\alpha(t)$ to the axis R and the cosine of the angle made by α at the time t with the parallel through $\alpha(t)$. Prove that $r(t)p(t)$ is a constant function.

(15) Show that, if a straight line R is contained in a surface S, then there is a geodesic of S having R as its trace.

(16) Let $\alpha : I \to S$ be a geodesic in an oriented surface S such that

$$\sigma_{\alpha(t)}(\alpha'(t), \alpha'(t)) = 0$$

for each $t \in I$. Prove that α is a straight line.

(17) ↑ Let $B_\varepsilon(p) = \exp_p(B_\varepsilon(0_p))$ be a geodesic ball on a surface S. Let $\alpha : I \to \mathbb{S}^1 \subset T_pS$ be a parametrization b.a.l. of the unit circle of T_pS. We define a map $F : (-\varepsilon, \varepsilon) \times I \to S$ by

$$F(t, s) = \exp_p(t\alpha(s)).$$

Show that—see Exercise (12) of Chapter 3—the Gauss curvature K of S satisfies

$$\left| \frac{\partial F}{\partial s} \right|_{tt} + (K \circ F) \left| \frac{\partial F}{\partial s} \right| = 0,$$

for all $0 < |t| < \varepsilon$ and all $s \in I$.

(18) ↑ *Minding's theorem.* Let S and S' be two surfaces with the same constant Gauss curvature, and let $p \in S$ and $p' \in S'$ be any two points such that $B_\delta(p) = \exp_p(B_\delta(0_p))$ and $B_\delta(p') = \exp'_{p'}(B_\delta(0_{p'}))$ are geodesic balls. Prove that

$$f = \exp'_{p'} \circ \phi \circ \exp_p^{-1},$$

is an isometry between the two geodesic balls, where $\phi : T_pS \to T_{p'}S'$ is any linear isometry. Hence, *two surfaces with the same constant Gauss curvature are locally isometric.*

(19) ↑ *Bertrand-Puiseux and Diquet formulas (1848).* Let $B_\varepsilon(p)$ be a geodesic ball on a surface S. For each $t \in (0, \varepsilon)$, we call the curve $\exp_p(\mathbb{S}^1(t))$ the *geodesic circle* of radius t and centre p. Analogously we call the subset of S given by $\exp_p(\overline{B_t(0_p)})$ the *geodesic disc* of radius t and centre p. Represent by L_t and A_t, respectively, the length and the area of the geodesic circle and the geodesic disc of radius t centred at p. Show that

$$\begin{cases} K(p) = \lim_{t \to 0} 3\dfrac{2\pi t - L_t}{\pi t^3}, \\[4mm] K(p) = \lim_{t \to 0} 12\dfrac{\pi t^2 - A_t}{\pi t^4}. \end{cases}$$

From this, compute the limit

$$\lim_{t \to 0} \frac{4\pi A_t - L_t^2}{\pi^2 t^4}$$

and give an interpretation of its sign.

(20) ↑ Consider a surface S like that studied in Exercise (7) of Chapter 3, that is, a right cylinder over a simple curve C. Calculate all its geodesics and see that the exponential map of this surface is defined on the whole of TS, provided that C is compact or has infinite length. Check that, however, for some curves C the resulting surface S is not closed. (This shows that the converse of Corollary 7.41 is not true.)

(21) Show that the right cylinder constructed over a logarithmic spiral—see Exercise 1.13—is not a complete surface.

(22) Let $\mathbb{S}^2$ be the unit sphere centred at the origin and d the distance defined in Exercise (5) above. Prove that $d(p, q) \leq \pi$ for all $p, q \in \mathbb{S}^2$ and that equality holds if and only if p and q are antipodal.

(23) ↑ Given two points $p, q \in \mathbb{S}^2$ non-antipodal of the unit sphere, prove that there exists a unique point $m \in \mathbb{S}^2$ such that the 180 degrees rotation around m takes p to q and q to p. It is called the *middle point* of p and q. Show that, moreover, if d is the intrinsic distance on the sphere, then $2d(p, m) = 2d(q, m) = d(p, q)$.

Hints for solving the exercises

Exercise 7.6: For each $x \in O$, the map $(dF)_x : \mathbb{R}^3 \to \mathbb{R}^3$ is orthogonal. That is,

$$\langle (dF)_x(u), (dF)_x(v) \rangle = \langle u, v \rangle, \qquad \forall x \in O \quad \forall u, v \in \mathbb{R}^3.$$

Choosing u, v as vectors of the canonical basis of $\mathbb{R}^3$, one has the equality of functions

$$\left\langle \frac{\partial F}{\partial x_i}, \frac{\partial F}{\partial x_j} \right\rangle = \delta_{ij}, \qquad i, j = 1, 2, 3,$$

where x_1, x_2, x_3 represent the coordinates of $\mathbb{R}^3$. Taking derivatives with respect to x_k, $k = 1, 2, 3$, one obtains

$$\left\langle \frac{\partial^2 F}{\partial x_k \partial x_i}, \frac{\partial F}{\partial x_j} \right\rangle + \left\langle \frac{\partial^2 F}{\partial x_k \partial x_j}, \frac{\partial F}{\partial x_i} \right\rangle = 0, \qquad i, j, k = 1, 2, 3.$$

This equality and the Schwarz theorem imply that

$$G_{ijk} = \left\langle \frac{\partial^2 F}{\partial x_i \partial x_j}, \frac{\partial F}{\partial x_k} \right\rangle$$

is symmetric in i, j and skew-symmetric in j, k. Hence

$$G_{ijk} = -G_{ikj} = -G_{kij} = G_{kji} = G_{jki} = -G_{jik} = -G_{ijk}.$$

Therefore

$$\left\langle \frac{\partial^2 F}{\partial x_i \partial x_j}, \frac{\partial F}{\partial x_k} \right\rangle \equiv 0$$

on O, for all $i, j, k = 1, 2, 3$. But the partial derivatives $\partial F/\partial x_1$, $\partial F/\partial x_2$, and $\partial F/\partial x_3$ form an orthonormal basis of $\mathbb{R}^3$ at each point of O. Consequently

$$\frac{\partial^2 F}{\partial x_i \partial x_j} \equiv 0$$

on O, for each $i, j = 1, 2, 3$. Then the three components of F are linear functions. Thus, $F(x) = Ax + b$ for all $x \in O$, some square matrix A of order three, and some $b \in \mathbb{R}^3$, because O is connected. So, $(dF)_x = A$ at each $x \in O$ and A must be orthogonal. Hence F is the restriction to O of a rigid motion of space $\mathbb{R}^3$.

Exercise 7.16: Since ϕ is self-adjoint, there exists an orthonormal basis of the Euclidean vector plane V where it is defined, which diagonalizes it. If λ_1 and λ_2 are its eigenvalues, since ψ is also self-adjoint, the two endomorphisms, with respect to this basis, will be represented by two matrices

$$\phi \equiv \begin{pmatrix} \lambda_1 & 0 \\ 0 & \lambda_2 \end{pmatrix}, \qquad \psi \equiv \begin{pmatrix} a & b \\ b & c \end{pmatrix}$$

for some $a, b, c \in \mathbb{R}$. The required formula is easily proved replacing each endomorphism by the matrix which represents it.

Exercise 7.17: If $\det(\phi + \psi) > 0$, the endomorphism $\phi + \psi$ would be definite, say, positive definite. If $\{e_1, e_2\}$ is a basis of the plane diagonalizing ϕ, one has

$$\langle \phi(e_i), e_i \rangle + \langle \psi(e_i), e_i \rangle > 0, \qquad i = 1, 2.$$

Consequently

$$\begin{aligned} \det \phi &= \langle \phi(e_1), e_1 \rangle \langle \phi(e_2), e_2 \rangle > \langle \psi(e_1), e_1 \rangle \langle \psi(e_2), e_2 \rangle \\ &\geq \langle \psi(e_1), e_1 \rangle \langle \psi(e_2), e_2 \rangle - \langle \psi(e_1), e_2 \rangle^2 = \det \psi, \end{aligned}$$

which gives a contradiction to our hypothesis. Therefore $\det(\phi + \psi) \leq 0$. Moreover, if equality holds, there would be at least a non-null vector in the kernel of $\phi + \psi$. Let $\{u_1, u_2\}$ be a basis diagonalizing $\phi + \psi$, that is, such that

$$\phi(u_1) + \psi(u_1) = 0 \quad \text{and} \quad \phi(u_2) + \psi(u_2) = \lambda u_2, \qquad \lambda \in \mathbb{R}.$$

From the first equality we deduce that

$$\langle \phi(u_1), u_1 \rangle = -\langle \psi(u_1), u_1 \rangle \quad \text{and} \quad \langle \phi(u_1), u_2 \rangle = -\langle \psi(u_1), u_2 \rangle,$$

which, together with the facts that $\det \phi = \det \psi$ and that ϕ and ψ are definite, gives the equality

$$\langle \phi(u_2), u_2 \rangle = -\langle \psi(u_2), u_2 \rangle,$$

implying $\lambda = 0$. Thus, in this case, $\phi = -\psi$.

Exercise 7.38: Given $p \in \mathbb{S}^2$, if $u \in T_p\mathbb{S}^2$ is a unit vector, Example 7.27 implies that

$$\gamma(p, u, t) = p \cos t + u \sin t, \qquad \forall t \in \mathbb{R}.$$

This is why, if $v \in T_p\mathbb{S}^2 - \{0\}$, it follows from the homogeneity of geodesics that

$$\gamma(p, v, t) = p \cos t + \frac{v}{|v|} \sin t, \qquad \forall t \in \mathbb{R}.$$

Thus, according to the definition,

$$\exp_p v = \begin{cases} p \cos |v| + \dfrac{v}{|v|} \sin |v| & \text{if } v \neq 0, \\ p & \text{if } v = 0_p. \end{cases}$$

Now, it only remains to check that the map $\phi : \mathbb{S}^2 - \{p, -p\} \to T_p\mathbb{S}^2$, given by

$$\phi(q) = \frac{\arccos \langle p, q \rangle}{\sqrt{1 - \langle p, q \rangle^2}}(q - \langle p, q \rangle p),$$

is a differentiable inverse map of $\exp_p$ restricted to $B_\pi(0_p) - \{0_p\} \subset T_p\mathbb{S}^2$. Thus, $\mathbb{S}^2 - \{-p\} = B_\pi(p)$ is a geodesic ball of radius π centred at p.

Then, taking $p = (0, 0, 1)$, the north pole of the sphere, the map $X : (0, \pi) \times (0, 2\pi) \to \mathbb{S}^2$ given by

$$X(\rho, \theta) = \cos \rho \, (0, 0, 1) + \sin \rho \, (\cos \theta, \sin \theta, 0)$$

is a parametrization covering the whole sphere except a half-meridian and moreover $|X_\rho \wedge X_\theta| = \sin \rho$. Then, if $f : \mathbb{S}^2 \to \mathbb{R}$ is integrable,

$$\int_{\mathbb{S}^2} f(p) \, dp = \int_0^\pi \int_0^{2\pi} f(\rho, \theta) \sin \rho \, d\theta \, d\rho.$$

Exercise (1): Let ϕ be a rigid motion of $\mathbb{R}^3$ fixing the point $a \in \mathbb{R}^3$. If $p \in \mathbb{S}_r^2(a)$, with $r > 0$, one has

$$|\phi(p) - a| = |\phi(p) - \phi(a)| = |p - a| = r.$$

Therefore $\phi(\mathbb{S}_r^2(a)) \subset \mathbb{S}_r^2(a)$. This also occurs for ϕ^{-1} and, so, ϕ keeps the sphere $\mathbb{S}_r^2(a)$ invariant. The restriction of ϕ to this sphere belongs, according to Proposition 7.1, to the group of isometries of $\mathbb{S}_r^2(a)$. But there are no other isometries of this sphere, since each isometry f of $\mathbb{S}_r^2(a)$ is the restriction of a rigid motion ϕ of $\mathbb{R}^3$, as seen in Example 7.9. It suffices now to show that ϕ must fix the centre of the sphere. This follows from, for example, the fact that any affine map takes the centre of a quadric to the centre of the quadric image.

Exercise (2): We know that the helicoid $\mathcal{H}$ is covered by the image of a parametrization $X : \mathbb{R}^2 \to \mathcal{H}$ given by

$$X(u, v) = (v \cos u, v \sin u, u), \qquad \forall (u, v) \in \mathbb{R}^2$$

and that the map $Y : \mathbb{R}^2 \to \mathcal{C}$ given by

$$Y(u, v) = (\cosh v \cos u, \cosh v \sin u, v),$$

restricted to each open interval of the form $(\theta, \theta + 2\pi) \times \mathbb{R}$, with $\theta \in \mathbb{R}$, is a parametrization of the catenoid $\mathcal{C}$. We define a map $\phi : \mathcal{H} \to \mathcal{C}$ by

$$\phi = Y \circ F \circ X^{-1},$$

where $F : \mathbb{R}^2 \to \mathbb{R}^2$ is the map defined by

$$F(u, v) = (u, \arg \sinh v),$$

which is clearly differentiable. Differentiating with respect to u and v in the equality $\phi \circ X = Y \circ F$, one has

$$(d\phi)_{X(u,v)}(-v \sin u, v \cos u, 1) = \sqrt{1 + v^2}(-\sin u, \cos u, 0)$$

$$(d\phi)_{X(u,v)}(\cos u, \sin u, 0) = \frac{1}{\sqrt{1 + v^2}}(v \cos u, v \sin u, 1).$$

Since $X_u = (-v \sin u, v \cos u, 1)$ and $X_v = (\cos u, \sin u, 0)$ form a basis of $T_X \mathcal{H}$, this shows that $(d\phi)_{X(u,v)}$ preserves the scalar product of vectors.

Exercise (5): If $p, q \in S$, there is one-to-one correspondence between $\Gamma_{p,q}$ and $\Gamma_{q,p}$, given as follows: if $\alpha : [a, b] \to S$ is a piecewise differentiable curve with $\alpha(a) = p$ and $\alpha(b) = q$, we consider another curve $\alpha^{-1} : [a, b] \to S$ defined by $\alpha^{-1}(t) = \alpha(-t + b + a)$. Since, moreover,

$$L_a^b(\alpha^{-1}) = L_a^b(\alpha),$$

we have $d(p, q) = d(q, p)$. That is, d is symmetric. It is also clearly non-negative. Let us see that d satisfies the triangular inequality. Let $p, q, r \in S$, $\alpha \in \Gamma_{p,q}$ and $\beta \in \Gamma_{q,r}$. Since lengths of the curves remain invariant through reparametrizations, we may consider that the two curves α and β are defined on $[0, 1]$. Then $\alpha * \beta : [0, 1] \to S$ belongs to $\Gamma_{p,r}$ and, so,

$$d(p, r) \leq L_0^1(\alpha * \beta) = L_0^1(\alpha) + L_0^1(\beta).$$

Since this happens for every $\alpha \in \Gamma_{p,q}$ and $\beta \in \Gamma_{q,r}$, taking infima on the right-hand side, we get

$$d(p, r) \leq d(p, q) + d(q, r), \qquad \forall p, q, r \in S.$$

In order to see that, in fact, d is a distance, we will relate it to the Euclidean distance. If $p, q \in S$, let $\alpha \in \Gamma_{p,q}$. Using Exercise 1.9,

$$L(\alpha) \geq |p - q|$$

and, so, taking infimum in α, we have

$$d(p, q) \geq |p - q|.$$

This inequality implies that, if $d(p, q) = 0$ for some $p, q \in S$, then $p = q$. Hence, d is a distance. But, furthermore, it implies that the metric space (S, d) is complete if S is closed in $\mathbb{R}^3$. In fact, each Cauchy sequence in S must be a Cauchy sequence in $\mathbb{R}^3$ as well and, so, it converges. Since S is closed, the limit has to be in S. Finally, if $f : S \to S$ is a diffeomorphism, it is clear that

$$\Gamma_{f(p), f(q)} = f \circ \Gamma_{p,q}, \qquad \forall p, q \in S.$$

If, moreover, f is an isometry, f preserves the lengths of curves. Thus

$$d(f(p), f(q)) = \inf_{\gamma \in \Gamma_{p,q}} L(f \circ \gamma) = \inf_{\gamma \in \Gamma_{p,q}} L(\gamma) = d(p, q),$$

for $p, q \in S$, and we conclude.

Exercise (6): Let us check first that σ^t is definite at any point $f_t(p)$, for each $t \in [0, 1]$. In fact,

$$\det \sigma^t_{f_t(p)} = K^t(f_t(p)) = (K^t \circ f_t)(p) = K(p) > 0,$$

where we have used the Theorema Egregium 7.10. Now, let $X : U \to S$ be a parametrization with $X(0, 0) = p$. Each $X^t = f_t \circ X$ is a parametrization of S_t with $X^t(0, 0) = f_t(p)$ and each X^t is obtained by fixing t in the map $G : U \times [0, 1] \to \mathbb{R}^3$ given by

$$G(u, v, t) = f(X(u, v), t), \qquad \forall (u, v) \in U \quad \forall t \in [0, 1].$$

As a consequence, if we have choosen X so that the inner normal of S is given by $\det(X_u, X_v, N \circ X) > 0$, using the expressions in (3.10), we have, if $w = X_u(0, 0) \in T_pS$,

$$\sigma^t_{f_t(p)}((df_t)_p(w), (df_t)_p(w)) = \frac{1}{|G_u \wedge G_v|} \det(G_u, G_v, G_{uu})(0, 0, t).$$

The point is that this expression is continuous in t and, since it cannot vanish because $\sigma^t_{f_t(p)}$ is definite, it does not change sign. But, when $t = 0$, this sign is positive. This finishes the exercise.

Exercise (8): Let $\{p_n\}_{n \in \mathbb{N}}$ be a sequence of points of S such that

$$f(p_n) = g(p_n) \quad \text{and} \quad (df)_{p_n} = (dg)_{p_n},$$

for each $n \in \mathbb{N}$, converging to a point $p \in S$. Since f and g are continuous, it is clear that $f(p) = g(p)$. Let $X : U \to S$ be a parametrization with

$X(u, v) = p$ for some $(u, v) \in U$. We may consider that $p_n = X(u_n, v_n)$ with $(u_n, v_n) \in U$ for each $n \in \mathbb{N}$. Then

$$(f \circ X)_u(u_n, v_n) = (df)_{p_n} X_u(u_n, v_n), \quad (f \circ X)_v(u_n, v_n) = (df)_{p_n} X_v(u_n, v_n),$$

and the corresponding equalities for g hold. Thus

$$(f \circ X)_u(u_n, v_n) = (g \circ X)_u(u_n, v_n), \quad (f \circ X)_v(u_n, v_n) = (g \circ X)_v(u_n, v_n).$$

Since the partial derivatives of $f \circ X$ and of $g \circ X$ are continuous, we take limits in the previous equalities and we are finished.

Exercise (9): From the solution to Exercise (8), the set

$$A = \{p \in S \mid f(p) = g(p) \text{ and } (df)_p = (dg)_p\}$$

is a closed subset of S that we are supposing to be non-empty. Let $\varepsilon > 0$ be a number such that $B_\varepsilon(p)$ is a geodesic ball. By Theorem 7.25, if $v \in T_pS$ and $t \in (-\varepsilon(p, v), \varepsilon(p, v))$, we have

$$f(\gamma(p, v, t)) = \gamma(f(p), (df)_p(v), t) \quad \text{and} \quad g(\gamma(p, v, t)) = \gamma(g(p), (dg)_p(v), t).$$

Therefore, if $v \in T_pS$ with $|v| < \varepsilon$,

$$f(\exp_p v) = f(\gamma(p, v, 1)) = g(\gamma(p, v, 1)) = g(\exp_p v)$$

and, so, f and g coincide on $B_\varepsilon(p)$. Hence, their differentials also coincide on the points of this open set. Thus A is an open subset of S. Since the surface is connected, $A = S$ and $f = g$.

Exercise (11): In Examples 7.26, 7.27, and 7.28, we saw that the traces of geodesics in a plane, in a sphere, and in a right circular cylinder are, respectively, straight lines, great circles, and circular helices (including generating lines and their orthogonal circles). Then, in the case of the plane, there is exactly one geodesic joining two different points: the corresponding segment. In the case of the sphere, if the two points are not antipodal, they determine a unique great circle and, so, there are exactly two geodesics joining them: the two segments determined by the two points in this circle, one of them with length less than π and the other one with length greater than π (provided that the sphere has unit radius). Then, by the Hopf-Rinow Theorem 7.43, a segment of the great circle of length less than π minimizes the length between its end points. If the two points are antipodal, there are infinitely many great circles passing through them. Each of them determines two geodesics with length π. In the case of the cylinder, it is a question of showing how many circular helices join two given different points. Cutting the cylinder along a generating line and unwinding (this process is a local isometry, according to Exercise 7.4 and, hence, maps geodesics to geodesics, by Theorem 7.25), the problem is reduced to finding, in a plane strip, finite sequences of parallel lines at a constant distance, joining two given points.

If these two points lie in the same circle, there are two geodesics: the two arcs of this circle. In any other case, there are countably many geodesics: two helices whose pitch is the difference of the height of the points divided by any natural number, where one of them is right-handed and the other one is left-handed.

Exercise (12): Let $p \in S$ be a point and $v \in T_pS$ any unit vector. The geodesic γ with $\gamma(0) = p$ and $\gamma'(0) = v$ lies in a plane with unit normal, for example, $a \in \mathbb{R}^3$. Then $\langle \gamma'(t), a \rangle = 0$ and $\langle \gamma''(t), a \rangle = 0$ for each t. A direct computation, or Exercise (10), permits us to rewrite the last equality as

$$\sigma_{\gamma(t)}(\gamma'(t), \gamma'(t))\langle (N \circ \gamma)(t), a \rangle = 0,$$

where N is a local Gauss map for S and σ is its second fundamental form. If $\sigma_p(v, v) \neq 0$, by continuity, we have $\langle (N \circ \gamma)(t), a \rangle = 0$. Hence, $\gamma' \wedge (N \circ \gamma)$ is always a or $-a$. Thus

$$0 = [\gamma' \wedge (N \circ \gamma)]' = \gamma' \wedge (N \circ \gamma)',$$

because γ is geodesic. This equality, at $t = 0$, tells us that $(dN)_p(v)\|v$, or that $\sigma_p(v, u) = 0$, for each $u \in T_pS$ such that $\langle u, v \rangle = 0$. Therefore, the two quadratic forms σ_p and $\langle \, , \, \rangle$ are proportional. So, S is totally umbilical.

Exercise (13): In Exercise (3) of Chapter 2, we saw that, if S is a surface of revolution whose axis is the straight line R passing through the origin with direction given by the unit vector $a \in \mathbb{R}^3$, we have, for each $p \in S$, that $p \wedge a \in T_pS$ is a non-null vector tangent to the parallel passing through p. Hence, the normal of S at p belongs to the plane spanned by p and a. Thus, if $\alpha : I \to S$ is a meridian p.b.a.l., we have $\alpha' \perp \alpha \wedge a$ or, in another form,

$$\det(\alpha', \alpha, a) = 0$$

along I. Differentiating, one obtains that $\det(\alpha'', \alpha, a) = 0$, or, equivalently, α'' is perpendicular to $\alpha \wedge a$ at any time. But α'' is also perpendicular to the unit vector α'. Since α' and $\alpha \wedge a$ are perpendicular, they generate $T_\alpha S$. Therefore, $\alpha'' \perp T_\alpha S$ and so α is a geodesic. Suppose now that $\alpha : I \to S$ is a parallel. The trace of α is, in this case, a circle of radius, say $r > 0$, in a plane perpendicular to the vector a. Let us put provisionally the origin at the point of intersection of this plane and the axis R. Then, if α is p.b.a.l., $\alpha'' = -(1/r)\alpha$ and, so, for a suitable choice on the normal of S, we have

$$N \circ \alpha = -\frac{1}{r}\alpha.$$

Let p be a point in the trace of α. We know that $|p| = r$, that $N(p) = -(1/r)p$, and that $\langle p, a \rangle = 0$. Consequently, if $\beta : J \to S$ is a meridian with

$\beta(s_0) = p$ for some $s_0 \in J$, one has

$$0 = \langle \beta'(s_0), N(\beta(s_0)) \rangle = -\frac{1}{r} \langle \beta'(s_0), p \rangle = -\frac{1}{r} \langle \beta'(s_0), \beta(s_0) \rangle.$$

Thus, $|\beta|^2$ has a critical point at s_0. In general, if a parallel is a geodesic, its distance to the axis of revolution is critical, or the tangent plane of S along the parallel is parallel to the axis of revolution. The converse is easy.

Exercise (14): Let γ be a geodesic p.b.a.l. on the surface of revolution S whose axis is R, the straight line through the origin with direction $a \in \mathbb{R}^3$, $|a| = 1$. Since γ'' is normal to S at the point γ, from the arguments in the solution to Exercise (13), γ'' is perpendicular to $\gamma \wedge a$. So

$$\det(\gamma'', \gamma, a) = 0.$$

Hence, it follows that

$$\det(\gamma', \gamma, a) = c,$$

where $c \in \mathbb{R}$ is a constant. This equality can also be written as

$$c = \langle \gamma', \gamma \wedge a \rangle = \left\langle \gamma', \frac{\gamma \wedge a}{|\gamma \wedge a|} \right\rangle |\gamma \wedge a|,$$

which is the Clairaut relation, because the scalar product on the left-hand side is the cosine of the angle made by γ with the parallel passing through γ and the length on the right-hand side is the distance from γ to the axis of revolution.

Exercise (17): Since $\exp_p : B_\varepsilon(0_p) \to B_\varepsilon(p)$ is a diffeomorphism, F is a parametrization of S, restricted to $(0, \varepsilon) \times I$. Furthermore

$$F_t(t, s) = (d\exp_p)_{t\alpha(s)} \alpha(s) \quad \text{and} \quad F_s(t, s) = t(d\exp_p)_{t\alpha(s)} \alpha'(s).$$

In this way, the Gauss Lemma 7.36 tells us that F is an orthogonal parametrization, that is, that $\langle F_t, F_s \rangle = 0$ and that, moreover, $E = |F_t|^2 = 1$. Then, we may apply Exercise (12) at the end of Chapter 3 and obtain

$$K \circ F = -\frac{1}{2\sqrt{G}} \left(\frac{G_t}{\sqrt{G}} \right)_t,$$

where $G = |F_s|^2$. Thus

$$|F_s|_{tt} + (K \circ F)|F_s| = 0,$$

as required. This expression is valid on the domain of the polar coordinates (t, s) of $B_\varepsilon(0_p)$, that is, on $B_\varepsilon(0_p) - \{0_p\}$.

Exercise (18): We have to show that $(df)_q$ is a linear isometry for each $q \in B_\varepsilon(p)$. But we know, by definition and by the chain rule, that

$$(df)_{\exp_p v} \circ (d\exp_p)_v = (d\exp_{p'}')_{\phi(v)} \circ \phi.$$

If $v = 0_p$, this equality is nothing more than $(df)_p = \phi$ and, so, $(df)_p$ is a linear isometry. Then, we have to prove that $(df)_{\exp_p v}$ is an isometry as well, for each $v \in B_\varepsilon(0_p) - \{0_p\}$ in $T_p S$. By the Gauss Lemma 7.36, it suffices to see that

$$|(d\exp_p)_v(w)| = |(d\exp'_{p'})_{v'}(w')|,$$

for $v, w \in T_p S$ and $v', w' \in T_{p'} S'$ such that

$$0 < |v| = |w| < \varepsilon \quad \text{and} \quad \langle v, w \rangle = \langle v', w' \rangle = 0.$$

Using the parametrization F, and the corresponding F' for S', constructed in Exercise (17) above with $t\alpha(s) = v$ and $\alpha'(s) = w$, the equality that we are looking for will be true if we show that $|F_s(t, s)|$ does not depend on s and its dependence on t is determined by the constant value c of the Gauss curvature K. Exercise (17) implies that $|F_s|$ is a solution of

$$x''(t) + cx(t) = 0, \qquad 0 < t < \varepsilon,$$

for each s. But

$$\lim_{t \to 0} F_s(t, s) = \lim_{t \to 0} (d\exp_p)_{t\alpha(s)} \alpha'(s) = 0,$$

where, in order to see that $(d\exp_p)_{t\alpha(s)} \alpha'(s)$ depends continuously on t, one can refer to Exercise (8). On the other hand, this same expression tells us that

$$\lim_{t \to 0} \frac{1}{t} F_s(t, s) = (d\exp_p)_{0_p} \alpha'(s) = \alpha'(s),$$

and, so, there exists $F_{ts}(0, s) = \alpha'(s)$. Therefore

$$\lim_{t \to 0} |F_s(t, s)|_t = |\alpha'(s)|.$$

Then, one obtains

$$|F_s(t, s)| = \begin{cases} |w|t & \text{if } c = 0, \\ |w| \sin \sqrt{c}t & \text{if } c > 0, \\ |w| \sinh \sqrt{-c}t & \text{if } c < 0. \end{cases}$$

We finish here, because one finds analogous expressions for $F'_s(t, s)$ by replacing $|w|$ by $|\phi(w)| = |w|$.

Exercise (19): It is clear that, again using the parametrization F from the above exercises,

$$L_t = \int_0^{2\pi} |F_s(t, s)|\, ds, \qquad 0 < t < \varepsilon,$$

and, so, $\lim_{t \to 0} L_t = 0$. Moreover

$$\lim_{t \to 0} \frac{d}{dt} L_t = \lim_{t \to 0} \int_0^{2\pi} |F_s(t, s)|_t\, ds = \int_0^{2\pi} |\alpha'(s)|\, ds = 2\pi.$$

The limit of the second derivative is

$$\lim_{t \to 0} \frac{d^2}{dt^2} L_t = \lim_{t \to 0} \int_0^{2\pi} |F_s(t,s)|_{tt}\, ds = -\lim_{t \to 0} \int_0^{2\pi} (K \circ F)(t,s)|F_s(t,s)|\, ds = 0.$$

Finally, the limit of the third derivative can be computed by applying the recurrence given by the second order equation. In fact

$$\lim_{t \to 0} \frac{d^3}{dt^3} L_t = -\lim_{t \to 0} \int_0^{2\pi} (K \circ F)(t,s)|(\mathrm{dexp}_p)_{t\alpha(s)}\alpha'(s)|\, ds = -2\pi K(p).$$

To compute the required limit, it only remains to use the L'Hôpital rule three consecutive times, taking into account the above equalities. For the second limit, we may argue in a similar way, bearing in mind that, by Proposition 5.8,

$$A_t = \int_0^t \int_0^{2\pi} |F_s(t,s)|\, ds\, dt = \int_0^t L_t\, dt.$$

To conclude, consider the equality

$$\frac{4\pi A_t - L_t^2}{\pi^2 t^4} = \frac{1}{3}\left(3\frac{2\pi t - L_t}{\pi t^3}\frac{2\pi t + L_t}{\pi t} - 12\frac{A_t - \pi t^2}{\pi t^4}\right).$$

Now, take limits in this expression as $t \to 0$, using the previous computations and the fact that

$$\lim_{t \to 0} \frac{2\pi t + L_t}{\pi t} = \lim_{t \to 0} \frac{2\pi + L_t'}{\pi} = 4,$$

by L'Hôpital rule. Therefore

$$\lim_{t \to 0} \frac{4\pi A_t - L_t^2}{\pi^2 t^4} = K(p).$$

This means that, near an elliptic point, the geodesic circles satisfy an isoperimetric inequality opposite to that shown for the plane. However, near a hyperbolic point we have a strict inequality in the same sense that that found in the plane.

Exercise (20): From Exercise (15), we know that the lines of S parametrized in an affine way are geodesics of S. Any other geodesic has to be, then, by uniqueness, transverse to all the straight lines that it meets. Then, it admits a one-to-one projection onto the generating curve C. Let $\alpha : \mathbb{R} \to C \subset P$ be a parametrization b.a.l. (see Example 2.18), where P is the plane passing through the origin and perpendicular to $a \in \mathbb{R}^3$, $|a| = 1$. Any geodesic γ of S is of the form

$$\gamma(s) = \alpha(s) + f(s)\, a,$$

for some differentiable function f. Moreover γ will be defined exactly where f is. But then

$$\gamma''(s) = k(s)n(s) + f''(s)\, a,$$

where n and k are the unit normal of α in P and its curvature, respectively. Hence γ is geodesic if and only if $\gamma''(s) \perp T_{\gamma(s)}S$, that is, if and only if $f'' \equiv 0$. Therefore

$$\gamma(s) = \alpha(s) + (\lambda s + \mu)\, a, \qquad \forall s \in \mathbb{R},$$

for some real numbers λ and μ. Hence, S is complete when C is compact or when, even being non-compact, it has infinite length. To complete the exercise, it suffices to present a simple plane curve with inifinite length which is not closed in its plane, for example, a spiral asymptotic to a circle.

Exercise (23): We know, from Exercise (11) above, that there is a unique arc of a great circle whose length is $d(p, q)$, joining p and q. Let $\gamma :$ $[0, d(p, q)] \to \mathbb{S}^2$ be a parametrization of γ b.a.l. Let $m = \gamma(d(p, q)/2) \in \mathbb{S}^2$. If ϕ is the rotation of π radians around m, the composition $\phi \circ \gamma$ is another geodesic with

$$(\phi \circ \gamma)(d(p, q)/2) = m \quad \text{and} \quad (\phi \circ \gamma)'(d(p, q)/2) = -\gamma'(d(p, q)/2).$$

By uniqueness, it follows that

$$(\phi \circ \gamma)(t) = \gamma(-t + d(p, q)), \qquad \forall t \in [0, d(p, q)].$$

Then, $\phi(p) = q$ and $\phi(q) = p$. Moreover, $\gamma : [0, d(p, q)/2] \to \mathbb{S}^2$ is a geodesic p.b.a.l. joining p and m and its length is less than π. Therefore

$$d(p, m) = L_0^{d(p,q)/2}(\gamma) = \frac{1}{2} d(p, q).$$

Analogously for $d(q, m)$.

7.7. Appendix: Other results of an intrinsic type

In the final section of this chapter, we will leave room for some important results of the intrinsic geometry of surfaces, such that the classification of surfaces with parallel second fundamental form—with constant principal curvatures—or the Hartmann-Nirenberg-Massey theorem, whose statement and proof do not require any new tools, but a more involved use of the tools that we have availed ourselves of until now. We include them separately in this appendix to point out clearly that they are not necessary for getting, in a first reading, a wide and non-superficial idea about the subject of the geometry of surfaces.

7.7.1. Positively curved surfaces. We will start by paying attention to one of the results which gave rise to the so-called differential geometry at large, originally *im Grossen*, a German translation of the Latin expression *in toto* due to Gauss, that is, to the global aspects which, as we consider currently, reveal in a very clear way the objectives pursued in the study of the relationship between the curvature and the topology. We will deal with positively curved surfaces. In fact, we already know that, if the surface is compact, its topology is necessarily spherical—this is a part of the contents of the Hadamard-Stoker Theorem 6.5. This result shows how the sign of the curvature influences the global shape of the surface. But it is still more surprising that, without assuming the surface to be compact, the fact that it is positively curved—far away from zero—forces it to become compact, as seen in Theorem 6.8. Now, we will sharpen this result by Bonnet a bit. To do this, we need the following lemma.

Lemma 7.44. *Let $\gamma : [a, b] \to S$ be a geodesic p.b.a.l. on an orientable surface closed as a subset of Euclidean space. The variation $F : [a, b] \times \mathbb{R} \to S$ given by*

$$F(t, s) = \exp_{\gamma(t)} sf(t)B(t), \qquad \forall (t, s) \in [a, b] \times \mathbb{R},$$

where $B(t) = \gamma'(t) \wedge (N \circ \gamma)(t)$ and N is a Gauss map of S, is a well-defined map for all differentiable function $f : [a, b] \to \mathbb{R}$ and, moreover, its associated length function L_F satisfies

$$L_F''(0) = \int_a^b [f'(t)^2 - (K \circ \gamma)(t)f(t)^2] \, dt.$$

Proof. Since S is closed, we may apply Corollary 7.41 and assume that the exponential map of S is defined on the whole of TS. We then define a map

$$F : [a, b] \times \mathbb{R} \longrightarrow S$$

by the relation

$$F(t, s) = \exp_{\gamma(t)} sf(t)B(t), \qquad \forall (t, s) \in [a, b] \times \mathbb{R},$$

where $B : [a, b] \to \mathbb{R}^3$ is given by

$$B(t) = \gamma'(t) \wedge N(t) \quad \text{and} \quad N(t) = (N \circ \gamma)(t),$$

where $f : [a, b] \to \mathbb{R}$ is any differentiable function and N is the Gauss map of S. So this F is a variation of the geodesic γ whose variational field is $f(t)B(t)$, and this B is the third vector of the Darboux trihedron associated to γ; see Exercise (10) in the previous section. If $L_F : \mathbb{R} \to \mathbb{R}$ is the length function of this variation, we know that, by definition,

$$(7.14) \quad L_F''(0) = \int_a^b \left\{ \left| \frac{\partial^2 F}{\partial s \partial t} \right|^2 + \left\langle \frac{\partial F}{\partial t}, \frac{\partial^3 F}{\partial s^2 \partial t} \right\rangle - \left\langle \frac{\partial F}{\partial t}, \frac{\partial^2 F}{\partial s \partial t} \right\rangle^2 \right\} (t, 0) \, dt.$$

But, for this specific variation, since $F(t, 0) = \exp_{\gamma(t)}(0_{\gamma(t)}) = \gamma(t)$ and taking Exercise (10) into account, we have

$$(7.15) \quad \frac{\partial F}{\partial t}(t, 0) = \gamma'(t) \quad \text{and} \quad \frac{\partial^2 F}{\partial t^2}(t, 0) = \gamma''(t) = \sigma_{\gamma(t)}(\gamma'(t), \gamma'(t)) \, N(t),$$

where σ is the second fundamental form of S relative to the Gauss map N. On the other hand, recalling the definition of the exponential map and the homogeneity of geodesics,

$$F(t,s) = \exp_{\gamma(t)} sf(t)B(t) = \gamma(\gamma(t), sf(t)B(t), 1) = \gamma(\gamma(t), f(t)B(t), s),$$

for each $t \in [a,b]$ and each $s \in \mathbb{R}$. Therefore

$$\frac{\partial F}{\partial s}(t,0) = \gamma'(\gamma(t), f(t)B(t), 0) = f(t)B(t).$$

Taking derivatives here with respect to t and again using the Darboux equations in Exercise (10), we obtain

$$(7.16) \qquad \frac{\partial^2 F}{\partial t \partial s}(t,0) = f'(t)B(t) + f(t)\sigma_{\gamma(t)}(\gamma'(t), B(t))\, N(t),$$

$$\frac{\partial^2 F}{\partial s^2}(t,0) = \gamma''(\gamma(t), f(t)B(t), 0) = f(t)^2 \sigma_{\gamma(t)}(B(t), B(t))\, N(t).$$

Then, it follows from (7.16) that

$$\left| \frac{\partial^2 F}{\partial s \partial t}(t,0) \right|^2 = f'(t)^2 + f(t)^2 \sigma_{\gamma(t)}(\gamma'(t), B(t))^2.$$

Also from (7.15) and (7.16), we get

$$\left\langle \frac{\partial F}{\partial t}, \frac{\partial^3 F}{\partial s^2 \partial t} \right\rangle (t,0) = \frac{\partial}{\partial t}\left\langle \frac{\partial F}{\partial t}, \frac{\partial^2 F}{\partial s^2} \right\rangle (t,0) - \left\langle \frac{\partial^2 F}{\partial t^2}, \frac{\partial^2 F}{\partial s^2} \right\rangle (t,0)$$

$$= -f(t)^2\, \sigma_{\gamma(t)}(B(t), B(t))\, \sigma_{\gamma(t)}(\gamma'(t), \gamma'(t)).$$

Finally, again using (7.15) and (7.16), it follows that

$$\left\langle \frac{\partial F}{\partial t}, \frac{\partial^2 F}{\partial s \partial t} \right\rangle (t,0) = 0.$$

Substituting the three last equalities in (7.14), we obtain

$$L_F''(0) = \int_a^b [f'(t)^2 - f(t)^2\{\sigma_{\gamma(t)}(\gamma'(t), \gamma'(t))\sigma_{\gamma(t)}(B(t), B(t))$$

$$-\sigma_{\gamma(t)}(\gamma'(t), B(t))^2\}]\, dt.$$

Now, it only remains to show that the $\{\gamma'(t), B(t)\}$ form an orthonormal basis of the tangent plane $T_{\gamma(t)}S$ for each $t \in [a,b]$ and to take the definition (3.5) of the Gauss curvature into account. $\qquad\square$

Theorem 7.45 (Bonnet's theorem). *Let S be a surface closed as a subset of Euclidean space $\mathbb{R}^3$ such that its Gauss curvature satisfies $\inf_{p \in S} K(p) > 0$. Then S is an ovaloid whose diameter, as a subset of $\mathbb{R}^3$, obeys*

$$\mathrm{diam}\, S \le \frac{\pi}{\sqrt{\inf_{p \in S} K(p)}}.$$

Proof. Let us represent by $k^2 > 0$ the infimum of the Gauss curvature K of the surface. Then, the mean curvature of S, relative to any local normal field, satisfies $H^2 \geq K \geq k^2 > 0$, as a consequence of (3.6). We choose at each point of the surface the local determination of the unit normal field for the function H to be positive. Then, one can see, by Exercise 3.9, that the surface S is orientable. On the other hand, let $p, q \in S$ be any two points of the surface. From the Hopf-Rinow Theorem 7.43, which can be applied because S is closed, we dispose of a geodesic p.b.a.l. $\gamma : [0, \ell] \to S$ which minimizes the length and such that $\gamma(0) = p$ and $\gamma(\ell) = q$, where ℓ is the length of γ. We apply Lemma 7.44 to this geodesic and conclude that

$$L_F''(0) = \int_0^\ell [f'(t)^2 - (K \circ \gamma)(t) f(t)^2] \, dt \leq \int_0^\ell [f'(t)^2 - k^2 f(t)^2] \, dt,$$

where F is the variation of γ given by

$$F(t, s) = \exp_{\gamma(t)} s f(t) B(t).$$

Moreover, if the function $f : [0, \ell] \to \mathbb{R}$ vanishes at the two ends of the interval, then all the longitudinal curves of the variation $F_s : [0, \ell] \to S$ join the points p and q because

$$F_s(0) = \exp_{\gamma(0)} 0_{\gamma(0)} = p \quad \text{and} \quad F_s(\ell) = \exp_{\gamma(\ell)} 0_{\gamma(\ell)} = q.$$

For this reason, the corresponding length function L_F has a minimum at $s = 0$. Thus $L_F''(0) \geq 0$. Consequently

$$\int_0^\ell [f'(t)^2 - k^2 f(t)^2] \, dt \geq 0$$

for all differentiable function $f : [0, \ell] \to \mathbb{R}$ such that $f(0) = f(\ell) = 0$. Now choose

$$f(t) = \sin \frac{\pi}{\ell} t \qquad \forall t \in [0, \ell],$$

which clearly obeys these conditions. Then

$$0 \leq \int_0^\ell \left[\frac{\pi^2}{\ell^2} \cos^2 \frac{\pi}{\ell} t - k^2 \sin^2 \frac{\pi}{\ell} t \right] dt = \frac{\pi^2}{2\ell} - \frac{k^2 \ell}{2}.$$

Therefore

$$|p - q| \leq L_0^\ell(\gamma) = \ell \leq \frac{\pi}{k}.$$

This is true for each pair of points $p, q \in S$ and so

$$\text{diam } S \leq \frac{\pi}{k} = \frac{\pi}{\sqrt{\inf_{p \in S} K(p)}}.$$

As a consequence, S is a bounded subset of $\mathbb{R}^3$ and, since it is closed, it is compact. $\qquad \square$

Figure 7.5. *Elliptic paraboloid*

Remark 7.46. The upper bound provided by the Bonnet Theorem 7.45 for the diameter of a closed surface with Gauss curvature greater than or equal to a positive constant cannot be improved. In fact, this is shown by the spheres $\mathbb{S}^2(r)$ with arbitrary radius $r > 0$. In this case, the Gauss curvature is constant and its value is $1/r^2$. Hence $k^2 = \inf_{p \in S} K(p) = 1/r^2$ and the diameter of a sphere with radius r in $\mathbb{R}^3$ is just π/r. It is important to note also that this upper bound is valid, not only for the extrinsic diameter of the surface, that is, the diameter resulting from using the Euclidean distance, but also for the intrinsic diameter which appears if we use the distance defined on the surface in Exercise (5) of this chapter.

Remark 7.47. For a closed surface, it is impossible to deduce compactness from the mere positivity of the Gauss curvature K. Indeed, let us recall the elliptic paraboloid S given by

$$S = \{(x, y, z) \in \mathbb{R}^3 \mid 2z = x^2 + y^2\},$$

which, since it is the graph of the differentiable function

$$(x, y) \in \mathbb{R}^2 \longmapsto \frac{1}{2}(x^2 + y^2),$$

is a closed surface diffeomorphic to the plane. According to Example 3.45, we have that the Gauss curvature of S obeys

$$K(x, y, z) = \frac{1}{(1 + 2z^2)^2}, \qquad \forall (x, y, z) \in S.$$

Thus, K is positive everywhere. However this paraboloid is not compact. Of course, $\inf_{p \in S} K(p) = 0$ because $\lim_{z \to +\infty} K(x, y, z) = 0$.

7.7.2. Tangent fields and integral curves. Let us consider a differentiable tangent field V on a given, not necessarily closed, surface S. That is, according to Definition 3.1, $V : S \to \mathbb{R}^3$ is a differentiable map such that

$$V(p) \in T_p S, \qquad \forall p \in S.$$

We will work, as in the proof of Theorem 7.30 of the existence and uniqueness of geodesics, on orientable open subsets U of S and we will denote by $N(U)$ the tubular neighbourhoods constructed on them, if possible. So, on each of the $N(U)$, we dispose of a projection onto S and of an oriented distance with the properties pointed out in Exercise (1) of Chapter 4. The field V may be extended to a field $\mathcal{V}$ on the open subset $N(U)$ of three-space where it is constant on the normal segments of the surface, that is, $\mathcal{V} : N(S) \to \mathbb{R}^3$ is defined as

$$\mathcal{V} = V \circ \pi,$$

as we did in (6.1) in order to prove the divergence Theorem 6.10 on a compact surface. Now, we set the following system of first order differential equations on $N(U)$:

(7.17) $$x' = \mathcal{V}(x).$$

The theorem of existence, uniqueness, and dependence on the initial conditions gives, for each point $a \in N(U)$, a real number $0 < \varepsilon(a) \le +\infty$ and a differentiable curve $\alpha_a : (-\varepsilon(a), \varepsilon(a)) \to N(U)$ such that

$$\alpha_a(0) = a \quad \text{and} \quad \alpha_a'(t) = \mathcal{V}(\alpha_a(t)), \qquad \forall t \in (-\varepsilon(a), \varepsilon(a)),$$

and this so-called *integral curve* of the field $\mathcal{V}$ is maximal under these conditions. Moreover, if we put

$$\mathcal{D} = \{(a, t) \in N(U) \times \mathbb{R} \,|\, |t| < \varepsilon(a)\},$$

this set is open in $N(U) \times \mathbb{R} \subset \mathbb{R}^3 \times \mathbb{R}$ and the map

$$F : \mathcal{D} \longrightarrow N(U)$$

determined by

$$F(a, t) = \alpha_a(t), \qquad \forall (a, t) \in \mathcal{D},$$

is differentiable.

Take as an initial condition for solving (7.17) a point $p \in U \subset N(U)$ and let $\alpha = \alpha_p$ be the corresponding solution—or integral curve—of $\mathcal{V}$ defined on $(-\varepsilon(p), \varepsilon(p))$. We denote by f the composition $r \circ \alpha : (-\varepsilon(a), \varepsilon(a)) \to \mathbb{R}$. Then $f(0) = r(\alpha(0)) = r(p) = 0$ and, using the properties given in Exercise (1) of Chapter 4,

$$\begin{aligned} f'(t) &= (dr)_{\alpha(t)}(\alpha'(t)) = \langle (N \circ \pi)(\alpha(t)), \alpha'(t) \rangle \\ &= \langle (N \circ \pi)(\alpha(t)), \mathcal{V}(\alpha(t)) \rangle = \langle (N \circ \pi)(\alpha(t)), (V \circ \pi)(\alpha(t)) \rangle = 0, \end{aligned}$$

for all $t \in (-\varepsilon(p), \varepsilon(p))$. Hence, $f = r \circ \alpha$ is identically zero, i.e., the trace of the curve α is contained in the surface S and, so, we have

$$\alpha'(t) = \mathcal{V}(\alpha(t)) = (V \circ \pi)(\alpha(t)) = V(\alpha(t)).$$

Conversely, each curve on the surface satisfying this last equality is an integral curve of the field $\mathcal{V}$ on the open subset $N(U)$ of $\mathbb{R}^3$. Bearing all this in mind and working as in the proof of Theorem 7.30, we may infer the following result.

Theorem 7.48 (Integral curves and flow of a field). *Let V be a differentiable tangent field on a surface S of Euclidean space. Given $p \in S$, there exist a number $0 < \varepsilon(p) \leq +\infty$ and a curve $\alpha_p : (-\varepsilon(p), \varepsilon(p)) \to S$ such that*

$$\alpha_p(0) = p \quad and \quad \alpha'_p = V \circ \alpha_p,$$

which is maximal under these conditions. Moreover, the set

$$\mathcal{F} = \{(p,t) \in S \times \mathbb{R} \,|\, |t| < \varepsilon(p)\}$$

is an open subset of $S \times \mathbb{R}$ and the map

$$F : \mathcal{F} \longrightarrow S$$

defined by $F(p,t) = \alpha_p(t)$ is differentiable. The curves α_p are called the integral curves *of the field V and the map F is called the* flow *of the field V.*

Remark 7.49. For all $p \in S$, we have $0 < \varepsilon(p)$. Then $(p,0) \in \mathcal{F}$ for any $p \in S$. Since the subset $\mathcal{F}$ of $S \times \mathbb{R}$ is open, there exist an open subset W of S containing p and a number $\delta > 0$ such that $W \times (-\delta, \delta) \subset \mathcal{F}$. That is, the flow F of the field V is defined on at least a set of the form $W \times (-\delta, \delta)$ with $\delta > 0$ and where W is an open neighbourhood of p in S, and this occurs for all $p \in S$.

Remark 7.50. Suppose that the surface S is closed in $\mathbb{R}^3$ and that the field V that we are dealing with is bounded, i.e.,

$$|V(p)| \leq M, \qquad \forall p \in S,$$

for some positive constant M. Let $\alpha : (-\varepsilon, \varepsilon) \to S$ be an integral curve of V which is maximal relative to the initial condition $\alpha(0) = p \in S$ and such that $0 < \varepsilon < +\infty$. We take a sequence $\{t_n\}_{n \in \mathbb{N}}$ of numbers in the domain of α such that $\lim_{n \to \infty} t_n = \varepsilon$. For each pair of natural numbers n and m, we have

$$|\alpha(t_n) - \alpha(t_m)| = \left| \int_{t_n}^{t_m} \alpha'(t)\, dt \right| \leq \int_{t_n}^{t_m} |\alpha'(t)|\, dt.$$

Thus, since $|\alpha'(t)| = |V(\alpha(t))| \leq M$,

$$|\alpha(t_n) - \alpha(t_m)| \leq M\, |t_n - t_m|.$$

Then, the sequence of images $\{\alpha(t_n)\}_{n \in \mathbb{N}}$ is a Cauchy sequence which must converge to some point $p \in \mathbb{R}^3$. This limit has to be on S, since it is closed. Furthermore, since V is a continuous map, we obtain

$$\lim_{n \to \infty} \alpha'(t_n) = \lim_{n \to \infty} V(\alpha(t_n)) = V(q).$$

Then α could be prolonged on the right-hand side beyond the interval $(-\varepsilon, \varepsilon)$ gluing the integral curve of V passing through q. Analogously, it could be extended on the left-hand side and we would obtain in this way a contradiction. Therefore, under our hypotheses, $\varepsilon(p) = +\infty$ for each $p \in S$. In other words, the following is true.

> *The integral curves of a bounded tangent field on a closed surface are defined on the whole of $\mathbb{R}$. In particular, this happens, if S is a compact surface, for any tangent field.*

As a consequence of this, in any of these situations, the flow F of the field will be defined on the whole of the product $S \times \mathbb{R}$.

7.7.3. Special parametrizations. Even though the very definition of a surface required the notion of a parametrization, we believe that throughout this text it has already been pointed out that the results that we are interested in do not depend on these parametrizations. Until now, almost every time we used such parametrizations, the choice of the parametrization did not matter. However, we also saw that, to solve some problems, it was convenient to dispose of parametrizations with special properties—see, for example, the Hilbert Theorem 3.47 and the Herglotz integral formula in Proposition 7.15. Now, we intend to show that, under some assumptions, there are parametrizations particularly useful on the surfaces. To do this, we will first obtain the so-called *first integrals* for the integral curves of a tangent field on a surface, that is, we will prove that these integral curves are, locally, level curves for some suitable functions defined on the surface. This will occur, of course, where the integral curves are not constant, that is, far away from the field zeroes.

Proposition 7.51 (First integrals of tangent fields). *Let V be a tangent vector field on a surface S, and let $p \in S$ be a point where $V(p) \neq 0$. Then there exist an open neighbourhood W of p in S and a differentiable function $f : W \to \mathbb{R}$ without critical points such that, if $\alpha : I \to W$ is an integral curve of V defined on an open interval I of $\mathbb{R}$, then $f \circ \alpha$ is constant. That is, the kernel of $(df)_q$ is exactly the line spanned by $V(q)$ for each $q \in W$.*

Proof. Using Remark 7.49, we know that the flow ϕ of the field V is defined on $E \times (-\varepsilon, \varepsilon)$, where E is an open neighbourhood of p in S and for some $\varepsilon > 0$. On the other hand, let $u \in T_pS$ be a vector that, together with $V(p)$, forms a basis of the tangent plane T_pS, and let $\gamma : (-\delta, \delta) \to E \subset S$, $\delta > 0$, be a regular curve with $\gamma(0) = p$ and $\gamma'(0) = u$. We now define a differentiable map $G : (-\delta, \delta) \times (-\varepsilon, \varepsilon) \to S$ by

$$G(s,t) = \phi(\gamma(s), t), \qquad \forall (s,t) \in (-\delta, \delta) \times (-\varepsilon, \varepsilon).$$

Then, $G(0,0) = \phi(\gamma(0),0) = p$ and, differentiating with respect to s at $s = 0$ and $t = 0$,

$$G_s(0,0) = \frac{d}{ds}\bigg|_{s=0} \phi(\gamma(s),0) = \frac{d}{ds}\bigg|_{s=0} \gamma(s) = w.$$

Now differentiating with respect to t at an arbitrary time,

$$G_t(t,s) = \frac{d}{dt}\phi(\gamma(s),t) = (V \circ \phi)(\gamma(s),t) = V(G(t,s)).$$

In particular, since $V(p)$ and w are linearly independent, the map $(dG)_{(0,0)}$ has a trivial kernel. Thus, in this situation, we may apply the inverse function theorem and find an open neighbourhood U of the origin in $\mathbb{R}^2$ and an open neighbourhood $W \subset E$ of p in S such that $G : U \to W$ is a diffeomorphism. We define the required function $f : W \to \mathbb{R}$ as $f = p_1 \circ G^{-1}$, where p_1 is the first projection of $\mathbb{R}^2$. Then

$$(df)_{G(s,t)} \circ (dG)_{(s,t)} = (dp_1)_{(s,t)} = p_1,$$

for each $(s,t) \in U$. Thus $(df)_{G(s,t)} \neq 0$ at each point of U, that is, f has no critical points in $G(U) = W$. Now let $\alpha : I \to W \subset S$ be an integral curve of the field V defined on an open interval I of the real line. If we put $g = p_2 \circ G^{-1}$, then we may write

$$\alpha(u) = G(f(u),g(u)), \qquad \forall u \in I,$$

where $f(u)$ and $g(u)$ denote $f(\alpha(u))$ and $g(\alpha(u))$, respectively. Taking derivatives with respect to u we obtain

$$\alpha'(u) = f'(u)G_s(f(u),g(u)) + g'(u)G_t(f(u),g(u)).$$

Using the facts that α is an integral curve of V and that

$$G_t(f(u),g(u)) = V(G(f(u),g(u))),$$

we infer that

$$V(G(f(u),g(u))) = f'(u)G_s(f(u),g(u)) + g'(u)V(f(u),g(u)),$$

for all $u \in I$. Since G_s and G_t are independent at each point of U, we have that $f'(u) = (f \circ \alpha)'(u)$ vanishes everywhere on I. Therefore, $f \circ \alpha$ is constant, as we wanted. The final observation in the statement follows immediately. $\qquad \square$

Theorem 7.52. *Let V_1 and V_2 be two tangent vector fields on a surface S. Suppose that, at some point $p \in S$, the values taken by the two fields $V_1(p)$ and $V_2(p)$ are linearly independent. Then, one can find a parametrization $X : U \to S$ of the surface, defined on an open subset U of $\mathbb{R}^2$, such that $p \in X(U)$ and such that the vectors X_u and X_v are proportional to $V_1 \circ X$ and $V_2 \circ X$, that is, the parametric lines of X are reparametrizations of integral curves of V_1 and V_2, respectively.*

Proof. Let W_1 and W_2 be the two open subsets of S provided by Proposition 7.51 applied to each of the fields V_1 and V_2, respectively, and let f_1 and f_2, the corresponding functions. So, on the open subset $W = W_1 \cap W_2$ of S containing p, we may define a differentiable map $G : W \to \mathbb{R}^2$ by

$$G(q) = (f_2(q), f_1(q)), \qquad \forall q \in W \subset S.$$

Then, we have $\ker (dG)_q = \ker (df_2)_q \cap \ker (df_1)_q$. Since V_1 and V_2 are linearly independent at the point p, one has that $(dG)_p : T_pS \to \mathbb{R}^2$ is an isomorphism. By the inverse function theorem, we find an open neighbourhood V of p in S and an open subset U of $\mathbb{R}^2$ such that $G : V \to U$ is a diffeomorphism. The required parametrization is nothing more than $X = G^{-1} : U \to V \subset S$, because, with this definition, one gets

$$f_2(X(u,v)) = u \quad \text{and} \quad f_1(X(u,v)) = v, \qquad \forall (u,v) \in U.$$

Taking partial derivatives in these two equalities with respect to v and u, one has

$$(df_2)_X(X_v) = 0 \quad \text{and} \quad (df_1)_X(X_u) = 0,$$

and we conclude because the kernels of $(df_2)_q$ and of $(df_1)_q$ at each point of W are spanned by $V_2(q)$ and $V_1(q)$, respectively, as seen in Proposition 7.51. $\square$

Example 7.53 (Orthogonal parametrizations). Let $Y : U \to S$ be any parametrization of the surface S. Then, the partial derivatives Y_u and Y_v give us at each point of U a basis of the corresponding tangent plane. Thus, the maps $V_1, V_2 : X(U) \to \mathbb{R}^3$ given by

$$V_1 = Y_u \circ Y^{-1} \quad \text{and} \quad V_2 = [-|Y_u|^2 Y_v + \langle Y_u, Y_v \rangle Y_u] \circ Y^{-1}$$

are clearly differentiable and determine tangent fields on the open subset $Y(U)$ of S which are non-null and orthogonal at each point. Applying Theorem 7.52, there exists a parametrization covering an open subset of S inside $Y(U)$ whose parametric lines have the same traces as the integral curves of V_1 and V_2. In conclusion, we have the following statement.

> *Given a point p of a surface S, there exists a parametrization $X : U \to S$ such that $p \in X(U)$ and the function $\langle X_u, X_v \rangle$ vanishes identically on U.*

This type of parametrization will be called an *orthogonal parametrization*.

Example 7.54 (Parametrizations by curvature lines). Let S be an oriented surface with no umbilical points; see Remark 3.29. By Proposition 3.43, the principal curvatures k_1 and k_2 of S are differentiable functions. Let $p \in S$, and let $X : U \to S$ be an orthogonal parametrization of S defined on an open subset U of $\mathbb{R}^2$ such that $p = X(q)$ for some $q \in U$. If we had that

$$(dN)_p(X_u(q)) = -k_1(p)X_u(q),$$

that is, that $X_u(q)$ is in the principal direction corresponding to k_1, since $X_v(q)$ is in the perpendicular direction and $k_1(p) \neq k_2(p)$, then we would obtain

$$(dN)_p(X_v(q)) \neq -k_1(p)X_v(q),$$

and so, relabeling, if necessary, the u and v variables, we may assume that

$$(dN)_p(X_u(q)) \neq -k_1(p)X_u(q),$$

and, hence, that

$$(dN)_p(X_v(q)) \neq -k_2(p)X_v(q).$$

Therefore, by continuity, on an open set smaller than the open set U where the parametrization X is defined, we may suppose that

$$W_1 = N^X{}_u + (k_1 \circ X)X_u = -A_X X_u + (k_1 \circ X)X_u \neq 0,$$

$$W_2 = N^X{}_v + (k_2 \circ X)X_v = -A_X X_v + (k_2 \circ X)X_v \neq 0$$

are satisfied on U. Since the endomorphism $A_p = -(dN)_p$ satisfies at each point

$$(A_p - k_1(p)I_{T_pS})(A_p - k_2(p)I_{T_pS}) = 0,$$

we have

$$A_X W_1 = (k_2 \circ X)W_1 \quad \text{and} \quad A_X W_2 = (k_1 \circ X)W_2,$$

that is, the differentiable fields $V_1 = W_1 \circ X^{-1}$ and $V_2 = W_2 \circ X^{-1}$ defined on the open set $X(U)$ give at each point a pair of non-trivial vectors in the principal directions corresponding to k_2 and k_1, respectively. Therefore, they are also perpendicular. Applying Theorem 7.52 to this pair of fields, we obtain the following statement.

> *If $p \in S$ is a non-umbilical point of a surface, there exists a parametrization $X : U \to S$ of the surface S such that $p \in X(U)$ and the vectors $\{X_u(q), X_v(q)\}$, at each point q of U, are principal directions at $X(q) \in S$.*

We will call this type of parametrization a *parametrization by the curvature lines*, since, in general, a regular curve on a surface is called a curvature line when its tangent vector is an eigenvector of the differential of the Gauss map at each time.

A first application of the existence of these classes of special parametrizations is the following local classification result, which works for those orientable surfaces of $\mathbb{R}^3$ whose—Gauss, mean, principal—curvatures are constant. They are usually called *surfaces with parallel second fundamental form*.

Theorem 7.55 (Classification of surfaces with parallel second fundamental form). *An orientable connected surface whose principal curvatures are constant, or, equivalently, whose Gauss and mean curvatures are constant, has to be an open subset of a plane, of a sphere, or of a right circular cylinder.*

Proof. Let k_1 and k_2 be the two principal curvatures, assumed to be constant, of a connected surface S oriented by means of a Gauss map N. If $k_1 = k_2$, then the surface is totally umbilical and, by Theorem 3.30, it must be an open subset of a plane or of a sphere. Thus, we may assume that the two constants k_1 and k_2 are different. Hence, by Example 7.54, for each point of S, there exists a parametrization by curvature lines whose image covers this point. Let $X : U \to S$ be a parametrization of this type defined on a connected open subset U of $\mathbb{R}^2$. We have

(7.18)
$$(dN)_X(X_u) = (N \circ X)_u = N_u^X = -k_1 X_u,$$
$$(dN)_X(X_v) = (N \circ X)_v = N_v^X = -k_2 X_v,$$
$$\langle X_u, X_v \rangle = 0,$$

equalities between functions defined on U. Differentiating in the first equality of (7.18) with respect to v and in the second one with respect to u, one obtains

$$N_{uv}^X = -k_1 X_{uv} \quad \text{and} \quad N_{vu}^X = -k_2 X_{vu}.$$

By the Schwarz theorem of elementary calculus, the two mixed partial derivatives are equal and, since we are supposing that the principal curvatures are different, we have

(7.19)
$$X_{uv} = 0$$

on U. Now, we take partial derivatives relative to v in the third equality of (7.18) and use the equation above. So, we have

$$\langle X_{vv}, X_u \rangle = 0$$

on U. Therefore, if the superscript T indicates, as usually, the component tangent to the surface, we may write

$$(X_{vv})^T = h X_v,$$

for some differentiable function $h : U \to \mathbb{R}$. Thus

$$\langle (X_{vv}^T)_u - (X_{uv}^T)_v, X_u \rangle = \langle h_u X_v + h X_{uv}, X_u \rangle = 0,$$

and, taking Lemma 7.11 into account, we conclude that

$$K \circ X \equiv 0.$$

Consequently, under our hypotheses, the surface S, provided that it is not totally umbilical, is flat, that is, it has vanishing Gauss curvature. Let us recall that K is the product of the principal curvatures k_1 and k_2. Hence, reorienting the surface if necessary, one can imagine that

$$k_1 = 0 < k_2.$$

From this, the first equality in (7.18) becomes

(7.20) $$N_u^X = 0$$

on the open set U. Let us see now that all the computations that we made earlier lead us to prove that the tangent field $X_u/|X_u|$ is constant on U. In fact, (7.20) implies that

$$\left\langle \left(\frac{X_u}{|X_u|} \right)_u, N^X \right\rangle = \left\langle \frac{X_u}{|X_u|}, N^X \right\rangle_u - \left\langle \frac{X_u}{|X_u|}, N_u^X \right\rangle = 0.$$

Moreover, from the third equality of (7.18) and equality (7.19), it follows that

$$\left\langle \left(\frac{X_u}{|X_u|} \right)_u, X_v \right\rangle = \left\langle \frac{X_u}{|X_u|}, X_v \right\rangle_u - \left\langle \frac{X_u}{|X_u|}, X_{vu} \right\rangle = 0.$$

Finally

$$\left\langle \left(\frac{X_u}{|X_u|} \right)_u, X_u \right\rangle = |X_u| \left\langle \left(\frac{X_u}{|X_u|} \right)_u, \frac{X_u}{|X_u|} \right\rangle = 0,$$

since $X_u/|X_u|$ is a unit field. Since the $\{X_u, X_v, N^X\}$ supply at each point of U a basis of $\mathbb{R}^3$, the above three equations are equivalent to

$$\left(\frac{X_u}{|X_u|} \right)_u = 0 \quad \text{on the open set } U.$$

But, on the other hand, we deduce from (7.19) that

$$\left(\frac{X_u}{|X_u|} \right)_v = \frac{X_{uv}}{|X_u|} - \frac{\langle X_{uv}, X_u \rangle}{|X_u|^3} X_u = 0.$$

Therefore, since the open set U is connected, there exists a vector $a \in \mathbb{S}^2$ such that

(7.21) $$\frac{X_u}{|X_u|} = a \quad \text{and, so,} \quad \langle X_v, a \rangle = 0 \quad \text{and} \quad \langle N, a \rangle = 0.$$

Now consider the differentiable map $G : U \to \mathbb{R}^3$ given by

$$G = X - \langle X, a \rangle a + \frac{1}{k_2} N^X.$$

The partial derivative of G relative to the u variable is

$$G_u = X_u - \langle X_u, a \rangle a + \frac{1}{k_2} N_u^X = 0,$$

where we have used (7.20) and the first equation of (7.21). Furthermore, the second equations of both (7.21) and (7.18) permit to infer that

$$G_v = X_v - \langle X_v, a \rangle a + \frac{1}{k_2} N_v^X = 0.$$

Thus, the map G is also constant on the open set U. That is,

$$X - \langle X, a \rangle a + \frac{1}{k_2} N^X = c,$$

for some $c \in \mathbb{R}^3$. If we perform scalar multiplication by the vector a on this last equation and take the third equality of (7.21) into account, we can see that, moreover, $\langle a, c \rangle = 0$. In short, we have found a unit vector a and another vector c of $\mathbb{R}^3$ perpendicular to the first one such that

$$X - \langle X, a \rangle a - c = -\frac{1}{k_2} N^X.$$

Taking lengths in this equality, we finally arrive at

$$|X - c|^2 - \langle X - c, a \rangle^2 = \frac{1}{k_2^2}.$$

In other words, the distance from the points of $X(U) \subset S$ to the straight line passing through c in the direction of a is constant and equal to $1/k_2$. This means that the open subset $X(U)$ of S is contained in a right circular cylinder of radius $1/k_2$ whose axis is the straight line that we referred to before. So, we have proved that each point of the surface has an open neighbourhood included in a right circular cylinder of radius $1/k_2$. Since the surface is connected, this cylinder is the same for all its points and we finish the proof. $\qquad\square$

If we recall now that any surface which is closed as a subset of $\mathbb{R}^3$ is orientable, as seen in Exercise (6) of Chapter 4, the following characterization follows immediately from the classification Theorem 7.55 that we have just proved.

Corollary 7.56. *The only connected surfaces closed as subsets of $\mathbb{R}^3$ having constant principal curvatures are planes, spheres, and right circular cylinders.*

7.7.4. Flat surfaces. In this subsection, we will undertake the main subject of this Appendix: the study of surfaces closed in $\mathbb{R}^3$ with vanishing Gauss curvature everywhere. Until now, all the examples of this type of surface that we know of are part of the so-called right cylinders over plane curves—see Exercise (7) of Chapter 3—and, in fact, the result that we pursue confirms that they are the unique examples. These right cylinders over simple plane curves can be viewed as the physical result of folding up a piece of paper with infinite extension—winding up infinitely many times if

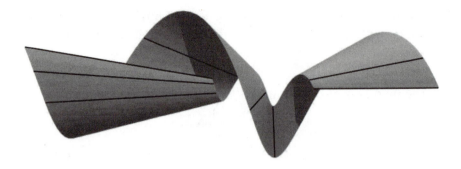

Figure 7.6. *Flat surface*

the curve is compact. Indeed, by the Gauss Theorema Egregium 7.10, any open piece of paper folded up smoothly originates a surface with vanishing Gauss curvature everywhere, that is, according to all the notation that we introduced in Remark 3.28, having only parabolic or flat points. We will refer to this situation by saying that the surface is *flat*. If the piece of paper that we take to build a flat surface, by folding up, is not the whole of the infinite sheet—which is always the realistic case—it is not true that the resulting surface has to be an open subset of a right cylinder. One can smoothly deform a piece of paper in a way such as is shown in Figure 7.6, without obtaining a piece of a cylinder. This is why it is surprising that, if we add the hypothesis that the surface is closed or, equivalently, that the piece of paper is infinite, we necessarily get a right cylinder.

In spite of all the above, it is always true that, even locally, any flat surface contains plenty of segments of straight lines. This is the subject of the following lemma, at least when the surface has no flat points, but all its points are parabolic.

Lemma 7.57. *Let S be an oriented flat surface whose mean curvature H is positive everywhere. Then, for each point $p \in S$, there exists a unique straight line R_p of $\mathbb{R}^3$ such that $p \in R_p$ and $R_p \cap S$ is a neighbourhood of p in R_p. Moreover, the connected component of $R_p \cap S$ containing p is an open interval of R_p on which $1/H$ is an affine function of the arc length.*

Proof. We have, with respect to the considered orientation, that $H > 0$ everywhere. Since $K = 0$, the principal curvatures k_1 and k_2 of S must satisfy

$$0 = k_1 < k_2.$$

Hence, there are no umbilical points on S and, so, the principal curvatures are differentiable functions on the whole of S and we have, at each point, exactly two orthogonal principal directions corresponding to the two curvatures. Given a point $p \in S$, let R_p be the straight line passing through

p and having as its direction the principal direction $e_1 \in T_pS$. Let us see now that this line satisfies the requirements on the lemma. In fact, let $X : (-\delta, \delta) \times (-\varepsilon, \varepsilon) \to S$ be a parametrization of S by the curvature lines, whose existence was assured by Example 7.54, with $X(0,0) = p$. The features of this X are

$$N_u^X = 0, \quad N_v^X = -k_2\, X_v, \quad \text{and} \quad \langle X_u, X_v \rangle = 0.$$

Differentiating with respect to v in the first equality and respect to u in the second one, we get

$$(7.22) \qquad\qquad (k_2)_u\, X_v + k_2\, X_{uv} = 0.$$

After scalar multiplication by X_u, we obtain

$$\langle X_{uv}, X_u \rangle = 0.$$

Hence

$$\langle X_{uu}, X_v \rangle = \langle X_u, X_v \rangle_u - \langle X_u, X_{uv} \rangle = 0.$$

Furthermore, again using that $N_u^X = 0$, we have

$$\langle X_{uu}, N^X \rangle = \langle X_u, N^X \rangle_u - \langle X_u, N_u^X \rangle = 0.$$

Consequently, the vector X_{uu} is, at each point of the domain of X, proportional to X_u. That is,

$$(7.23) \qquad\qquad X_{uu} = g\, X_u,$$

where $g : (-\delta, \delta) \times (-\varepsilon, \varepsilon) \to \mathbb{R}$ is some differentiable function.

We now define a differentiable curve $\alpha : (-\delta, \delta) \to S$ on the surface by

$$\alpha(u) = X(u, 0), \qquad \forall u \in (-\delta, \delta).$$

This curve is regular and clearly satisfies

$$\alpha(0) = p, \quad \alpha'(u) = X_u(u, 0), \quad \text{and} \quad \alpha''(0) = X_{uu}(u, 0).$$

Then, from (7.23), we have that the acceleration α'' of α is proportional to the velocity α', at each point. Thus, the trace of $\alpha((-\delta, \delta))$ is contained in a straight line. Indeed, in the straight line passing through $\alpha(0) = p$ with direction $\alpha'(0) = X_u(0,0) \| e_1 \in T_pS$, that is,

$$\alpha((-\delta, \delta)) \subset R_p \cap S.$$

Since $\alpha((-\delta, \delta))$ is a neighbourhood of p in R_p, the same happens with $R_p \cap S$, as required.

With respect to uniqueness, if R were another straight line of $\mathbb{R}^3$ such that $p \in R$ and $R \cap S$ were a neighbourhood of p in R, each vector $v \neq 0$ in the direction of R would satisfy $\sigma_p(v, v) = 0$. But the only isotropic direction of the quadratic form σ_p on T_pS is the principal direction e_1. Hence, v and e_1 should be proportional, and so $R = R_p$.

Let C_p be the connected component of $R_p \cap S$ containing the point p. Since C_p is a connected subset of R_p containing p in its interior, we know that C_p is an interval of the line R_p with p in its interior. If this interval had q as an end point, then there would be a curve $\beta : [0, \rho) \to R_p \cap S$, for some $\rho > 0$, with $\beta(0) = q$ and $\beta'(0) = v \neq 0$. Then, differentiating $N \circ \beta$ on the right at 0, one would obtain that $\sigma_q(v, v) = 0$ and, hence, v would be a direction for R_q. We would conclude that $R_p = R_q$ and, since q is in the interior of $R_q \cap S$, it would also be in the interior of $R_p \cap S$ and we would have a connected set greater than C_p in $R_p \cap S$. This is impossible and, thus, C_p is an interval of R_p without end points, that is, an open interval.

To finish the proof, it only remains to study the mean curvature function H on each straight line R_p, $p \in S$. By uniqueness, if $q \in C_p \subset R_p \cap S$, one has that $R_q = R_p$ and so $C_p = C_q$. Therefore, each point $q \in C_p$ can be covered by a parametrization $X : (-\delta, \delta) \times (-\varepsilon, \varepsilon) \to S$ of S by the curvature lines with $X(0, 0) = q$ and such that $\alpha(u) = X(u, 0)$ is a parametrization of $C_q = C_p$ on a neighbourhood of q. For this X, equation (7.22) and all equalities below it are valid. Differentiating (7.22) with respect to u and (7.23) with respect to v and taking into account that $2H = k_2$, one gets

$$H_{uu} X_v + 2 H_u X_{uv} + H X_{uuv} = 0,$$

$$X_{uuv} = g_v X_u + g X_{uv}.$$

Substituting, it turns out that

$$2 H_{uu} X_v - 4 \frac{H_u^2}{H} X_v + 2 H g_v X_u - \frac{H_u}{H} g X_v = 0.$$

Therefore, extracting only the X_v-component, we have

(7.24) $H H_{uu} - 2 H_u^2 - 2 g H_u = 0,$

where the function g is determined by (7.23) and, so, it is

(7.25) $g = \dfrac{(|X_u|^2)_u}{2|X_u|^2} = \dfrac{|X_u|_u}{|X_u|}.$

On the other hand, if s is the arc length function along C_p, relative to any choice of origin, we know that

$$\frac{ds}{du} = |\alpha'| = |X_u|.$$

Substituting this expression in (7.24) and in (7.25), one sees that

$$\frac{d^2}{ds^2} \left(\frac{1}{H} \right) = 0.$$

Therefore, the inverse function of the mean curvature H is affine with respect to the arc length along $C_p \subset R_p \cap S$. $\square$

When the flat surface S of Lema 7.57 is assumed to be closed, one can obtain stronger consequences. In fact, we will get good control on the parabolic points of the surface and of the flat points on which the parabolic points accumulate.

Lemma 7.58. *Let S be a flat surface closed in $\mathbb{R}^3$. Let $\mathcal{P}$ be the open subset of S consisting of all its parabolic points, and let $\mathcal{F}$ be the closed subset formed by its flat points. For each $p \in \overline{\mathcal{P}}$, there exists a unique straight line R_p such that $p \in R_p \subset S$. Moreover*

$$p \in \mathcal{P} \Longrightarrow R_p \subset \mathcal{P},$$
$$p \in \mathrm{Bdry}\,\mathcal{P} = \mathrm{Bdry}\,\mathcal{F} \Longrightarrow R_p \subset \mathrm{Bdry}\,\mathcal{P}.$$

Proof. Let us notice that $\mathcal{P}$ is an open subset of S because S is closed and, so, it is orientable. Thus, if H is its mean curvature for a given orientation, the set $\mathcal{P}$ is formed by the points where $H \neq 0$. Hence, $\mathcal{F}$ is closed and we have that

$$S = \mathcal{P} \cup \mathcal{F} = \mathcal{P} \cup \mathrm{Bdry}\,\mathcal{F} \cup \mathrm{int}\,\mathcal{F}$$

are disjoint unions and so $\mathrm{Bdry}\,\mathcal{F} = \mathrm{Bdry}\,\mathcal{P}$.

First, suppose that $p \in \mathcal{P}$ is a parabolic point of the surface. From Lemma 7.57, there is a unique straight line R_p of $\mathbb{R}^3$ such that $p \in R_p$ and the connected component C_p of $R_p \cap \mathcal{P}$ containing p is an open interval of R_p where $1/H$ is an affine function of the arc length. Let D_p be the connected component of $R_p \cap S$ containing p. Clearly, D_p is an interval of R_p including C_p. If this inclusion is not an equality, C_p cannot be the whole of the line R_p and, hence, C_p has an end point q which, since C_p is open, lies in $D_p - C_p \subset \mathcal{F}$. Therefore, there exists an end point q of the interval C_p where $H(q) = 0$. This contradicts the fact that $1/H$ is an affine function of the arc length of C_p and q is an end point at finite distance. It follows that $D_p = C_p$ is the connected component of $R_p \cap S$ containing p, which is, as already mentioned, an open interval of the line R_p. Since we are assuming the surface S to be closed in $\mathbb{R}^3$, the intersection $R_p \cap S$ is a closed subset of the line R_p and the same is true for each of its connected components. Then C_p is a closed subset of R_p. By connectedness, it turns out that $C_p = R_p$, that is, $R_p \subset \mathcal{P} \subset S$, as we wanted. The uniqueness of the line R_p among those passing through p and which are contained in S was already a consequence of Lemma 7.57.

Second, now take a point $p \in \mathrm{Bdry}\,\mathcal{P} \subset \mathcal{F}$, that is, a flat point which can be described as the limit of a sequence $\{p_n\}_{n \in \mathbb{N}}$ of parabolic points. Consider the sequence of straight lines $R_{p_n} \subset S$ supplied by the above reasoning. A subsequence of this sequence, which will be denoted in the same way, converges to a straight line R of $\mathbb{R}^3$; see Exercise 4.12. Then, since S is closed, we have $p \in R \subset S$. It remains to see that this line R

belongs to $\operatorname{Bdry} \mathcal{P} = \mathcal{F} \cap \overline{\mathcal{P}}$. Each point of R is the limit of a sequence of points in the lines $R_{p_n} \subset \mathcal{P}$ and, so, $R \subset \overline{\mathcal{P}}$. Moreover, if there were a point $q \in R \cap \mathcal{P}$, one would have $p \in R = R_q \subset \mathcal{P}$, which is impossible. We conclude, then, that $R \subset \operatorname{Bdry} \mathcal{P}$.   □

It is obvious that we cannot expect to generalize this result to each point of a flat closed surface. Indeed, through each point of a plane we find infinitely many straight lines entirely contained in the surface. But, in fact, we will see that this is the only example of this behaviour.

Lemma 7.59. *Let S be a flat connected surface closed as a subset of $\mathbb{R}^3$ different from a plane. Then, through each point $p \in S$ a unique straight line R_p passes which is entirely contained in the surface.*

Proof. After having proved Lemma 7.58, it only remains to show that the assertion is also true for the points of $\operatorname{int} \mathcal{F} \subset S$. We know that $\operatorname{int} \mathcal{F}$ is an orientable surface whose Gauss and mean curvatures vanish, that is, with trivial second fundamental form at each point. Thus, Exercise 3.23 assures us that each of its connected components is contained in a plane. Then, let C be a connected component of $\operatorname{int} \mathcal{F}$ and P the plane of $\mathbb{R}^3$ including it. Thus C, being a surface, is a non-empty open subset of the plane P, different from the whole plane P. Indeed, if $C = P$, then $P = C \subset \operatorname{int} \mathcal{F} \subset S$ and, since S is connected, one would have that $P = S$, and this is forbidden under our assumptions. Once we have established that C is an open subset of the plane P different from P itself, we may accept that the topological boundary $\operatorname{Bdry}^P C$ of C in P is non-empty. Choose a point $q \in \operatorname{Bdry}^P C$. Then, there is a sequence $\{q_n\}_{n \in \mathbb{N}}$ of points of $C \subset \operatorname{int} \mathcal{F}$ converging to q. Since S is closed, $q \in S$ and, since $\mathcal{F}$ is closed in S, we have that $q \in \mathcal{F} \cap P$. If q were an interior point of $\mathcal{F}$, it would be in a connected component C' of $\operatorname{int} \mathcal{F} \cap P$ different from C. But, since $q \in P$ and C' has to be contained in some plane of $\mathbb{R}^3$, we deduce that $C' \subset P$. In conclusion, we would have two non-empty open disjoint subsets C and C' of a plane with $q \in C' \cap \overline{C}$, which is impossible. Hence

$$q \in \operatorname{Bdry}^P C \implies q \in \operatorname{Bdry} \mathcal{F} \cap P.$$

On the other hand, the affine tangent plane of the surface S at the point q is P itself, because $T_{q_n} S = T_{q_n} \operatorname{int} \mathcal{F} = T_{q_n} P$ for each $n \in \mathbb{N}$. Using Lemma 7.58, there is a unique straight line R_q such that $q \in R_q \subset S$ which, for this reason, is tangent to the surface at q. This implies that $R_q \subset P$ and, moreover, we know that $R_q \subset \operatorname{Bdry} \mathcal{F}$. Thus, for each $q \in \operatorname{Bdry}^P C$, there is a unique straight line R_q such that $q \in R_q \subset \operatorname{Bdry} \mathcal{F} \cap P$. All these R_q with $q \in \operatorname{Bdry}^P C$ are parallel because they are disjoint and coplanar. Then each R_q is entirely contained in $\operatorname{Bdry}^P C$, because, if $p \in C$ and $r \in R_q$

and if the segment $[p, r[$ is not contained in C, there would be a point $q' \in [p, r[\cap \text{Bdry}^P C$. From this, it would follow that $C \subset P - R_{q'}$ and it would be impossible that $q \in \text{Bdry}^P C$. Consequently, C must be either a half-plane of P or a strip between two parallel lines. Anyway, through each point of C there exists a unique straight line contained in S. □

In order to complete our study of the shape of the flat surfaces which are closed in $\mathbb{R}^3$, it will useful to introduce a property of the exponential map which distinguishes surfaces with vanishing Gauss curvature.

Proposition 7.60. *If S is a surface closed in Euclidean space with zero Gauss curvature everywhere, then the exponential map $\exp_p : T_pS \to S$ is a local isometry at any point $p \in S$.*

Proof. If $p \in S$, since S is closed in $\mathbb{R}^3$, Corollary 7.41 says that $\exp_p$ is defined on the whole of the tangent plane T_pS. Moreover, from the Gauss Lemma 7.36, in order to prove that $\exp_p$ is a local isometry, it only remains to check that, if $v, w \in T_pS$ and $\langle v, w \rangle = 0$, then

$$|(d\exp_p)_v(w)| = |w|.$$

We may also assume that $w \neq 0_p$ and that $v \neq 0_p$. The latter, due to $(d\exp_p)_{0_p}$, is the identity map, as seen in (7.13). Then, let $\alpha : \mathbb{R} \to T_pS$ be a parametrization by the arc length of the circle centred at $0_p \in T_pS$ with radius $|v|$, such that $\alpha(0) = v$ and $\alpha'(0) = w/|w|$. Consider the differentiable map $G : \mathbb{R}^2 \to S$ given by

$$G(s, t) = \exp_p t\alpha(s).$$

Either from a direct computation or from looking at the proof of the Gauss Lemma 7.36, we see that the equalities

$$\langle G_s, G_t \rangle = 0 \quad \text{and} \quad |G_t|^2 = |v|^2$$

are true on $\mathbb{R}^2$. Taking partial derivatives relative to the two variables, one obtains

$$\langle G_{ss}, G_t \rangle + \langle G_s, G_{ts} \rangle = 0, \quad \langle G_{tt}, G_t \rangle = 0,$$

$$\langle G_{st}, G_t \rangle + \langle G_s, G_{tt} \rangle = 0, \quad \langle G_{ts}, G_t \rangle = 0.$$

Combining these expressions and denoting by the superscript T the component tangent to the surface, it turns out that

(7.26) $$G_{tt}^T = 0 \quad \text{and} \quad G_{st}^T \| G_s.$$

Hence, since

$$\frac{\partial}{\partial t} |G_s| = \frac{\langle G_{st}^T, G_s \rangle}{|G_s|},$$

we have, differentiating again with respect to t,

$$\frac{\partial^2}{\partial t^2}|G_s| = \frac{\langle (G_{st}^T)_t, G_s \rangle + \langle G_{st}^T, G_{st} \rangle}{|G_s|} - \frac{\langle G_{st}^T, G_s \rangle^2}{|G_s|^3},$$

and, since, from (7.26), $\langle G_{st}^T, G_s \rangle^2 = |G_s|^2 |G_{st}^T|^2$, we deduce that

$$\frac{\partial^2}{\partial t^2}|G_s| = \frac{\langle (G_{st}^T)_t, G_s \rangle}{|G_s|^3} = \frac{\langle (G_{st}^T)_t - (G_{tt}^T)_s, G_s \rangle}{|G_s|^3},$$

where we have used the first equality of (7.26). Finally, Lemma 7.11 implies

$$\frac{\partial^2}{\partial t^2}|G_s| = 0,$$

since the surface has vanishing Gauss curvature. As a consequence, $|G_s|$ is an affine function in t which, on the other hand, satisfies

$$|G_s|(s,0) = \left| \frac{d}{ds} G(s,0) \right| = 0.$$

Thus $|G_s|(s,t) = a(s)t$ for some function $a(s)$ and all $t \in \mathbb{R}$. But

$$G_s(s,t) = \frac{d}{ds} \exp_p t\alpha(s) = t (d\exp_p)_{t\alpha(s)} \alpha'(s)$$

and, from this, it follows that, for any t,

$$a(s) = |(d\exp_p)_{t\alpha(s)} \alpha'(s)|.$$

Putting $t = 0$, since $(d\exp_p)_{0_p}$ is the identity map,

$$a(s) = |(d\exp_p)_{0_p} \alpha'(s)| = |\alpha'(s)| = 1,$$

because the curve α was taken to be p.b.a.l. Finally

$$\left| (d\exp_p)_v \left(\frac{w}{|w|} \right) \right| = |(d\exp_p)_{\alpha(0)} \alpha'(0)| = |G_s|(0,1) = 1.$$

This equality shows the property that we were looking for. $\square$

Exercise 7.61. Revise Exercise (17) in the list of exercises in this chapter, bearing in mind the above proof.

Exercise 7.62. If S is a surface closed in $\mathbb{R}^3$ with negative Gauss curvature everywhere, show that, for each $p \in S$, the map $\exp_p : T_p S \to S$ diminishes the length of curves.

Exercise 7.63. Let P be a plane, S a surface, and $f : P \to S$ a local isometry. Prove that

$$|f(x) - f(y)| \leq |x - y|$$

for each pair $x, y \in P$.

Now, we are ready to announce and prove the result which classifies all flat surfaces closed in $\mathbb{R}^3$. This result was shown the first time in 1959, as a part of a much wider theorem, by Hartman and Nirenberg and later, in 1961 and 1962, by Stoker and Massey, respectively. These are surprisingly late dates, which can be understood only because the geometers of earlier centuries studied exclusively the local version of the problem—that is, they do not assume the surface to be closed in Euclidean space.

Theorem 7.64 (Hartman-Nirenberg, Stoker, Massey). *Let S be a surface closed in $\mathbb{R}^3$ with zero Gauss curvature everywhere. Then there exists a simple curve C closed in a plane P of $\mathbb{R}^3$ such that*

$$S = \bigcup_{p \in C} R(p, a),$$

where a is a unit vector normal to P and $R(p, a)$ denotes the straight line passing through p with direction a. That is, S is a right cylinder over a simple curve closed in a plane.

Proof. We may consider separately each connected component of S, or, equivalently, suppose from now on that S itself is connected. If S is a plane, it is clear that S is a right cylinder over any of its straight lines. Thus, assume that S is not a plane. By Lemma 7.59, we know that, for each $p \in S$, there exists a unique straight line R_p such that $p \in R_p \subset S$ and, by Proposition 7.60, there exists a plane $P \subset \mathbb{R}^3$ passing through the origin and a local isometry $f : P \to S$ which is surjective; take $P = T_{p_0}$ and $f = \exp_{p_0}$ for any $p_0 \in S$. We will show first that one can find a non-null vector $u \in P$ such that

(7.27) $$f(R(x, u)) = R_{f(x)}, \qquad \forall x \in P,$$

where $R(x, u)$ is the straight line of P passing through x with direction u. In fact, let $p = f(0) \in S$ and denote by v a unit vector in the direction of the line $R_p \subset S$. We have $v \in T_p S$. Let us choose

$$u = (df)_0^{-1}(v) \in P.$$

This u is a unit vector because $(df)_0$ is a linear isometry. Moreover, since $R(0, u)$ is a geodesic of P—see Example 7.26—its image through f must be, by Theorem 7.25, a geodesic of S passing through $f(0) = p$ with velocity at this point $(df)_0(u) = v$. But this geodesic has to be, from the uniqueness of Theorem 7.30, the straight line $R_{f(0)} = R_p$, since, according to Exercise (15) of this chapter, any straight line contained in a surface is a geodesic. That is,

$$f(R(0, u)) = R_p = R_{f(0)},$$

which is the required relation (7.27) for the origin $0 \in P$. Now, let $x \in P$ be an arbitrary point of the plane P, and let $R_{f(x)}$ be the straight line corresponding to its image $f(x) \in S$. Denote by w a unit vector in the direction of this line, and represent by y the vector $(df)_x^{-1}(w) \in P$. Suppose that y and u have different directions. Thus, the two lines $R(x, y)$ and $R(0, u)$ of the plane P intersect at a point z. But, reasoning exactly as we did for $R(0, u)$, it turns out that

$$f(R(x, y)) = R_{f(x)}.$$

Hence we have a point $f(z)$ in $R_{f(x)} \cap R_{f(0)}$ and this is not possible, by the uniqueness of Lemma 7.59, unless $R_{f(x)} = R_{f(0)} = R_p$. If the latter were the case, the vectors v and w would have the same direction and, consequently, since the vectors $(df)_z(u)$ and $(df)_z(y)$ would be in the direction of $R_p = f(x)$, they would be proportional. Since $(df)_z$ is an isomorphism, u and y would be proportional as well, and this is a contradiction. In short, the vector y is parallel to u and so

$$f(R(x, u)) = f(R(x, y)) = R_{f(x)},$$

which is the required equality (7.27) shown to be true for all $x \in P$.

Now, take $p, q \in S$ such that $R_p \neq R_q$, and let R_1 and R_2 be two parallel lines of the plane P such that

$$f(R_1) = R_p \quad \text{and} \quad f(R_2) = R_q,$$

whose existence follows from (7.27), which we have just seen, and the fact that f is surjective. Using Exercise 7.63 above, we have

$$\text{dist}\,(f(x), R_q) \leq \text{dist}\,(x, R_2), \qquad \forall x \in R_1.$$

Since R_1 and R_2 are parallel, $\text{dist}\,(x, R_2) = \text{dist}\,(R_1, R_2)$, and so

$$\text{dist}\,(z, R_q) \leq \text{dist}\,(R_1, R_2), \qquad \forall z \in R_p.$$

Therefore, R_p and R_q are two straight lines of $\mathbb{R}^3$ which do not intersect and such that the distance from the points of one of them to the other one is bounded. Hence, they are parallel. So, there exists a vector $v \in \mathbb{S}^2$ such that, for each $p \in S$, the straight line $R(p, v)$ is entirely contained in S. Consequently, if Q is a plane perpendicular to v, Q and S cut transversely and their intersection, which is non-empty, is a simple curve—see Example 2.64—which is closed in Q because S is closed in $\mathbb{R}^3$.   □

Chapter 8

The Gauss-Bonnet Theorem

8.1. Introduction

The problems that we dealt with in Chapters 6 and 7 will most likely have given enough information to convince the reader that global differential geometry is one of the most attractive branches of mathematics. In this discipline, we can see how general, algebraic, and differential topology and analysis closely interact. One of the most brilliant and famous examples of this topic is the Gauss-Bonnet theorem—or formula—which relates the integral of the Gauss curvature over a compact surface, its total Gauss curvature, with an integer number naturally associated to it, its Euler characteristic, which coincide for any two diffeomorphic surfaces.

The first version of this result—still a local version—was obtained by GAUSS in his *Disquisitiones*. There, he discovered a relation between the integral of the Gauss curvature over a geodesic triangle and the sum of the angles at its vertices. We state it here concisely.

> *The total Gauss curvature of a geodesic triangle is equal to the excess over π of the sum of its angles.*

GAUSS commented on this statement: *This theorem, if I am not wrong, should be counted among the most elegant ones of the theory of skew surfaces.* Indeed, the theorem due to GAUSS generalized another well-known theorem by LEGENDRE about spherical trigonometry: if we denote by α_1, α_2, and α_3 the three angles at the vertices of a spherical triangle—whose edges are by definition arcs of great circles—then *the area of a spherical triangle is equal to the product of its spherical excess* $\alpha_1 + \alpha_2 + \alpha_3 - \pi$ *by the square of its radius.* Later, in 1848, BONNET generalized the result by GAUSS to any simply connected region of a surface, originating in this way the well-known Gauss-Bonnet formula, which still has a local nature.

We will consider only the global version of the theorem, which matches better with the goals of this text and is more relevant from the point of view of modern geometry. Moreover, since we are assuming that our reader has no more knowledge of topology than an introductory course, we cannot prove that the total Gauss curvature of a compact surface is equal to 2π times its Euler characteristic, arising from triangulations or from homology groups or from the fundamental group of the surface. That is, we assume that the reader is not familiar with this topological number and its properties. Thus, we will show that this total curvature is 2π times a certain integer which, at first, will only mean the sum of the indices of a tangent field on the surface with isolated zeroes, that is, a hybrid between the Gauss-Bonnet formula and the famous results by POINCARÉ and H. HOPF. As a consequence, this integer will be an invariant of the diffeomorphisms class of the surface—and will be called, of course, the Euler characteristic. After this, we will prove that this integer is twice the degree of the Gauss map N, that is, twice the number of preimages—counted with their orientations—of a point of $\mathbb{S}^2$ which is a regular value of N. Hence, it will always be an even number, as we will prove at the end of the chapter.

There appears in this way, with not too much effort, a theorem which deserves a chapter to itself. It reveals clearly what the spirit of modern geometry is, connecting in a strange and exciting way, several quantities which, at first, might appear to be very different: the total Gauss curvature of a compact surface, the number of zeroes of a tangent field on the surface, and the degree of the Gauss map. In fact, the result is not only *very elegant*, as GAUSS claimed about his local version, but, from a modern point of view, it is more *egregium* than the Theorema Egregium that we studied in Chapter 7. It is the illustrious precedent of a series of results relating the global topological structure of surfaces—in general, of Riemann manifolds—to the properties of their curvatures. Furthermore, like this Gauss-Bonnet theorem, most of the results of global differential geometry have something to do with some integral formula or, using more modern language, with some cohomological system naturally associated with differential calculus.

8.2. Degree of maps between compact surfaces

Let S and S' be two surfaces of Euclidean space $\mathbb{R}^3$ that we will assume to be oriented by means of two Gauss maps N and N', respectively. In this situation, we already defined—see Exercise 3.17—for a given differentiable map $\phi : S \to S'$ the Jacobian of ϕ as the function

$$\text{Jac}\,\phi : S \longrightarrow \mathbb{R}$$

determined by the equality

$$(\text{Jac}\,\phi)(p) = \det((d\phi)_p(e_1), (d\phi)_p(e_2), N'(\phi(p))),$$

where $p \in S$ and $\{e_1, e_2\}$ is a positively oriented orthonormal basis of T_pS, that is, such that

$$\det(e_1, e_2, N(p)) = 1.$$

It is worth noting that this function would not be a well-defined object if the surfaces S and S' were not oriented and that it clearly changes sign when one of the orientations of S or S' is reversed. Notice that, instead, its absolute value

$$|\text{Jac}\,\phi|(p) = |(d\phi)_p(e_1) \wedge (d\phi)_p(e_2)|, \qquad \forall p \in S,$$

does not depend of the chosen orientations and, so, it can also be defined for non-orientable surfaces. In fact, we defined this absolute value of the Jacobian, for a given differentiable map between surfaces, in Exercise (13) of Chapter 2 and it played a central role in the theory of integration developed in Chapter 5.

Definition 8.1 (Degree of a map between compact surfaces). If $\phi : S \to S'$ is a differentiable map between compact oriented surfaces, we define the *degree* of the map ϕ to be the *real* number

$$\deg \phi = \frac{1}{A(S')} \int_S \text{Jac}\,\phi.$$

This degree is clearly additive relative to the connected components of S. This is why we will assume that both S and S' are connected. Moreover, it is also immediate that, if one of the two orientations on S or S' is reversed, this degree changes sign.

Now, in the following remarks, we will gather some of the more important properties and particular cases relative to the degree of maps between compact surfaces. Throughout this chapter, unless otherwise specified and for reasons which will be clear later, each compact connected surface will be oriented by means of its *outer* Gauss map.

Remark 8.2. If the map ϕ is a diffeomorphism, then, either from Definition 8.1 or from Exercise 3.17, its Jacobian $\text{Jac}\,\phi$ has a constant sign, either positive or negative according to whether it preserves or reverses the orientation. If ϕ preserves the orientation, we have

$$\deg \phi = \frac{1}{A(S')} \int_S |\text{Jac}\,\phi| = \frac{1}{A(S')} \int_{S'} 1 = 1,$$

by Theorem 5.14 of the change of variables. Similarly, if ϕ is a diffeomorphism which reverses the orientation, then $\deg \phi = -1$.

Remark 8.3. Consider a Gauss map $N : S \to \mathbb{S}^2$ of a compact surface S. Wen we talk about the degree of N, it seems natural to consider on S the orientation determined by N itself. Once we have fixed this orientation, if $p \in S$ and $\{e_1, e_2\} \subset T_p S$ is a positively oriented orthonormal basis, consisting of principal directions of S at p, we have

$$(\text{Jac } N)(p) = \det((dN)_p(e_1), (dN)_p(e_2), N(p))$$

$$= k_1(p) k_2(p) \det(e_1, e_2, N(p)) = K(p).$$

That is, the previous computation shows that the Jacobian of the Gauss map—either of the two possible ones—is the Gauss curvature, provided that we consider on the surface the orientation compatible with this normal field and on the sphere the orientation corresponding to the outer normal field; see Exercise (5) in Chapter 3 and Remark 6.6. This is one of the reasons for why we are choosing, for compact surfaces, unless otherwise indicated, the outer orientation. As a consequence

$$\deg N = \frac{1}{4\pi} \int_S K(p) \, dp,$$

that is, *the degree of the Gauss map of a compact surface is its total Gauss curvature divided by the area of the unit sphere.* For the present, the above formula is not significant, because the degree of N is nothing more than a real number without any specific property.

Remark 8.4. Suppose that S is a compact surface in $\mathbb{R}^3$ and $V : S \to \mathbb{R}^3$ is a nowhere vanishing differentiable map, or, in other words, a differentiable vector field, not necessarily tangent to the surface—see Definition 3.1— without zeroes. We can see that

$$\frac{V}{|V|} : S \longrightarrow \mathbb{S}^2$$

is a differentiable map between compact surfaces and, so, it makes sense to talk about its degree. To compute it, take $p \in S$ and $v \in T_p S$. We have

$$\left(d\frac{V}{|V|} \right)_p (v) = \frac{1}{|V|} (dV)_p(v) + \text{multiple of } V(p).$$

Consequently

$$(\text{Jac } \frac{V}{|V|})(p) = \frac{1}{|V(p)|^3} \det((dV)_p(e_1), (dV)_p(e_2), V(p)),$$

where $e_1, e_2 \in T_pS$ form a positively oriented orthonormal basis. As a consequence, the degree of the normalized field associated to V is

$$\deg \frac{V}{|V|} = \frac{1}{4\pi} \int_S \frac{1}{|V(p)|^3} \det((dV)_p(e_1), (dV)_p(e_2), V(p)) \, dp.$$

Remark 8.5. A particular case included in Remark 8.4 is when the field V on the surface S is given by

$$V(p) = Ap, \qquad \forall p \in S,$$

where A is a regular square matrix of order three or, equivalently, a linear automorphism of $\mathbb{R}^3$. Suppose that $0 \notin S$. In this situation, the differential of V coincides with A at each point and, if $p \in S$ and $e_1, e_2 \in T_pS$ form a positively oriented orthonormal basis, one has

$$\det(Ae_1, Ae_2, Ap) = \langle p, N(p) \rangle \det A,$$

where N is the Gauss map compatible with the orientation chosen on S. Therefore

$$(8.1) \qquad \deg \frac{V}{|V|} = \frac{\det A}{4\pi} \int_S \frac{\langle p, N(p) \rangle}{|Ap|^3} \, dp$$

and this is true whenever V does not vanish on S, that is, when $0 \notin S$.

We now define a new vector field X on the open subset $\mathbb{R}^3 - \{0\}$ by

$$X(x) = \frac{x}{|Ax|^3}, \qquad \forall x \in \mathbb{R}^3 - \{0\},$$

which is clearly differentiable. Moreover

$$(dX)_x(v) = \frac{1}{|Ax|^3} v - 3 \frac{\langle Ax, Av \rangle}{|Ax|^5} x$$

for each $x \in \mathbb{R}^3 - \{0\}$ and each $v \in \mathbb{R}^3$. Thus, X is a field with vanishing divergence on the whole of its domain. If the origin does not lie in the inner domain Ω determined by the surface S, the field X is differentiable on $\overline{\Omega}$ and we may apply the divergence Theorem 5.31 to conclude that

$$\int_S \langle X, N \rangle = 0.$$

Then, from (8.1) and the definition of X, we have

$$\deg \left(p \in S \longmapsto \frac{Ap}{|Ap|} \in \mathbb{S}^2 \right) = 0 \qquad \text{if } 0 \notin \overline{\Omega}.$$

If, instead, the origin of $\mathbb{R}^3$ lies in Ω, we choose $\varepsilon > 0$ so that the ball B_ε centred at the origin, along with its closure, is contained in Ω. Thus, we may again apply the divergence Theorem 5.31 to the same field X on the regular domain $\Omega' = \Omega - \overline{B}_\varepsilon$—see Remark 5.33—and get

$$\int_S \langle X, N \rangle = -\int_{\mathbb{S}^2_\varepsilon} \left\langle X(p), \frac{p}{\varepsilon} \right\rangle \, dp,$$

where N is the inner unit normal field. Substituting here the precise value of X and taking (8.1) into account, we have

$$\deg \frac{V}{|V|} = \pm \frac{\varepsilon \det A}{4\pi} \int_{\mathbb{S}^2_\varepsilon} \frac{1}{|Ap|^3} \, dp,$$

and the sign is either positive or negative depending upon whether the orientation fixed on S is the inner or the outer orientation. We now apply Theorem 5.14 of the change of variables to the diffeomorphism

$$p \in \mathbb{S}^2 \longmapsto \varepsilon p \in \mathbb{S}^2_\varepsilon$$

and transform the integral on the right-hand side into an integral over the unit sphere, that is,

$$\deg \frac{V}{|V|} = \pm \frac{\det A}{4\pi} \int_{\mathbb{S}^2} \frac{1}{|Ap|^3} \, dp.$$

The sign depends again whether the orientation of S is either the inner or the outer orientation. Suppose that we choose the inner orientation. Then, from Exercise (4) of Chapter 5, which computes the integral on the right-hand side, it follows that

$$\deg \left(p \in S \longmapsto \frac{Ap}{|Ap|} \in \mathbb{S}^2 \right) = \begin{cases} +1 & \text{if } \det A > 0, \\ \\ -1 & \text{if } \det A < 0, \end{cases}$$

which is true when $0 \in \Omega$.

Remark 8.6 (Degree and composition with diffeomorphisms). Consider two differentiable maps

$$\phi : S \to S' \quad \text{and} \quad \psi : S' \to S''$$

between oriented compact surfaces, and suppose that the first of them, ϕ, is a diffeomorphism preserving the orientation; see Exercise 3.17. Thus, since the Jacobian of a map between surfaces is multiplicative—compare with Exercise 5.2—we have

$$\text{Jac}\,(\psi \circ \phi) = [(\text{Jac}\,\psi) \circ \phi]|\text{Jac}\,\phi|.$$

As a consequence of Definition 8.1, we obtain

$$\deg(\psi \circ \phi) = \frac{1}{A(S'')} \int_S [(\text{Jac}\,\psi) \circ \phi]|\text{Jac}\,\phi| = \frac{1}{A(S'')} \int_{S'} \text{Jac}\,\psi = \deg \psi,$$

where we have again used Theorem 5.14 of the change of variables. Analogously

$$\deg(\psi \circ \phi) = - \deg \psi$$

if the diffeomorphism ϕ reverses the orientation.

Now, once we have stated the elementary properties of the degree introduced in Definition 8.1 for differentiable maps between oriented compact surfaces, we are going to obtain a geometrical interpretation of this degree, as a sort of mean value of the number of preimages of a point through the map. To do this, we will use the area formula of Theorem 5.24. In fact, given a map $\phi : S \to S'$ of this type, let $q \in S'$ be one of its regular values, that is—see Definition 4.5—a point such that

$$p \in \phi^{-1}(\{q\}) \Longrightarrow (\operatorname{Jac} \phi)(p) \neq 0.$$

We know, by the inverse function Theorem 2.75, that the set of the preimages of q through ϕ is finite. We define the *local degree* of the map ϕ at the point q as the integer number given by

$$\deg(\phi, q) = \sum_{p \in \phi^{-1}(\{q\})} \begin{cases} +1 & \text{if } (\operatorname{Jac} \phi)(p) > 0, \\ -1 & \text{if } (\operatorname{Jac} \phi)(p) < 0. \end{cases}$$

Hence, this local degree of ϕ at q defines a map $\deg(\phi, -)$ by

$$q \in S' \text{ regular value of } \phi \longmapsto \deg(\phi, q) \in \mathbb{Z}$$

on the subset $\mathcal{R}$ of S' of regular values of ϕ, which we know, by the Sard Theorem 4.33, has a measure zero complement. Notice that, *grosso modo*, $\deg(\phi, q)$ measures the number of preimages of q through ϕ taking into account the local behaviour of ϕ relative to the orientation and that, moreover, in general, it is impossible to extend the definition of $\deg(\phi, -)$ to the whole of S', because the non-regular values can have infinitely many preimages. The relation between the degree of ϕ and this local degree will be established in the following result.

Proposition 8.7. *Let* $\phi : S \to S'$ *be a differentiable map between two oriented compact surfaces, and let* $\mathcal{R}$ *be the subset of* S' *consisting of the regular values of* ϕ, *whose complement is a measure zero set. Then the function*

$$\deg(\phi, -) : \mathcal{R} \subset S' \longrightarrow \mathbb{Z},$$

defined above, is constant on each connected component of the open subset $\mathcal{R}$ *of* S' *and is integrable on* S'. *Moreover*

$$\deg \phi = \frac{1}{A(S')} \int_{S'} \deg(\phi, q) \, dq.$$

Proof. It is clear that the function $\deg(\phi, -)$ is integrable over S' because it follows from its definition that

$$|\deg(\phi, -)| \leq n(\phi, 1)$$

on $\mathcal{R}$, where the integrable function $n(\phi, 1)$ was introduced in Section 5.6 to prove the area formula. Likewise, it follows that

$$\deg(\phi, q) = n(\phi, \mathrm{sgJac})(q), \qquad \forall q \in \mathcal{R},$$

where the function sign of the Jacobian, denoted by sgJac, is given by

$$p \in S \longmapsto \mathrm{sgJac}(p) = \begin{cases} +1 & \text{if } (\mathrm{Jac}\,\phi)(p) > 0, \\[2mm] -1 & \text{if } (\mathrm{Jac}\,\phi)(p) < 0, \\[2mm] 0 & \text{if } (\mathrm{Jac}\,\phi)(p) = 0. \end{cases}$$

Consequently, if we apply the above-mentioned area formula—the version for surfaces contained in Theorem 5.28—to the map $\phi : S \to S'$ and the function $\mathrm{sgJac} : S \to \mathbb{R}$, then we obtain

$$\int_{S'} \deg(\phi, q)\, dq = \int_S \mathrm{sgJac}(p)\, |\mathrm{Jac}\,\phi|(p)\, dp.$$

Now, notice that the integrand on the right-hand side is nothing more than the function $\mathrm{Jac}\,\phi$. This, together Definition 8.1, gives the required equality. Thus, it remains to show that $\deg(\phi, -)$ is constant on each connected component of $\mathcal{R}$, which is an open subset of S' because $\mathcal{R} = S' - \phi(\mathcal{N})$, where $\mathcal{N} \subset S$ is the set of the points of S where the Jacobian of ϕ vanishes. In other words, it remains to show that $\deg(\phi, -)$ is locally constant. To prove this, take $q \in S' - \phi(\mathcal{N})$ a regular value of ϕ and let

$$\phi^{-1}(\{q\}) = \{p_1, \ldots, p_k\}$$

be the set of its preimages. Since $(\mathrm{Jac}\,\phi)(p_i) \neq 0$ for $i = 1, \ldots, k$, each linear map

$$(d\phi)_{p_i} : T_{p_i}S \longrightarrow T_q S', \qquad i = 1, \ldots, k,$$

is regular. Using the inverse function Theorem 2.75, we can find open neighbourhoods that we may choose to be disjoint, $U_1, \ldots, U_k$ of $p_1, \ldots, p_k$ in S and $V_1, \ldots, V_k$ of q in S', respectively, such that each restriction $\phi_{|U_i}$ is a diffeomorphism from U_i to V_i, $i = 1, \ldots, k$. Then, there exists an open subset V of S' such that

$$q \in V \subset V_1 \cap \cdots \cap V_k \quad \text{and} \quad \phi^{-1}(V) \subset U_1 \cup \cdots \cup U_k.$$

If not, there would exist a sequence $\{q_n\}_{n \in \mathbb{N}}$ of points of S' converging to q and a sequence $\{a_n\}_{n \in \mathbb{N}}$ of points of S such that

$$\phi(a_n) = q_n \quad \text{and} \quad a_n \in S - (U_1 \cup \cdots \cup U_k), \qquad n \in \mathbb{N}.$$

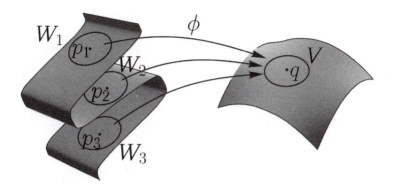

Figure 8.1. *Neighbourhood of q where $\deg(\phi, -)$ is constant*

Since $S - (U_1 \cup \cdots \cup U_k)$ is a closed subset of S, it is compact. Thus, a subsequence of $\{a_n\}_{n \in \mathbb{N}}$, which we will keep denoting by the same symbol, would converge to a point $a \in S - (U_1 \cup \cdots \cup U_k)$. This leads to a contradiction because, by continuity, $\phi(a) = q$ and, however, a has to be different from the $p_1, \ldots, p_k$.

In conclusion, if we put $W_i = U_i \cap \phi^{-1}(V)$, with $i = 1, \ldots, k$, we have shown (see Figure 8.1) that, for each $q \in S' - \phi(\mathcal{N})$, there exists an open neighbourhood V, which can be taken to be connected, such that

$$\phi^{-1}(V) = W_1 \cup \cdots \cup W_k,$$

where the W_i, $i = 1, \ldots, k$, are disjoint connected open subsets of S and where each $\phi_{|W_i}$ is a diffeomorphism from W_i to V. Hence, each element of V has exactly k preimages through ϕ—as the point q has—and at all the points of each W_i the map preserves or reverses orientation simultaneously because its Jacobian does not change sign. This proves that $\deg(\phi, -)$ is constant on V. $\qquad\square$

Remark 8.8. Proposition 8.7 describes the *real* number $\deg \phi$ associated to a differentiable map between oriented compact surfaces as the mean of the number of preimages through ϕ, counted with positive or negative sign depending on whether the orientation is preserved or not. Then, it is the mean of an integer valued function which, furthermore, is locally constant. On the other hand, all the computations that we have made until now have supplied only integer degrees. In fact, as one might already suspect, the degree of such a map ϕ is always an integer number and the function $\deg(\phi, -)$ is not only locally constant on $\mathcal{R}$, but it coincides identically with the degree of ϕ. In any case, we will not need to know that much to obtain our version of the Gauss-Bonnet theorem. We will need to show that result only when the second surface is the unit sphere, in order to get an improvement of this version.

Exercise 8.9. Consider the surface of $\mathbb{R}^3$ given by

$$S = \{(x, y, z) \in \mathbb{R}^3 \mid (\sqrt{x^2 + y^2} - a)^2 + z^2 = r^2\},$$

where $0 < r < a$, that is, the torus of revolution of Example 2.17. Show that, if K is its Gauss curvature and $(x, y, z) \in S$, then

$$K(x, y, z) > 0 \quad \text{if and only if} \quad x^2 + y^2 > a^2,$$

$$K(x, y, z) < 0 \quad \text{if and only if} \quad x^2 + y^2 < a^2$$

and that the image of the set of the points of S with vanishing Gauss curvature through a Gauss map N of S consists only of the north and south poles of the unit sphere. Show also that each point of the sphere, except these two poles, has exactly two preimages through N, one of them with positive Gauss curvature and the other one with negative Gauss curvature. From this, determine the function $\deg(N, -)$ defined on $\mathbb{S}^2 - \{\text{north}, \text{south}\}$ and compute $\deg N$.

Exercise 8.10. Let $\phi : S \to S'$ be a differentiable map. Deduce, from the area formula in Theorem 5.28 and Definition 8.1 of $\deg(\phi, -)$, that

$$\int_S (f \circ \phi)(p)(\operatorname{Jac} \phi)(p) \, dp = \int_{S'} f(q) \deg(\phi, q) \, dq,$$

for each integrable function $f : S' \to \mathbb{R}$.

8.3. Degree and surfaces bounding the same domain

The main property of the degree of maps between oriented compact surfaces of $\mathbb{R}^3$ will be given now: when maps defined on different surfaces $S_1, \ldots, S_k$ taking values on the same surface S can be extended to a regular domain— see Remark 5.33—whose (topological) boundary consists of the union $S_1 \cup \cdots \cup S_k$, the sum of their degrees vanishes, provided that we have chosen a suitable orientation on each of them. From this property, the well-known invariance under homotopies and also our first version of the Gauss-Bonnet theorem follow.

Theorem 8.11. *Let $\Omega \subset \mathbb{R}^3$ be a regular domain whose boundary $\partial\Omega = S_1 \cup \cdots \cup S_k$ is a union of k disjoint compact connected surfaces. If S' is another compact surface and $\phi : \overline{\Omega} \to S'$ is a differentiable map, then*

$$\deg \phi_{|S_1} + \cdots + \deg \phi_{|S_k} = 0,$$

where we have taken any orientation of S' and, simultaneously on all the S_i, $i = 1, \ldots, k$, either the inner orientation or the outer orientation relative to the domain Ω.

Proof. We pick up on the surface S' any Gauss map $N' : S' \to \mathbb{S}^2$ and the corresponding orientation and we define a differentiable vector field X_ϕ on $\overline{\Omega}$ as

$$X_\phi = (\det(N' \circ \phi, \phi_y, \phi_z), -\det(N' \circ \phi, \phi_x, \phi_z), \det(N' \circ \phi, \phi_x, \phi_y)),$$

where the subscripts x, y, z stand for partial derivatives. Then, if we recall the definition of the divergence of a field—see Definition 5.30—we have

$$\operatorname{div} X_\phi = \frac{\partial}{\partial x} \det(N' \circ \phi, \phi_y, \phi_z) - \frac{\partial}{\partial y} \det(N' \circ \phi, \phi_x, \phi_z) + \frac{\partial}{\partial z} \det(N' \circ \phi, \phi_x, \phi_y).$$

All the summands on the right-hand side containing partial derivatives of second order cancel each other by the Schwarz theorem of calculus and, so,

$$\operatorname{div} X_\phi = \det((N' \circ \phi)_x, \phi_y, \phi_z) - \det((N' \circ \phi)_y, \phi_x, \phi_z) + \det((N' \circ \phi)_z, \phi_x, \phi_y).$$

Now then, if $p \in \overline{\Omega}$, one has that

$$\phi_y(p) = (d\phi)_p(0, 1, 0) \in T_{\phi(p)}S',$$

$$\phi_z(p) = (d\phi)_p(0, 0, 1) \in T_{\phi(p)}S',$$

$$(N' \circ \phi)_x(p) = d(N' \circ \phi)_p(1, 0, 0) = (dN')_{\phi(p)}[(d\phi)_p(1, 0, 0)] \in T_{\phi(p)}S',$$

because $(dN')_{\phi(p)}$ is an endomorphism of $T_{\phi(p)}S'$. Therefore, the three vectors

$$(N' \circ \phi)_x(p), \quad \phi_y(p), \quad \text{and} \quad \phi_z(p)$$

lie in a plane. Thus

$$\det((N' \circ \phi)_x, \phi_y, \phi_z) = 0$$

on $\overline{\Omega}$. By similar reasonings

$$\det((N' \circ \phi)_y, \phi_x, \phi_z) = \det((N' \circ \phi)_z, \phi_x, \phi_y) = 0,$$

and, so, X_ϕ is a vector field with vanishing divergence.

On the other hand, if $\{u_1, u_2, u_3\}$ is the canonical basis of $\mathbb{R}^3$, the very definition of the field X_ϕ implies that

$$\det(X_\phi, u_1, u_2) = \langle X_\phi, u_3 \rangle = \det(N' \circ \phi, \phi_x, \phi_y),$$

$$\det(X_\phi, u_1, u_3) = -\langle X_\phi, u_2 \rangle = \det(N' \circ \phi, \phi_x, \phi_z),$$

$$\det(X_\phi, u_2, u_3) = \langle X_\phi, u_1 \rangle = \det(N' \circ \phi, \phi_y, \phi_z).$$

Consequently, for each $p \in \overline{\Omega}$, the two skew-symmetric bilinear forms

$$(u, v) \in \mathbb{R}^3 \times \mathbb{R}^3 \longmapsto \det(X_\phi(p), u, v),$$

$$(u, v) \in \mathbb{R}^3 \times \mathbb{R}^3 \longmapsto \det((N' \circ \phi)(p), (d\phi)_p(u), (d\phi)_p(v))$$

coincide on the canonical basis u_1, u_2, u_3. Hence, they both must be equal, that is,

$$(8.2) \qquad \det(X_\phi(p), u, v) = \det((N' \circ \phi)(p), (d\phi)_p(u), (d\phi)_p(v))$$

for all $p \in \overline{\Omega}$ and all $u, v \in \mathbb{R}^3$.

Now, we apply the divergence Theorem 5.31 to the field X_ϕ whose divergence vanishes on the regular domain Ω whose boundary is $S_1 \cup \cdots \cup S_k$ and we have

$$\int_{S_1} \langle X_\phi, N_1 \rangle + \cdots + \int_{S_k} \langle X_\phi, N_k \rangle = 0,$$

where each N_i, $i = 1, \ldots, k$, is the Gauss map of the surface S_i inner with respect to the domain Ω. But, if $p \in S_i$ and $\{e_1^i, e_2^i\} \subset T_p S_i$ is an orthonormal basis positively oriented with respect to N_i, for $i = 1, \ldots, k$, it turns out, from (8.2), that

$$\langle X_\phi, N_i \rangle(p) = \langle X_\phi(p), e_1^i \wedge e_2^i \rangle = \det(X_\phi(p), e_1^i, e_2^i)$$

$$= \det((N' \circ \phi)(p), (d\phi)_p(e_1^i), (d\phi)_p(e_2^i)) = (\mathrm{Jac}\, \phi_{|S_i})(p).$$

Finally, we obtain the equality

$$\int_{S_1} \mathrm{Jac}\, \phi_{|S_1} + \cdots + \int_{S_k} \mathrm{Jac}\, \phi_{|S_k} = 0,$$

that is, the sum of the degrees of the restrictions $\phi_{|S_i}$, with $i = 1, \ldots, k$, vanishes, as required. $\square$

Corollary 8.12. *If a differentiable map $\phi : S \to S'$ between compact connected surfaces extends smoothly to the closure of the inner domain determined by S, then $\deg \phi = 0$.*

Proof. It suffices to apply Theorem 8.11 above to the inner domain Ω determined by the compact connected surface S. $\square$

Remark 8.13. Let S_1 be a compact connected surface, Ω_1 the corresponding inner domain and suppose that S_2 is another surface of the same type such that $S_2 \subset \Omega_1$. Let Ω_2 be the inner domain of this second surface. We know, by Remark 5.33, that $\Omega = \Omega_1 - \overline{\Omega_2}$ is a regular domain with boundary $S_1 \cup S_2$. If $\phi : \overline{\Omega} \to S'$ is a differentiable map taking values on a compact surface S', we have

$$\deg \phi_{|S_1} = \deg \phi_{|S_2},$$

if, on S_1 and S_2, we choose simultaneously either the inner or the outer orientations.

Remark 8.14 (Invariance under homotopies). Suppose that S and S' are two oriented compact surfaces of $\mathbb{R}^3$ and that

$$\phi : S \times [a, b] \longrightarrow S'$$

is a differentiable map, where a and b are two real numbers with $a < b$. Take $\varepsilon > 0$ small enough so that $N_{2\varepsilon}(S)$ is a tubular neighbourhood of S, such as was defined in Definition 4.24. If Ω is the inner domain of S and Ω_ε the domain of the inner parallel surface at distance ε—see Remark 4.28—that we denote by S_ε, then $\Gamma = \Omega - \overline{\Omega_\varepsilon}$ is a regular domain of $\mathbb{R}^3$ with boundary $\partial\Gamma = S \cup S_\varepsilon$. We define a map $\psi : \overline{\Gamma} \longrightarrow S'$ as

$$\psi(F(p,t)) = \psi(p + tN(p)) = \phi(p, a + \frac{t}{\varepsilon}(b-a)),$$

where $p \in S$, $t \in [0, \varepsilon]$, and $F : S \times (-2\varepsilon, 2\varepsilon) \to N_{2\varepsilon}(S)$, as defined in (4.1), is a diffeomorphism. For this reason, ψ is a well-defined map and it is differentiable. Using Remark 8.13, we have

$$\deg \psi_{|S} = \deg \psi_{|S_\varepsilon},$$

where we have put the inner orientations on S and S_ε. Now then, it follows from the definition of the map ψ that

$$\psi_{|S} = \phi_a \quad \text{and} \quad \psi_{|S_\varepsilon} = \phi_b \circ (F_\varepsilon)^{-1}.$$

Since $F_\varepsilon : S \to S_\varepsilon$ is a diffeomorphism which preserves orientation—see, for instance, Example 5.17—we deduce from Remark 8.6 that

$$\deg \phi_a = \deg(\phi_b \circ F_\varepsilon^{-1}) = \deg \phi_b.$$

That is, *two differentiable maps which are homotopic through a differentiable homotopy have the same degree.*

Exercise 8.15. Let S be a compact connected surface. Suppose that its Gauss map $N : S \to \mathbb{S}^2$ extends to a differentiable map of $\overline{\Omega}$ to $\mathbb{S}^2$, where Ω is its inner domain. Show that

$$\int_S K(p)\, dp = 0.$$

Exercise 8.16. Let S be a compact surface and $V : S \times [a, b] \to \mathbb{R}^3$ a nowhere vanishing differentiable map. Demonstrate that

$$\deg \frac{V_a}{|V_a|} = \deg \frac{V_b}{|V_b|},$$

with respect to any orientations taken on S and on $\mathbb{S}^2$.

Exercise 8.17. Let $\phi : S \to \mathbb{S}^2$ be a differentiable map from an oriented compact surface to the unit sphere. Show that, if A is an orthogonal matrix, one has $\deg(A \circ \phi) = \pm \deg \phi$, depending on $\det A = \pm 1$.

Exercise 8.18. Check that the function

$$f(x, y, z) = (\sqrt{x^2 + y^2} - a)^2 + z^2 - r^2, \qquad 0 < r < a,$$

which is differentiable on $\mathbb{R}^3 - \{z\text{-axis}\}$ and which was used in Example 2.17 to define the torus of revolution $S = f^{-1}(\{0\})$, has no critical points on the open set

$$\Omega = f^{-1}((-\infty, 0)),$$

which is the domain determined by S. From this, deduce that the Gauss map of S can be smoothly extended to a map from $\overline{\Omega}$ to the unit sphere and confirm the assertion in Exercise 8.9 about the degree of this Gauss map.

Exercise 8.19. Let S be a compact connected surface and let Ω be its inner domain. If $a \in \mathbb{R}^3 - S$, we define a differentiable map $\phi_a : S \to \mathbb{S}^2$ by means of the equality

$$\phi_a(p) = \frac{p - a}{|p - a|}, \qquad p \in S.$$

Prove that $\deg \phi_a = 0$ if $a \notin \Omega$ and that $\deg \phi_a = 1$ when $a \in \Omega$. (Use Remark 5.34.)

An interesting application of the invariance of the degree under homotopies is the following well-known result on tangent vector fields defined on spheres, which is due to Brower.

Theorem 8.20 (Brower's hairy ball theorem). *A tangent vector field on the unit sphere $\mathbb{S}^2$ of $\mathbb{R}^3$ necessarily vanishes at some point.*

Proof. Suppose, on the contrary, that W is a differentiable field of tangent vectors on $\mathbb{S}^2$ with no zeroes. Then $V = W/|W|$ is a differentiable field of tangent unit vectors defined on $\mathbb{S}^2$, that is, a differentiable map $V : \mathbb{S}^2 \to \mathbb{S}^2$ such that $\langle V(p), p \rangle = 0$, for each $p \in \mathbb{S}^2$. We define, using this V, a map

$$\phi : \mathbb{S}^2 \times [0, \pi] \longrightarrow \mathbb{S}^2$$

as

$$\phi(p, \theta) = p \cos \theta + V(p) \sin \theta.$$

This map is differentiable and, clearly, supplies a homotopy between $\phi_0 = I_{\mathbb{S}^2}$ and $\phi_\pi = -I_{\mathbb{S}^2}$. From Remark 8.14, we deduce that

$$\deg I_{\mathbb{S}^2} = \deg(-I_{\mathbb{S}^2}).$$

However, Exercise 8.17 above, or equivalently Definition 8.1, implies that these degrees are $+1$ and -1, respectively. This is a contradiction, leading to the required conclusion.   □

8.4. Index of a field at an isolated zero

The degree of a differentiable map between compact surfaces, which measures a kind of mean of the number of preimages, will permit us to introduce the notion of the *index* of a field at an isolated zero of the field. We will do this first for fields defined on $\mathbb{R}^3$ and, later, for tangent fields defined on a given surface.

Definition 8.21 (Index of a field at an isolated zero). Let $O \subset \mathbb{R}^3$ be an open set and $X : O \to \mathbb{R}^3$ a differentiable vector field defined on O. Given a point $a \in O$, we will say that a is a—possible—isolated zero of X if there exists a neighbourhood U of a in O such that X does not vanishes on $U - \{a\}$. In this case, for $r > 0$ small enough, since the sphere $\mathbb{S}^2_r(a)$ centred at a with radius r will be contained in the punctured neighbourhood $U - \{a\}$, we may define the *index* of the field X at a as the number

$$i(X, a) = \deg \frac{X}{|X|}\bigg|_{\mathbb{S}^2_r(a)} \in \mathbb{R}.$$

It is important to note that the right-hand side does not depend, by Remark 8.13, on the radius r, and that, since we have not shown that the degree has to be an integer, the index of a field at an isolated zero is, for us, a real number. We will see that, however, under some assumptions, it is not difficult to prove that this index is indeed an integer number.

Lemma 8.22. *If the vector field X does not vanish at the point a of its domain, then $i(X, a) = 0$.*

Proof. In fact, in this case, the map $X/|X|$ extends differentiably to a closed ball centred at a and we finish by Definition 8.21 and Corollary 8.12. □

Another relevant case where we can prove that the index of a field at an isolated zero is an integer number is the following. Suppose that, at the point $a \in O$, there is an effective zero of X, that is, $X(a) = 0$. We will say that X is a *non-degenerate field at the point a* when the linear map

$$(dX)_a : \mathbb{R}^3 \longrightarrow \mathbb{R}^3$$

is regular or, equivalently, when $\det(dX)_a = (\operatorname{Jac} X)(a) \neq 0$. We must point out that, as a consequence of the inverse function theorem of calculus, if the field X is non-degenerate at the point a, then a is an isolated zero of X. In general, we will say that the field X itself is *non-degenerate* when it is non-degenerate at each of its zeroes. In this case, the discussion above shows that its set of zeroes is discrete.

Lemma 8.23. *Let X be a non-degenerate vector field. Then the index $i(X, a)$ of the field X at each of its zeroes $a \in O$ is either $+1$ or -1, depending on whether $(dX)_a$ preserves or reverses orientation in $\mathbb{R}^3$, that is, depending on whether $\det(dX)_a$ is positive or negative.*

Proof. In fact, by Definition 8.21 of the index of a field at an isolated zero and Remark 8.4, one has that

$$i(X, a) = \frac{1}{4\pi} \int_{\mathbb{S}^2_r(a)} \frac{1}{|X(p)|^3} \det((dX)_p(e_1), (dX)_p(e_2), X(p)) \, dp,$$

where, for each $p \in \mathbb{S}^2_r(a)$, the vectors $e_1, e_2 \in T_p\mathbb{S}^2_r(a)$ form an orthonormal basis positively oriented with respect to the outer normal, according to our convention throughout this chapter. We will now use Theorem 5.14 of the change of variables to transform the integral on the right-hand side into an integral over the unit sphere, by applying it to the diffeomorphism of $\mathbb{S}^2$ to $\mathbb{S}^2_r(a)$ given by $p \mapsto a + rp$. We have

$$i(X, a) = \frac{1}{4\pi} \int_{\mathbb{S}^2} \frac{r^2}{|X(a + rp)|^3} \det((dX)_{a+rp}(e_1), (dX)_{a+rp}(e_2), X(a+rp)) \, dp$$

for each $r > 0$ small enough. But, since X vanishes at a and is non-degenerate at this point,

$$\lim_{r \to 0} \frac{X(a + rp)}{r} = (dX)_a(p) \neq 0, \qquad \forall p \in \mathbb{S}^2.$$

From this, we infer that the integrand above converges pointwise, as r goes to zero, to the continuous function

$$p \in \mathbb{S}^2 \mapsto \frac{1}{|(dX)_a(p)|^3} \det((dX)_a(e_1), (dX)_a(e_2), (dX)_a(p)) = \frac{\det(dX)_a}{|(dX)_a(p)|^3}.$$

By using the continuous dependence of parameters, one obtains

$$i(X, a) = \frac{\det(dX)_a}{4\pi} \int_{\mathbb{S}^2} \frac{dp}{|(dX)_a(p)|^3}.$$

The conclusion that we are looking for can be obtained by recalling that

$$\int_{\mathbb{S}^2} \frac{dp}{|(dX)_a(p)|^3} = \frac{4\pi}{|\det(dX)_a|},$$

by Exercise (4) at the end of Chapter 5. $\square$

Now, we will deal with differentiable tangent vector fields defined on surfaces.

Definition 8.24 (Index of a field on a surface at an isolated zero). Let S be a surface oriented by means of a Gauss map $N : S \to \mathbb{S}^2$, and let V be a tangent vector field on S. In other words, V is a differentiable map from S to $\mathbb{R}^3$ such that $\langle V(p), N(p) \rangle = 0$ for all $p \in S$. A point $a \in S$ will be called a—possible—*isolated zero* of V when one can find a neighbourhood U of a in S such that V does not vanish on $U - \{a\}$. If this is the case, since, for $r > 0$ small enough, the open ball $B_r(a)$ centred at a with radius r is contained in a tubular neighbourhood of U, we may define a new vector field X_V on the open subset $B_r(a)$ of $\mathbb{R}^3$ by

$$X_V(F(p,t)) = X_V(p + tN(p)) = V(p) + tN(p), \qquad \forall p \in U \cap B_r(a),$$

where F is the map defined in (4.1). By this definition one sees that X is differentiable and that X does not vanish on $B_r(a)$, except maybe at the point a. Thus, a is a—possible—isolated zero of X_V. We define the *index* of the field V at the point a by means of the equality

$$i(V, a) = i(X_V, a).$$

This definition does not depend, as in the case of the index of vector fields defined on open domains of $\mathbb{R}^3$, either on U or on r, as a consequence of what we studied above. Also, we immediately have a version for surfaces like that Lemma 8.22. Now suppose that the point $a \in S$ is really a zero of V, that is, $V(a) = 0$. Then, if $u \in T_a S$ and $\alpha : (-\delta, \delta) \to S$ is a curve on the surface with $\alpha(0) = a$ and $\alpha'(0) = u$, one has

$$0 = \frac{d}{dt}\bigg|_{t=0} \langle (N \circ \alpha)(t), (V \circ \alpha)(t) \rangle = \langle N(a), (dV)_a(u) \rangle.$$

Hence,

$$(dV)_a(T_a S) \subset T_a S,$$

and we can think of the map $(dV)_a$ as an endomorphism of the tangent plane $T_a S$. We will say that the field V is *non-degenerate* at the point a when the map $(dV)_a$ is regular, that is, when $\det(dV)_a \neq 0$. Let us show that this is equivalent to the field X_V being non-degenerate at a, in the sense that we gave to this property for vector fields on $\mathbb{R}^3$. Since, by definition,

$$X_{V|B_r(a) \cap S} = V_{|B_r(a) \cap S},$$

one has that

$$(dX_V)_{a|T_a S} = (dV)_a.$$

On the other hand, since

$$X_V(a + tN(a)) = V(a) + tN(a)$$

for each t small, taking derivatives with respect to t at $t = 0$, one obtains

$$(dX_V)_a(N(a)) = N(a).$$

In conclusion, we have $\det (dX_V)_a = \det(dV)_a$ and, then, V is non-degenerate at a if and only if X_V is non-degenerate. Therefore, we may give the following result.

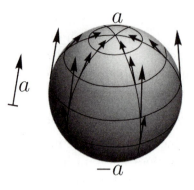

Figure 8.2. *Conformal field on a sphere*

Proposition 8.25. *Let V be a differentiable tangent vector field on a surface S, and let $p \in S$. If $V(p) \neq 0$, then $i(V, p) = 0$. If, instead, $V(p) = 0$, then $(dV)_p$ is an endomorphism of the tangent plane T_pS. If the field V is non-degenerate at p, that is, if this endomorphism is regular, p is an isolated zero of V and moreover*

$$i(V, p) = \begin{cases} +1 & \text{if } \det(dV)_p > 0, \\[2mm] -1 & \text{if } \det(dV)_p < 0. \end{cases}$$

Therefore, if the surface S is compact and V is non-degenerate, it has finitely many zeroes and

$$\sum_{p \in S} i(V, p) \in \mathbb{Z},$$

that is, the sum of the indices of the zeroes of V is an integer number.

Example 8.26 (Conformal fields on a sphere). Given a unit vector $a \in \mathbb{R}^3$, one can define a differentiable map a^T on the unit sphere by

$$p \in \mathbb{S}^2 \longmapsto a^T(p) = a - \langle a, p \rangle p.$$

In this way, $\langle a^T(p), p \rangle = 0$, for each $p \in \mathbb{S}^2$, that is, a^T is a tangent vector field on the unit sphere $\mathbb{S}^2$, which will be called, for reasons that we will not explain here, a *conformal* field. It is clear that the conformal field a^T corresponding to the unit vector a has exactly two zeroes at the antipodal points a and $-a$ of $\mathbb{S}^2$. On the other hand, if $p \in \mathbb{S}^2$ and $v \in T_p\mathbb{S}^2$, one has

$$(da^T)_p(v) = -\langle a, v \rangle p - \langle a, p \rangle v.$$

Thus, putting $p = a$ and $p = -a$ in this expression,

$$(da^T)_a(v) = -v \quad \text{and} \quad (da^T)_{-a}(v) = v, \qquad \forall v \in T_a\mathbb{S}^2 = T_{-a}\mathbb{S}^2.$$

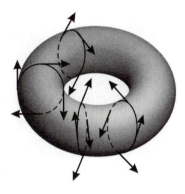

Figure 8.3. *Field without zeroes on a torus of revolution*

At these two zeroes, the corresponding linear maps $(da^T)_a$ and $(da^T)_{-a}$ preserve the orientation of the plane $T_a\mathbb{S}^2 = T_{-a}\mathbb{S}^2$. Hence, each conformal field a^T is a non-degenerate field on the sphere and, according to Proposition 8.25,

$$i(a^T, a) = 1 \qquad \text{and} \qquad i(a^T, -a) = 1.$$

Finally, we see that the sum of the indices of the zeroes of the field a^T on the sphere is exactly 2.

Example 8.27 (A non-vanishing tangent field on a torus). Consider again the torus of revolution

$$S = \{(x, y, z) \in \mathbb{R}^3 \mid (\sqrt{x^2 + y^2} - a)^2 + z^2 = r^2\},$$

where $0 < r < a$. Since this surface is given implicitly, one can see, for example, using Example 3.7, that

$$N(x, y, z) = \frac{1}{r\sqrt{x^2 + y^2}} \left(x(\sqrt{x^2 + y^2} - a), y(\sqrt{x^2 + y^2} - a), z\sqrt{x^2 + y^2} \right),$$

for each $(x, y, z) \in S$, is a Gauss map for the torus. From this, one can check without difficulty that the differentiable field $V : S \to \mathbb{R}^3$ defined on the torus by

$$V(x, y, z) = \frac{1}{r} \left(\frac{xz}{\sqrt{x^2 + y^2}}, \frac{yz}{\sqrt{x^2 + y^2}}, a - \sqrt{x^2 + y^2} \right)$$

is tangent to S at each point. In fact, V provides at any point of the torus a unit vector tangent to the meridian passing through this point. We have, then, a sample of a tangent field on the torus of revolution which does not vanish anywhere. This is in clear contrast to the case of the sphere where, according to the Brower Theorem 8.20, this situation cannot occur.

A fundamental property of the non-degenerate vector fields is that they can be transferred from a surface to another one through a diffeomorphism, and this process does not vary the indices at their zeroes.

Proposition 8.28. *Let $\phi : S \to S'$ be a diffeomorphism between two surfaces of $\mathbb{R}^3$ and suppose we have a non-degenerate tangent vector field V on S. Then, the map $V^\phi : S' \to \mathbb{R}^3$ given by*

$$(V^\phi \circ \phi)(p) = (d\phi)_p(V(p)), \qquad \forall p \in S,$$

defines a tangent vector field on S' which is non-degenerate as well. Moreover, $\phi(p) \in S'$ is a zero of V^ϕ if and only if $p \in S$ is a zero of V and, in this case, we have

$$i(V^\phi, \phi(p)) = i(V, p).$$

Proof. It is clear, by definition, that V^ϕ is a map from S' to $\mathbb{R}^3$ such that

$$V^\phi(q) \in T_q S', \qquad \forall q \in S'.$$

Let us see that V^ϕ is differentiable. Suppose that $X : U \to S$ is a parametrization of S defined on an open subset U of $\mathbb{R}^2$, whose inverse $X^{-1} : X(U) \subset \mathbb{R}^3 \to U \subset \mathbb{R}^2$ is the restriction of a coordinate projection $\pi : \mathbb{R}^3 \to \mathbb{R}^2$; see Proposition 2.23. Then $X' = \phi \circ X : U \to S'$ is a parametrization of S' and, moreover,

$$(V^\phi \circ X')(u, v) = (d\phi)_{X(u,v)}[(V \circ X)(u, v)],$$

for each $(u, v) \in U$. Since $(V \circ X)(u, v) \in T_{X(u,v)}S$, we have

$$(V \circ X)(u, v) = (dX)_{(u,v)}(W(u, v)), \qquad \forall (u, v) \in U,$$

where, using our precise choice of X, the map $W = (W^1, W^2) : U \to \mathbb{R}^2$ can be written as

(8.3) $$W = \pi \circ V \circ X$$

and, so, it is clearly differentiable. Thus, the expression above for $V^\phi \circ X'$ can be transformed into

(8.4) $$(V^\phi \circ X')(u, v) = d(\phi \circ X)_{(u,v)}(W(u, v)),$$

for each $(u, v) \in U$. This shows that $V^\phi \circ X'$ is differentiable on U because $d(\phi \circ X)_{(u,v)}$ is a linear map from $\mathbb{R}^2$ to $\mathbb{R}^3$ depending smoothly on $(u, v) \in U$. It is also clear by the definition of the field V^ϕ that, if $p \in S$, then

$$V^\phi(\phi(p)) = 0 \quad \text{if and only if} \quad V(p) = 0.$$

The reason is that $(d\phi)_p$ is an isomorphism. Now suppose that $p \in S$ is a zero of V and that, thus, $\phi(p)$ is a zero of V^ϕ. We already know that, in this case, $(dV)_p$ is an endomorphism of $T_p S$ and that $(dV^\phi)_{\phi(p)}$ is an endomorphism of $T_{\phi(p)}S'$. We are going to study the relation between these two linear maps. To do this, put $p = X(a, b)$ for some $(a, b) \in U$. Then, from (8.3), we have $W(a, b) = 0$. Taking this equality into account, we take partial derivatives of (8.4) and obtain

$$d(V^\phi \circ X')_{(a,b)} = d(\phi \circ X)_{(a,b)} \circ (dW)_{(a,b)}.$$

On the other hand, if we apply the chain rule to (8.3), we have

$$(dW)_{(a,b)} = \pi \circ (dV)_p \circ (dX)_{(a,b)}.$$

Putting this equality into the previous one, it turns out that

$$d(V^\phi \circ X')_{(a,b)} = (d\phi)_p \circ (dV)_p \circ (dX)_{(a,b)}.$$

Recalling now that $X' = \phi \circ X$ and applying the chain rule to the left-hand side, we finally have that

$$(dV^\phi)_{\phi(p)} \circ (d\phi)_p = (d\phi)_p \circ (dV)_p$$

or, equivalently,

$$(dV^\phi)_{\phi(p)} = (d\phi)_p \circ (dV)_p \circ (d\phi)_p^{-1}.$$

An immediate consequence is that V is a non-degenerate field at p if and only if V^ϕ is non-degenerate at $\phi(p)$ and, moreover,

$$\det(dV^\phi)_{\phi(p)} = \det(dV)_p.$$

From this and Proposition 8.25, it follows that

$$i(V^\phi, \phi(p)) = i(V, p)$$

as we wanted.   □

8.5. The Gauss-Bonnet formula

The global version of the celebrated Gauss-Bonnet theorem asserts that the total Gauss curvature of a compact surface of $\mathbb{R}^3$ is exactly 2π times a certain integer associated to the surface: its Euler characteristic. This integer has a topological nature and can be assigned to the surface in different ways, for instance, using triangulations or by means of the homology groups of the surface or through its fundamental group. In any case, two diffeomorphic surfaces have the same Euler characteristic. Furthermore, if two compact surfaces of $\mathbb{R}^3$ share this number, then they are diffeomorphic. A detailed study of all these features of the Euler characteristic of a surface, and even its rigorous definition, go beyond both the aim—remember that this is a text on differential geometry—and the possibility that we may attain with the prerequisites that we proposed for reading this book.

But, fortunately, there is another approach to the Euler characteristic. In 1885, Poincaré showed that the Euler characteristic of a compact surface also appears when one sums up the indices of a tangent vector field on the surface having only isolated zeroes. In 1926 this was generalized to higher

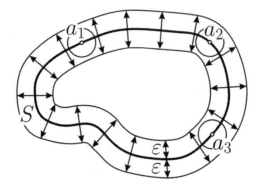

Figure 8.4. *The regular domain* Ω

dimensions by H. Hopf, after some attempts by Brower and Hadamard. The version of the Gauss-Bonnet theorem that we will state and prove now combines the two above-mentioned results and will connect the total Gauss curvature of a compact surface of $\mathbb{R}^3$ with the sum of the indices at the zeroes of a field of this class on the surface. This will allow us to see that the total Gauss curvatures of two diffeomorphic surfaces have to coincide.

Theorem 8.29 (First version of the Gauss-Bonnet theorem). *Let S be a compact surface of Euclidean space $\mathbb{R}^3$, and let V be a differentiable tangent vector field defined on S which has only isolated zeroes. Then*

$$\int_S K(p)\,dp = 2\pi \sum_{p \in S} i(V, p),$$

where K is the Gauss curvature function on the surface S.

Proof. Since the two sides of the required equality are additive with respect to the connected components of the surface S, we may suppose, as usual, that the surface is connected. Also, since S is compact, the number of isolated zeroes of the field V is finite. We will denote them by $\{a_1, \ldots, a_k\}$, for some $k \in \mathbb{N}$. We choose, using Theorem 4.26, a real number $\varepsilon > 0$ such that $N_{2\varepsilon}(S)$ is a tubular neighbourhood of S and, so, the open set $\Omega_{2\varepsilon} = \Omega_{-\varepsilon} - \overline{\Omega_\varepsilon}$, where $\Omega_{-\varepsilon}$ and Ω_ε are the inner domains determined by the parallel surfaces $S_{-\varepsilon}$ and S_ε, respectively, is a regular domain whose boundary is the union of these two parallel surfaces. In other words, we have

$$\Omega_{2\varepsilon} = F(S \times (-\varepsilon, \varepsilon)),$$

where $F : S \times (-2\varepsilon, 2\varepsilon) \to N_{2\varepsilon}(S)$ is the diffeomorphism introduced in (4.1), corresponding to the normal field N inner to S. Now, we take an open ball $B(a_i)$ of $\mathbb{R}^3$ centred at the zero a_i of V, for $i = 1, \ldots, k$, with radius small enough so that

$$\overline{B(a_i)} \cap \overline{B(a_j)} = \emptyset \quad \text{when } i \neq j$$

and also so that

$$\overline{B(a_i)} \subset \Omega_{2\varepsilon} \quad \text{for } i = 1, \ldots, k.$$

Then, we have the situation of Figure 8.4. Therefore, according to Remark 5.33, the open subset of $\mathbb{R}^3$ given by

$$\Omega = \Omega_{2\varepsilon} - (\overline{B(a_1)} \cup \cdots \cup \overline{B(a_k)})$$

is a regular domain contained in the tubular neighbourhood $N_{2\varepsilon}(S)$ whose boundary is the disjoint union

$$\partial\Omega = S_{-\varepsilon} \cup S_\varepsilon \cup \mathbb{S}^2(a_1) \cup \cdots \cup \mathbb{S}^2(a_k),$$

where each $\mathbb{S}^2(a_i)$ is the sphere with centre a_i, $i = 1, \ldots, k$, which is the boundary of $B(a_i)$.

Now, we consider, as in Definition 8.24, the vector field X on the tubular neighbourhood $N_{2\varepsilon}(S)$ of the surface S given by

$$X(F(p,t)) = X(p + tN(p)) = V(p) + tN(p), \qquad \forall(p,t) \in S \times (-2\varepsilon, 2\varepsilon).$$

Since F is a diffeomorphism, X is a well-defined differentiable map. Furthermore, since $V(p) \in T_pS$ for each $p \in S$, X vanishes only at the same points $a_1, \ldots, a_k$ where V does. As a consequence, the field X does not vanish on the regular domain Ω that we constructed in $\mathbb{R}^3$. Then, in this situation, we may apply Theorem 8.11 to the map

$$\phi = \frac{X}{|X|} : \overline{\Omega} \longrightarrow \mathbb{S}^2.$$

Consequently, we have

$$\sum_{i=1}^{k} \deg \frac{X}{|X|}\bigg|_{\mathbb{S}^2(a_i)} + \deg \frac{X}{|X|}\bigg|_{S_{-\varepsilon}} + \deg \frac{X}{|X|}\bigg|_{S_\varepsilon} = 0,$$

where, in order to compute these degrees, we have to put on S_ε, $S_{-\varepsilon}$ and on the spheres $\mathbb{S}^2(a_i)$, $i = 1, \ldots, k$, the inner orientations relative to Ω; see Figure 8.4. But, if we consider the same equation for the outer orientations that each of them has as a compact connected surface, we obtain

$$(8.5) \qquad \deg \frac{X}{|X|}\bigg|_{S_{-\varepsilon}} - \deg \frac{X}{|X|}\bigg|_{S_\varepsilon} = \sum_{i=1}^{k} \deg \frac{X}{|X|}\bigg|_{\mathbb{S}^2(a_i)}.$$

Now then, from the definition of the field X, it follows that, for each $\rho \in (-2\varepsilon, 2\varepsilon)$,

$$X_{|S_\rho} = V \circ F_\rho^{-1} + \rho N \circ F_\rho^{-1} = (V + \rho N) \circ F_\rho^{-1},$$

where we already know that $F_\rho : S \to S_\rho$ is a diffeomorphism preserving orientation. For this reason, taking Remark 8.6 into account, we conclude that

$$\deg \frac{X}{|X|}_{|S_\rho} = \deg \frac{V + \rho N}{|V + \rho N|},$$

for each $\rho \neq 0$ with $-2\varepsilon < \rho < 2\varepsilon$. Finally, the invariance of the degree under homotopies, which we saw in Remark 8.14, implies

$$\deg \frac{V + \rho N}{|V + \rho N|} = \begin{cases} \deg N & \text{when } \rho > 0, \\ \deg(-N) & \text{when } \rho < 0, \end{cases}$$

because it is not difficult to see that, if $\rho > 0$, $\phi : S \times [0, 1/\rho] \to \mathbb{S}^2$ given by

$$\phi(p, t) = \frac{N + tV}{|N + tV|}, \qquad \forall p \in S, \quad 0 \leq t \leq \frac{1}{\rho},$$

is a homotopy between $\phi_0 = N$ and $\phi_{1/\rho} = (V + \rho N)/|V + \rho N|$, and that, analogously, if $\rho < 0$, then $\psi : S \times [1/\rho, 0] \to \mathbb{S}^2$ given by

$$\psi(p, t) = -\frac{N + tV}{|N + tV|}, \qquad \forall p \in S, \quad 1/\rho \leq t \leq 0,$$

is a homotopy between $\psi_0 = -N$ and $\psi_{1/\rho} = (V + \rho N)/|V + \rho N|$. Consequently, since N is the inner normal field of S, from Remark 8.3 and Exercise 8.17, we have

$$\deg \frac{X}{|X|}_{|S_{-\varepsilon}} = -\deg \frac{X}{|X|}_{|S_\varepsilon} = \frac{1}{4\pi} \int_S K(p) \, dp.$$

Substituting this into (8.5), we deduce that

$$\frac{1}{2\pi} \int_S K(p) \, dp = \sum_{i=1}^k \deg \frac{X}{|X|}_{|\mathbb{S}^2(a_i)}.$$

To finish the proof it suffices to recall Definition 8.24 of index of a field on a surface at an isolated zero and the fact that, by Proposition 8.25, this index vanishes at the points of S where V does not. $\qquad\square$

We can derive a first immediate and important consequence of the Gauss-Bonnet Theorem 8.29.

Corollary 8.30. *All the differentiable tangent vector fields on a compact surface of $\mathbb{R}^3$ with isolated zeroes have the same sum of the indices at these zeroes.*

For example, if the surface is an ovaloid, in particular, a sphere $\mathbb{S}^2(r)$ of arbitrary radius $r > 0$, we saw in Remark 6.6 that its total Gauss curvature is exactly 4π. From this, we get the following consequence.

Corollary 8.31. *The sum of the indices at the zeroes of a differentiable tangent vector field which has only isolated zeroes on an ovaloid is exactly 2.*

This was the situation that we saw in Example 8.26 for the conformal fields on a sphere. As a consequence of this, we may generalize the well-known result by Brower that we already gave in Theorem 8.20.

Corollary 8.32. *On an ovaloid, each differentiable tangent vector field has at least one zero.*

Finally, recall that in Example 8.27 we exhibited a non-vanishing differentiable tangent field on the torus of revolution. From this fact, the following statement can be derived.

Corollary 8.33. *The sum of the indices of a differentiable tangent field with only isolated zeroes on a torus of revolution is zero.*

In order to strengthen the Gauss-Bonnet Theorem 8.29, we will deal with a particular type of tangent fields on surfaces: those fields which are the gradient, as defined in Exercise (17) of Chapter 6, of differentiable functions defined on them.

Proposition 8.34. *Suppose that S is a surface in $\mathbb{R}^3$ and that $f : S \to \mathbb{R}$ is a differentiable function defined on it. The following assertions are true.*

- **A:** *The gradient field ∇f has its zeroes just at the critical points of the function f.*
- **B:** *Let $p \in S$ be a zero of the gradient ∇f. Then the gradient is non-degenerate at p if and only if the Hessian $(d^2 f)_p$ of f at this critical point is a non-degenerate quadratic form on the plane $T_p S$.*
- **C:** *The index of ∇f at a critical point p of the function f with non-degenerate Hessian is $+1$ if, at this point p, f has a local extreme and is -1 if f has at p a saddle point.*

Proof. Part **A**. For $p \in S$ and $v \in T_p S$, we have

$$\langle (\nabla f)(p), v \rangle = (df)_p(v).$$

So the first assertion is obvious.

Part **B**. Suppose that, indeed, the point $p \in S$ is a zero of the field ∇f, that is, a critical point of the function f. If $v \in T_p S$, we take a curve $\alpha : (-\delta, \delta) \to S$ such that $\alpha(0) = p$ and such that $\alpha'(0) = v$. By the definition of gradient, one has

$$\langle (\nabla f)(\alpha(t)), \alpha'(t) \rangle = (df)_{\alpha(t)}(\alpha'(t)) = \frac{d}{dt}(f \circ \alpha)(t)$$

for all $t \in (-\delta, \delta)$. Taking derivatives at the time $t = 0$, without forgetting that $(\nabla f)(p) = 0$, one obtains

$$\left.\frac{d^2}{dt^2}\right|_{t=0} (f \circ \alpha)(t) = \langle (d\nabla f)_p(v), v \rangle.$$

The left-hand side of this equality is nothing more than the Hessian of the function f at the critical point p, as defined in Definition 3.34, applied to the vector v, that is,

$$\langle (d\nabla f)_p(v), v \rangle = (d^2 f)_p(v),$$

at each zero p of ∇f and each $v \in T_pS$. Therefore, the differential $(d\nabla f)_p$ is regular if and only if $(d^2 f)_p$ is a non-degenerate quadratic form.

Part **C**. If $p \in S$ is a critical point of f with non-degenerate Hessian, Proposition 8.25 implies, in this case, that

$$i(\nabla f, p) = \begin{cases} +1 & \text{if } \det(d^2 f)_p > 0, \\ -1 & \text{if } \det(d^2 f)_p < 0, \end{cases}$$

and one can finish by applying Proposition 3.35.

$$\square$$

Example 8.35. Consider the height function $f : \mathbb{S}^2 \to \mathbb{R}$ on the unit sphere, given by

$$f(p) = \langle p, a \rangle, \qquad \forall p \in \mathbb{R}^3,$$

corresponding to the unit vector a. It is clear, using for instance Example 2.59 and the definition of gradient, that the gradient of this height function f is just the conformal field a^T studied in Example 8.26. We saw that it was a non-degenerate field with two zeroes of index $+1$, that is, f has exactly two critical points which are extremes, specifically, the global maximum and the global minimum.

Example 8.36. In Example 8.27, we gave an example of a tangent field on the torus of revolution S, given by the equation $(\sqrt{x^2 + y^2} - a)^2 + z^2 = r^2$ with $0 < r < a$, with no zeroes. On such a torus S, consider the two height functions

$$f(x, y, z) = z \quad \text{and} \quad g(x, y, z) = x, \qquad \forall (x, y, z) \in S.$$

It is easy to see, for instance looking at Figure 8.5, that the field ∇f vanishes along the two circles where the planes $z = a$ and $z = -a$ touch the torus S. Then ∇f is not a field with isolated zeroes. However, the field ∇g is non-degenerate on S because g has exactly four critical points

$$(-a - r, 0, 0), \quad (-a + r, 0, 0), \quad (a - r, 0, 0), \quad \text{and} \quad (a + r, 0, 0)$$

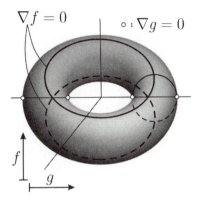

$$\nabla f = 0 \qquad \circ : \nabla g = 0$$

$$f \qquad g$$

Figure 8.5. *Two height functions on a torus*

and, among them, the first one is the global minimum of the function g, the fourth one is its global maximum, and the two remaining points are saddle points. We point out that the sum of the indices of ∇g at these four zeroes, according to Proposition 8.34, is zero, as predicted by Corollary 8.33.

With these preliminaries and taking Example 3.36 into account, where the Hessian of height functions on surfaces was studied, we have the following corollary of the Sard Theorem 4.33.

Proposition 8.37. *Let S be a compact surface. There exists a measure zero subset $\mathcal{S}$ of the unit sphere $\mathbb{S}^2$ such that, for each $a \in \mathbb{S}^2 - \mathcal{S}$, the gradient ∇h_a of the height function h_a, defined on S by $p \in S \mapsto h_a(p) = \langle p, a \rangle$, is a non-degenerate tangent field. Moreover, the sum of the indices at its zeroes satisfies*

$$\sum_{p \in S} i(\nabla h_a, p) = \deg(N, a) + \deg(N, -a),$$

that is, it is equal to the number of elliptic points of S with tangent plane perpendicular to the vector a minus the number of hyperbolic points of S with tangent plane perpendicular to a.

Proof. Let $N : S \to \mathbb{S}^2$ be a Gauss map for the surface S. We define the subset $\mathcal{S} \subset \mathbb{S}^2$ as the complement of the set consisting of the points of the sphere which are simultaneously regular values for the maps N and $-N$. By the Sard Theorem 4.33 and Remark 4.7, this set indeed has measure zero. Now let a be any point of $\mathbb{S}^2 - \mathcal{S}$ and consider its corresponding height function h_a on S. Its critical points are just the points of S with tangent plane perpendicular to the vector a, that is, the points lying in $N^{-1}(\{a, -a\})$. Since a and $-a$ are regular values for N, by our choice, the Jacobian of N at these points does not vanish. But this Jacobian, according to Remark 8.3, is the Gauss curvature, which is also the determinant of the Hessian of the height function h_a, from Proposition 3.37. Now, from Proposition 8.34, the vector field ∇h_a is non-degenerate and its indices at each zero are either $+1$ or -1, depending on the sign of the Gauss curvature at each of them. $\qquad\square$

From this result, we will be able to deduce a stronger version of the global Gauss-Bonnet formula than we got in Theorem 8.29.

Theorem 8.38 (The Gauss-Bonnet-Poincaré theorem). *For each compact surface S of Euclidean space $\mathbb{R}^3$, there exists an integer number $\chi(S)$, which we will call its Euler characteristic, with the following properties.*

A: *If S_1 and S_2 are diffeomorphic, then $\chi(S_1) = \chi(S_2)$.*

B: *Each differentiable tangent vector field V on S, with only isolated zeroes, satisfies*

$$\sum_{p \in S} i(V, p) = \chi(S).$$

C: *If K is the Gauss curvature function of S, we have*

$$\int_S K(p) \, dp = 2\pi \, \chi(S),$$

and, so, $2 \deg N = \chi(S)$, if N is a Gauss map of S.

Proof. Given the compact surface S, we know, from Proposition 8.37, that there exists a differentiable tangent vector field W on the surface which is non-degenerate. Proposition 8.25 implies that

$$\chi(S) = \sum_{p \in S} i(W, p)$$

is an integer and, as a consequence of the Gauss-Bonnet Theorem 8.29 and Corollary 8.30, it only depends on the surface S and satisfies requirements **B** and **C**. To prove assertion **A**, suppose that S_1 and S_2 are two compact surfaces such that there is a diffeomorphism $\phi : S_1 \to S_2$. We know that

$$\chi(S_1) = \sum_{p \in S_1} i(W_1, p) \quad \text{and} \quad \chi(S_2) = \sum_{p \in S_2} i(W_2, p),$$

where W_1 and W_2 are two non-degenerate tangent fields on S_1 and S_2, respectively. Now consider the vector field W_1^{ϕ} defined on S_2 from W_1 and ϕ. Proposition 8.28 asserts that W_1^{ϕ} is, like W_2, a non-degenerate tangent field on the surface S_2 and that, moreover,

$$\sum_{q \in S_2} i(W_1^{\phi}, q) = \sum_{p \in S_1} i(W_1, p) = \chi(S_1).$$

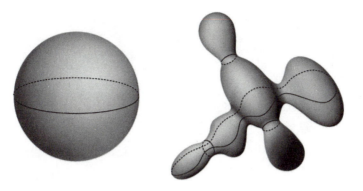

Figure 8.6. *Sphere and deformed sphere*

Applying statement **B**, which was already proved, to the surface S_2, we have

$$\sum_{q \in S_2} i(W_1^\phi, q) = \sum_{q \in S_2} i(W_2, q) = \chi(S_2),$$

and, so, **A** is true. $\qquad\square$

One of the most surprising and beautiful consequences of this Gauss-Bonnet-Poincaré Theorem 8.38 is that, when a compact surface is deformed, for example, if we *dent* a sphere, the values taken by the curvatures, say, the Gauss curvature, can be remarkably modified on those regions affected by the deformation. However, the total amount of Gauss curvature on the surface must remain invariant. For instance, the Gauss curvatures of the two surfaces in Figure 8.6 take considerably different values; note that one of them has positive constant curvature and the other one possesses some regions of negative curvature. In spite of this, they both must have total Gauss curvature equal to 4π.

Remark 8.39. We usually say that a surface S of $\mathbb{R}^3$ is a *topological sphere* if it is diffeomorphic to the unit sphere $\mathbb{S}^2$. For example, ovaloids are topological spheres, by the Hadamard-Stoker Theorem 6.5, although, of course, there are topological spheres with negative Gauss curvature regions. The Gauss-Bonnet-Poincaré Theorem 8.38 implies that the Euler characteristic $\chi(S)$ of such a topological sphere is just 2. Therefore, its total Gauss curvature has to be exactly 4π and the sum of the indices at the zeroes of a tangent field with isolated zeroes has to be exactly 2. In particular, on all these surfaces each tangent field must have at least one zero, as seen for ovaloids in Corollary 8.32.

Remark 8.40. A surface of $\mathbb{R}^3$ is called a *torus*, or a *toroidal surface*, if it is diffeomorphic to a torus of revolution as was introduced in Example 2.17. For example, Figure 8.7 shows two samples of toroidal surfaces with quite

Figure 8.7. *Two toroidal surfaces*

different properties. Example 8.27 and the Gauss-Bonnet-Poincaré Theorem 8.38 lead to the conclusion that the Euler characteristic $\chi(S)$ of a torus in $\mathbb{R}^3$ has to be 0. Therefore, its total Gauss curvature must also vanish. This means that the negatively curved and the positively curved regions must compensate for each other. Finally, each tangent vector field on a torus, with isolated zeroes, has null sum of indices, as we checked for the fields described in Examples 8.27 and 8.36 on the torus of revolution.

Remark 8.41. It is not difficult to see that there are compact surfaces in $\mathbb{R}^3$ with Euler characteristic different from 0 and 2. One can construct a compact surface S_k, with $k \in \mathbb{N}$, $k \geq 1$, by *gluing* k tori of revolution and smoothing the unions, as in Figure 8.8. The height function corresponding to the vector a in this figure has exactly two extremes—which are global—and $2k$ saddle points. Hence, $\chi(S_k) = 2(1 - k)$, according to the Gauss-Bonnet-Poincaré Theorem 8.38 and Proposition 8.34. In this way, we have surfaces whose Euler characteristic is any non-positive integer. In fact, any compact surface of $\mathbb{R}^3$ is a topological sphere, a torus, or is homeomorphic to some of these S_k. The proof of this assertion is beyond this text, but a consequence of this is that the only possible Euler characteristics have already appeared in our examples. In Section 8.6, we will sketch, in a list of exercises, a proof of the fact that a compact surface of $\mathbb{R}^3$ always has an *even* Euler characteristic. By the way, other interesting facts will appear.

8.6. Exercise: The Euler characteristic is even

If S is a compact surface of Euclidean space $\mathbb{R}^3$ and $N : S \to \mathbb{S}^2$ is a Gauss map, we know that

$$\chi(S) = 2 \deg N,$$

according to the Gauss-Bonnet-Poincaré Theorem 8.38. Now, we will give some exercises which lead to the following result.

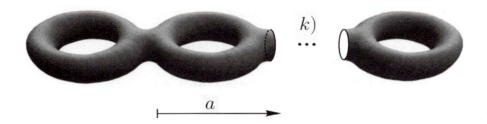

Figure 8.8. *A surface S_k*

Theorem 8.42. *The degree of a differentiable map $\phi : S \to \mathbb{S}^2$ from a compact surface to the unit sphere is always an integer. In fact, $\deg \phi = \deg(\phi, q)$ for each $q \in \mathcal{R} \subset \mathbb{S}^2$ which is a regular value of ϕ. As a consequence, the Euler characteristic of a compact surface is always an even integer.*

This result is also true for differentiable maps taking values not necessarily on the unit sphere but on any other compact surface.

Exercises: Steps of the proof

(1) Let S and S' be surfaces oriented by Gauss maps N and N', respectively, and let $\phi : S \to S'$ be a differentiable map between them. Given a differentiable tangent vector field W on S', we define, for each $p \in S$, a vector $V(p) \in T_pS$ by

$$\det(V(p), N(p), v) = \det((W \circ \phi)(p), (N' \circ \phi)(p), (d\phi)_p(v))$$

for all $v \in T_pS$. Demonstrate that the tangent field V defined on S in this way is differentiable by showing that, if $X : U \to S$ is a parametrization of S defined on an open subset U of $\mathbb{R}^2$, the following equalities are true:

$$\det(V \circ X, N \circ X, X_u) = \det(W \circ \phi \circ X, N' \circ \phi \circ X, (\phi \circ X)_u),$$

$$\det(V \circ X, N \circ X, X_v) = \det(W \circ \phi \circ X, N' \circ \phi \circ X, (\phi \circ X)_v).$$

Take the partial derivatives of these equations, alternatively with respect to v and to u, subtract, and finally obtain

$$\operatorname{div} V = (\operatorname{Jac} \phi)[(\operatorname{div} W) \circ \phi].$$

(2) Combining Exercise (1), Exercise 8.10, and the divergence Theorem 6.10 for surfaces, prove that, if $\phi : S \to S'$ is a differentiable map between

two oriented compact surfaces, then

$$\int_{S'} \deg(\phi, q)(\operatorname{div} W)(q)\, dq = 0,$$

for each differentiable tangent vector field W on S'.

(3) For each $t \in \mathbb{R}$ and each vector $a \in \mathbb{S}^2$, we define a differentiable map $\phi_t^a : \mathbb{S}^2 \to \mathbb{S}^2$ by

$$\phi_t^a(p) = \frac{p + [(\cosh t - 1)\langle p, a\rangle + \sinh t]a}{\cosh t + \langle p, a\rangle \sinh t}, \qquad \forall p \in \mathbb{S}^2.$$

Prove that the following assertions are true.

a: $\phi_t^a \circ \phi_s^a = \phi_s^a \circ \phi_t^a = \phi_{s+t}^a$, for all $s, t \in \mathbb{R}$ and $a \in \mathbb{S}^2$. Hence, each ϕ_t^a is a diffeomorphism of the sphere and ϕ_0^a is the identity map.

b: The Jacobian of the diffeomorphism ϕ_t^a is given by

$$(\operatorname{Jac} \phi_t^a)(p) = \frac{1}{(\cosh t + \langle p, a\rangle \sinh t)^2},$$

for each $t \in \mathbb{R}$, $a, p \in \mathbb{S}^2$.

c: If a^T is the conformal field on the sphere defined in Example 8.26, one has

$$\frac{d}{dt}\phi_t^a(p) = (d\phi_t^a)_p[a^T(p)] = a^T(\phi_t^a(p))$$

for each $t \in \mathbb{R}$ and each $a, p \in \mathbb{S}^2$.

d: For each pair of points $a, p \in \mathbb{S}^2$, one has

$$\phi_t^a(a) = a, \qquad \phi_t^a(-a) = -a, \qquad \phi_{-t}^a = \phi_t^{-a},$$

$$\lim_{t \to +\infty} \phi_t^a(p) = a \quad \text{if } p \neq -a.$$

(4) Given a differentiable function $h : \mathbb{S}^2 \to \mathbb{R}$ on the unit sphere, show that the integrals

$$f^h(p) = \int_0^\infty (h \circ \phi_t^a)(p)(\operatorname{Jac} \phi_t^a)(p)\, dt,$$

$$g^h(p) = \int_{-\infty}^0 (h \circ \phi_t^a)(p)(\operatorname{Jac} \phi_t^a)(p)\, dt$$

define two differentiable functions f^h and g^h on the open sets $\mathbb{S}^2 - \{-a\}$ and $\mathbb{S}^2 - \{a\}$, respectively, and that, moreover, we have

$$\operatorname{div}(f^h\, a^T) = -h \quad \text{and} \quad \operatorname{div}(g^h\, a^T) = h,$$

for each $a \in \mathbb{S}^2$.

(5) Suppose that $a \in \mathbb{S}^2$ and that $\alpha : \mathbb{R} \to \mathbb{S}^2$ is a 2π-periodic curve p.b.a.l. whose image is the equator of the sphere perpendicular to the vector a. We define $X : \mathbb{R}^2 \to \mathbb{S}^2$ by $X(s,t) = \phi_t^a(\alpha(s))$. Demonstrate that $X(\mathbb{R}^2) = \mathbb{S}^2 - \{a, -a\}$ and that X restricted to any open subset of the plane of the form $(a, a + 2\pi) \times \mathbb{R}$ is a parametrization of the sphere.

(6) Given a differentiable function $h : \mathbb{S}^2 \to \mathbb{R}$ on the sphere, one can consider another differentiable function $r^h : \mathbb{S}^2 - \{a, -a\} \to \mathbb{R}$ given by

$$r^h(p) = \int_{-\infty}^{\infty} (h \circ \phi_t^a)(p)(\operatorname{Jac} \phi_t^a)(p)\, dt.$$

If $\alpha : \mathbb{R} \to \mathbb{S}^2$ is a 2π-periodic curve p.b.a.l. whose image is the equator of $\mathbb{S}^2$ perpendicular to the vector a, use the parametrization in the exercise above to show that

$$\int_0^{2\pi} (r^h \circ \alpha)(s)\, ds = \int_{\mathbb{S}^2} h(p)\, dp.$$

As a consequence, $r^h \circ \alpha$ has a 2π-periodic primitive if and only if the integral of h over the sphere vanishes.

(7) Given a differentiable tangent vector field V defined on an open set $U \subset \mathbb{S}^2$ of the unit sphere, we can define a map $JV : U \to \mathbb{R}^3$ by

$$JV(p) = V(p) \wedge p, \qquad \forall p \in U.$$

Prove that JV is another tangent field on U. If $V = \nabla f$, where f is a differentiable function defined on U, show that

$$\operatorname{div} J\nabla f = 0.$$

(8) Let $h : \mathbb{S}^2 \to \mathbb{R}$ be a differentiable function with

$$\int_{\mathbb{S}^2} h(p)\, dp = 0,$$

and let $a \in \mathbb{S}^2$. Show that there exists a function $H : \mathbb{S}^2 - \{a, -a\} \to \mathbb{R}$ constant along the meridians of the sphere joining a with $-a$ and such that

$$(\nabla H)(p) = r^h(p)\, a \wedge p, \qquad \forall p \in \mathbb{S}^2 - \{a, -a\},$$

where r^h is the function defined in Exercise (6).

(9) Prove that, for each differentiable function $h : \mathbb{S}^2 \to \mathbb{R}$, with

$$\int_{\mathbb{S}^2} h(p)\, dp = 0$$

and each $a \in \mathbb{S}^2$, there exist two differentiable tangent vector fields V and W defined, respectively, on $\mathbb{S}^2 - \{-a\}$ and on $\mathbb{S}^2 - \{a\}$ and a differentiable function $H : \mathbb{S}^2 - \{a, -a\} \to \mathbb{R}$ such that

$$\operatorname{div} V = h_{|\mathbb{S}^2 - \{-a\}}, \quad \operatorname{div} W = h_{|\mathbb{S}^2 - \{a\}}, \quad \text{and} \quad J\nabla H = (V - W)_{|\mathbb{S}^2 - \{a, -a\}}.$$

(10) Let $h : \mathbb{S}^2 \to \mathbb{R}$ be a differentiable function. Then, the following statements are equivalent.

    a: The integral of h over $\mathbb{S}^2$ is zero.

    b: There exists a differentiable tangent vector field T on the sphere such that $\operatorname{div} T = h$.

(Hint for **a** $\Rightarrow$ **b**: Fix $a \in \mathbb{S}^2$ and define T as the field V of the previous exercise on a neighbourhood A of a and as $W + J\nabla \tilde{H}$ on the complement of a neighbourhood B of a such that $\overline{B} \subset A$, where the field W is also given by the previous exercise and the function $\tilde{H}$ is defined as fH, with H also given by the previous exercise and $f : \mathbb{S}^2 \to \mathbb{R}$ any differentiable function with $f \equiv 1$ on a neighbourhood of $\overline{A}$ and $f \equiv 0$ on a neighbourhood C of $-a$.)

(11) Combining Exercises (2) and (10), show that, if $\phi : S \to \mathbb{S}^2$ is a differentiable map from an oriented compact surface and the unit sphere, one has
$$\int_{\mathbb{S}^2} \deg(\phi, p) h(p)\, dp = 0,$$
for each differentiable function h defined on the sphere such that
$$\int_{\mathbb{S}^2} h(p)\, dp = 0.$$
Deduce from this that
$$\int_{\mathbb{S}^2} \deg(\phi, p) f(p)\, dp = \deg \phi \int_{\mathbb{S}^2} f(p)\, dp$$
if f is any differentiable function defined on the sphere.

(12) Using functions with arbitrarily small support on the sphere and the facts that the function $\deg(\phi, -)$ is locally constant and takes its values in $\mathbb{Z}$, where $\phi : S \to \mathbb{S}^2$ is a differentiable map and S is an oriented compact surface, prove that
$$\deg \phi = \deg(\phi, p)$$
for each $p \in \mathbb{S}^2$ which is a regular value of ϕ.

(13) Prove that the degree of a differentiable map from an oriented compact surface to the unit sphere has to be an integer. As a consequence, the Euler characteristic of such a surface of $\mathbb{R}^3$ is even.

Chapter 9

Global Geometry
of Curves

9.1. Introduction and historical notes

We finish our study of classical differential geometry by coming back to the topic of Chapter 1: geometric properties of curves. Curves were defined as differentiable maps from a real interval to a Euclidean plane or space. This definition allowed us to quickly approach the easiest geometric properties which can be obtained using methods of differential calculus. We already pointed out in Chapter 1 that we use the term geometric properties only for those properties depending on the image of the map, that is, on the trace of the curve.

After this first exposure to curves, we concentrated on the study of surfaces, which, on one hand, form the central block of classical differential geometry and, on the other hand, show more clearly the differences between extrinsic/intrinsic and local/global properties. We immediately observed an asymmetry between the definitions of a curve, as a map, and of a surface, as a set of points. This is why, in Chapter 2, after discussing the concept of a surface, we introduced the idea of a simple curve as an analogous one-dimensional object. At the beginning of this chapter, we will deal with the relation between the two classes of curves: parametrized and simple. We will show that the simple curves are just the traces of some parametrized curves and, moreover, if they are connected, they are topologically equal to either the real line or to the circle.

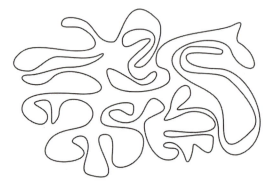

Figure 9.1. *Which are the inner points relative to this curve?*

Another reason for leaving the study of global properties of curves until the end of the book is that some of them are shared with surfaces. Furthermore, their proofs can be translated from one case to the other, but it is always easier to transfer results about surfaces to results about curves. This is why, once the problem of the connection between the two possible definitions of a curve is solved, we proceed to show which results of the theory of surfaces can be transferred, along with their proofs, except for minor changes, to the theory of curves.

First, we recover the JORDAN *curve theorem*, which asserts the following: a compact connected simple curve C of the plane, that is, homeomorphic to the circle, divides the plane into exactly two domains, one of them bounded. This means that, in the presence of the curve C, the remaining points of the plane separate into two classes: inner and outer points, in such a way that any two points in the same class can be joined by a curve which does not intersect C and any curve joining two points in two different classes must meet C. This result is obviously true for curves such as circles or ellipses, but its *evidence* grows weak depending upon how the curve complicates its figure; see Figure 9.1.

This theorem was announced for the first time by CAMILLE JORDAN (1838–1922) in his *Cours d'Analyse*, and the complexity of the proof of such an *evident* assertion surprised quite a number of mathematicians at that time. The surprise grew when some of them observed that the proof by JORDAN suffered from some gaps which demanded considerable effort to avoid them. The difficulty of the proof lies in the amplitude of the notion of the *compact simple curve* as a subset of the plane homeomorphic to the circle. The proof that we made in Chapter 4 for the JORDAN-BROWER theorem, corresponding to the case of surfaces, is also valid for differentiable curves.

Once we have accepted the JORDAN curve theorem, we may talk, for a given simple plane curve C which is compact and connected, about the inner domain Ω determined by C. The corresponding *divergence theorem* will relate, as in the higher dimensional case, the integrals of some functions defined on $\overline{\Omega}$ to integrals along the curve C. In the introduction of Chapter 5, we already talked about the statement and first proofs of this theorem due to STOKES and MAXWELL. In order for the proof that we made there to remain valid, it is enough to ensure the existence of a one-dimensional version of the SARD theorem. Verifying this point will allow us to establish some results about the total absolute value of the curvature for periodic plane curves.

Let us continue with the results that we proved for surfaces and that, in some way, also occur in the global theory of curves and which were even first solved in this context. For instance, we have the *isoperimetric question,* that is: *Among all compact simple curves with a given length, which enclose the greatest area?* This problem was raised as early as in the writings of the Greek mathematician PAPPUS (4th century B.C.), who asserted that the solution was the circle. PAPPUS wrote a complete treatise on isoperimetric problems, starting with the following motivation: why have bees *chosen* the hexagonal form for the cells of their honeycomb? It was only after twenty-three centuries that the first rigorous proof of the isoperimetric inequality in the plane was published in the book on the calculus of variations written by WEIERSTRASS in (1870). Other later proofs are due to HURWITZ (1902), using Fourier expansions, and to SCHMIDT (1939), which appears in almost all the texts on the differential geometry of curves. Recently GROMOV [1] gave a proof showing isoperimetric inequality in any dimension as a consequence of the divergence theorem. For us, the proof given in Chapter 6 of the three-dimensional case and based on the BRUNN-MINKOWSKI inequality remains valid in the two-dimensional case. In spite of this, we will sketch in some exercises a version of the proof by HURWITZ and another proof using only the divergence theorem, which is different from the GROMOV proof, and which has been outlined independently by F. HÉLEIN and the authors.

The last global results having an analogue in the theory of surfaces correspond to the theorems by HADAMARD and STOKER given in Chapter 6. There, we dealt with ovaloids. Here, we will consider *ovals*, that is, compact connected simple curves in the plane with positive curvature everywhere.

In our study of surfaces in $\mathbb{R}^3$, it was clear that different integrals of elementary functions of the Gauss and mean curvatures, such as

$$\int_S K \, dA, \quad \int_S |K| \, dA, \quad \text{and} \quad \int_S H^2 \, dA$$

are quantities with important geometrical properties of a global type. In the case of plane curves, all the local information is enclosed in the—oriented—curvature function. We will prove that the integral of this curvature along a periodic curve is an integer multiple of 2π—the length of the unit circle. The integer appearing here, called the *rotation index* of the curve, must contain, if things go in a way similar to the Gauss-Bonnet theorem, topological information about the curve. But the—intrinsic—topology of the curve is known in advance by the classification that we made at the beginning of this chapter: the curve is homeomorphic to the circle. We will give a theorem by WHITNEY and GRAUENSTEIN which states that this integer refers to the extrinsic topology of the curve. In fact, it characterizes the regular homotopy classes of periodic curves. A curve of this type can be deformed into another one, avoiding the formation of *cusps*, only when their rotation indices coincide. We will also show that, when the curve has no self-intersections, its rotation index is ± 1, a result which, in some sense, was cited by EULER, although it was correctly announced and understood for the first time by the first researcher in degree theory, mainly, H. HOPF.

When we studied the local theory of curves in Chapter 1, an important difference between the curvature functions of plane and space curves was that for the first ones we could define an oriented curvature and for the second ones only a non-negative curvature. For this reason, the total curvature of curves in $\mathbb{R}^3$ looks more like the total absolute curvature of plane curves. We knew that 2π was a lower bound for this integral. We will prove the W. FENCHEL theorem (1929), which says that this bound remains valid for periodic curves in $\mathbb{R}^3$ and that the equality characterizes the plane convex curves. We will give a proof due to B. SEGRE (1934), also discovered independently by H. RUSTIHAUSER and by H. SAMELSON. Later, in 1948, K. BORSUK generalized this proof to periodic curves of Euclidean space of arbitrary dimension and conjectured that the lower bound 2π should become 4π if the curve is a non-trivial *knot*, that is, if the curve is not the boundary of a disc embedded in $\mathbb{R}^3$. The proof of this assertion was achieved first by I. FARY and J.W. MILNOR in 1949 and 1950, respectively. The proof that we will present is based on an integral formula by M. W. CROFTON, a more or less immediate consequence of the area formula given in Chapter 5.

We will finish this chapter by proving a global theorem about the simple plane curves which are compact and connected: the so-called *four vertices theorem*, which establishes the existence of at least four local extremes for the curvature function of a curve of this type. It was first shown for ovals by the Indian geometer H. MUKHOPADHYAYA in 1909 and rediscovered in 1912 by A. KNESER. Since then, it has been one of the most *popular* theorems of the global theory of curves. There is quite a number of proofs for it, for example, by G. HERGLOTZ in 1930 and by S. B. JACKSON in 1944. Furthermore, H.

Figure 9.2. *Four-petal rose*

GLUCK (1966) showed that the theorem supplies in fact a sufficient condition for a positive periodic function to be the curvature function of an oval. The theorem is indeed true without the convexity assumption and we will prove it in its complete generality, following an idea due to R. OSSERMAN in [**14**].

9.2. Parametrized curves and simple curves

We began Chapter 1 with the local theory of curves. There we define the notion of a *curve*—see Definition 1.1—as a differentiable map of an interval of $\mathbb{R}$ either to $\mathbb{R}^2$ or to $\mathbb{R}^3$, depending on whether the curve is plane or not. In spite of this definition, the properties that we studied and the concepts, such as curvature or torsion, that we introduced were of a geometrical nature, that is, they depended only upon the image or trace of the curve. We will refer to them now as *parametrized curves*. Later, in Chapter 2, after the definition of a surface in $\mathbb{R}^3$, we defined a *simple curve* in $\mathbb{R}^2$ or in $\mathbb{R}^3$ as a one-dimensional analogue of a surface; see Example 2.18.

We must point out that the traces of parametrized curves, even when regular, and the simple curves are geometrical objects closely related, but not identical. This is more or less clear if we consider, for example, that the first ones can have self-intersection points and the second ones cannot have any. The following exercises can also shed light upon this point.

Exercise 9.1. ↑ Prove that straight lines and circles of $\mathbb{R}^3$ are simple curves.

Exercise 9.2. ↑ We call a *four-petal rose* the parametrized curve $\alpha : \mathbb{R} \to \mathbb{R}^2$ given by

$$\alpha(t) = (\sin 2t \, \cos t, \sin 2t \, \sin t)$$

for each $t \in \mathbb{R}$. Prove that its trace (see Figure 9.2) is not a simple curve.

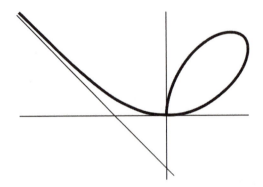

Figure 9.3. Descartes folium

Exercise 9.3. ↑ Show that the parametrized curve $\alpha : (-1, +\infty) \to \mathbb{R}^2$ given by

$$\alpha(t) = \left(\frac{3t}{1+t^3}, \frac{3t^2}{1+t^3} \right), \qquad \forall t \in \mathbb{R},$$

is regular and injective and, however, its trace (see Figure 9.3), which is called the *Descartes folium*, is not a simple curve.

Exercise 9.4. If $\alpha : I \to \mathbb{R}^3$ is a regular parametrized curve, which is also a homeomorphism onto its image $\alpha(I)$, prove that $\alpha(I)$ is a simple curve.

Now, let $\alpha : \mathbb{R} \to \mathbb{R}^3$ be a regular parametrized curve which is *periodic* as a map, that is, such that there exists a real number $a \in \mathbb{R}^+$ with $\alpha(t+a) = \alpha(t)$ for all $t \in \mathbb{R}$. If there is a sequence $\{a_n\}_{n \in \mathbb{N}}$ of positive numbers such that $\alpha(t+a_n) = \alpha(t)$ and $\lim_{n \to \infty} a_n = 0$ for some $t \in \mathbb{R}$, then

$$\alpha'(t) = \lim_{n \to \infty} \frac{\alpha(t+a_n) - \alpha(t)}{a_n} = 0,$$

which is not true because α is regular. So, if α is periodic, there exists $A \in \mathbb{R}^+$ such that $\alpha(t+A) = \alpha(t)$ for all $t \in \mathbb{R}$, where A is the least positive number satisfying this condition. We will say that A is the *period* of α and that α is an *A-periodic curve*.

Example 9.5. The circle $\alpha(s) = c + r \left(\cos(s/r), \sin(s/r) \right)$ with centre $c \in \mathbb{R}^2$ and radius $r > 0$ is a periodic curve p.b.a.l. with period $2\pi r$.

Example 9.6. Other examples of periodic curves with period 2π having self-intersections are, for instance, the curves $\alpha(t) = \sin 2t(\cos t, \sin t)$, which we called earlier the four-petal rose, and $\beta(t) = (\cos t, \sin 2t)$; see Figure 9.7.

Let $\alpha : \mathbb{R} \to \mathbb{R}^3$ be an A-periodic parametrized curve, with $A > 0$, and let $S : \mathbb{R} \to \mathbb{R}$ be its arc length function given by

$$S(t) = L_0^t(\alpha) = \int_0^t |\alpha'(u)|\, du.$$

Then we have

$$S(t + A) = \int_0^{t+A} |\alpha'(u)|\, du = \int_0^A |\alpha'(u)|\, du + \int_A^{t+A} |\alpha'(u)|\, du,$$

and, since α is A-periodic,

$$S(t + A) = \int_0^A |\alpha'(u)|\, du + \int_0^t |\alpha'(u)|\, du = L + S(t),$$

where we have put

$$L = \int_0^A |\alpha'(u)|\, du > 0.$$

This number will be called the *length* of the curve α. Thus, if $\phi : \mathbb{R} \to \mathbb{R}$ is the inverse diffeomorphism of S and $\beta = \alpha \circ \phi$ is the reparametrization of α by arc length, one obtains that

$$\beta(s + L) = \alpha\big(\phi(s + L)\big) = \alpha\big(\phi(s) + A\big) = \alpha\big(\phi(s)\big) = \beta(s),$$

and so β is L-periodic. Moreover, since A is the least period of α, then L is the least period of β, provided that $\alpha_{|[0,A)}$ is injective. As a consequence, from now on, when we say that $\alpha : \mathbb{R} \to \mathbb{R}^3$ is an L-periodic parametrized curve, it will be understood that it is parametrized by arc length and that $L > 0$ is its length, which coincides with its period.

Proposition 9.7. *Let $\alpha : \mathbb{R} \to \mathbb{R}^3$ be an L-periodic parametrized curve, $L > 0$, such that its trace has no self-intersections, that is, such that α is injective on $[0, L)$. Then the trace of α is a simple curve homeomorphic to a circle.*

Proof. We have $C = \alpha(\mathbb{R}) = \alpha([0, L])$ because α is L-periodic. On the other hand, the map $\alpha : [0, L] \to C$ is surjective, continuous, and closed—any continuous map defined on a compact space taking values in a Haussdorf space is closed. Hence, it is an identification map. Let R_α be the equivalence relation induced by α in $[0, L]$, given by

$$s_1 R_\alpha s_2 \iff \alpha(s_1) = \alpha(s_2).$$

Then, the map $\overline{\alpha} : [0, L]/R_\alpha \to \alpha(\mathbb{R})$ given by $\overline{\alpha}([s]) = \alpha(s)$ is a homeomorphism. If we observe now that

$$s_1 R_\alpha s_2 \iff \begin{cases} s_1 = s_2, \\ s_1 = 0,\, s_2 = L, \\ s_1 = L,\, s_2 = 0 \end{cases}$$

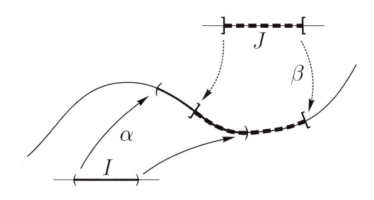

Figure 9.4. *Parametrizations of a simple curve*

because α is injective on $[0, L)$ and L-periodic, then C is homeomorphic to the quotient of $[0, L]$ identifying the end points 0 and L, that is, to a circle. But, moreover, $\alpha : (0, L) \to C$ is a homeomorphism onto its image, since it is an injective identification map. Therefore, all the maps

$$\alpha : (t, t + L) \longrightarrow C, \qquad \forall t \in \mathbb{R},$$

form a family of homeomorphisms whose images cover C endowing it with a structure of a simple curve as defined in Example 2.18. □

Remark 9.8. In general, we call any subset of $\mathbb{R}^3$ which is homeomorphic to a circle a *Jordan curve*. Then, the previous result shows that the trace of a periodic curve without self-intersections is a Jordan curve.

Exercise 9.4 and Proposition 9.7 above assert that, in at least two situations, the trace of a regular parametrized curve is a simple curve: when the curve is a homeomorphism onto its image and when it is L-periodic and injective on the subintervals with length less than L. In the first case, the resulting simple curve is topologically equal to $\mathbb{R}$ and, in the second case, it is homeomorphic to a circle, that is, it is a Jordan curve. Now, we will see that these two alternatives are the only possibilities for a connected simple curve in $\mathbb{R}^3$.

In fact, let $C \subset \mathbb{R}^3$ be a connected simple curve, and let

$$\alpha : I \to U \subset C \quad \text{and} \quad \beta : J \to V \subset C$$

be two parametrizations of C. That is, U and V are open subsets of C, I and J are open intervals of $\mathbb{R}$, and α and β are two regular curves— we can assume them to be p.b.a.l. without lost of generality—which are also homeomorphisms. Assume also that $U \cap V \neq \emptyset$ and that neither U is contained in V nor V is contained in U; see Figure 9.4. Then $U \cap V$ is a non-empty open subset of U and of V. Thus, $\alpha^{-1}(U \cap V)$ is a nonempty open subset of the interval I and $\beta^{-1}(U \cap V)$ is a non-empty open subset

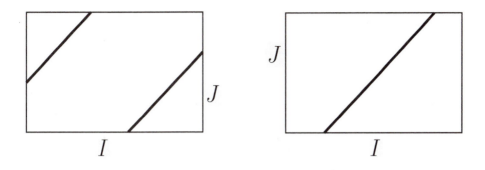

Figure 9.5. *The set X, cases* **a** *and* **b**

of the interval J. Hence, each connected component of any of these open subsets of $\mathbb{R}$ has to be an open interval contained in I or in J. Furthermore, the change of parameters

$$f = \beta^{-1} \circ \alpha : \alpha^{-1}(U \cap V) \subset I \longrightarrow \beta^{-1}(U \cap V) \subset J$$

is a diffeomorphism—recall Theorem 2.26—which satisfies $\alpha = \beta \circ f$ on $\alpha^{-1}(U \cap V) \subset I$. Therefore, by the chain rule,

$$\alpha'(s) = f'(s)\,\beta'(f(s)), \qquad \forall s \in \alpha^{-1}(U \cap V).$$

Since α and β are curves p.b.a.l., one has that $|f'(s)| = 1$, that is, either $f' \equiv 1$ or $f' \equiv -1$ on each connected component of $\alpha^{-1}(U \cap V)$. In other words, f is, up to a constant, either the identity map or its opposite on each of these connected components.

Let us consider the subset X of the rectangle $I \times J$ of $\mathbb{R}^2$ defined by

$$X = \{(s_1, s_2) \in I \times J \mid \alpha(s_1) = \beta(s_2)\}.$$

This set X is clearly closed in $I \times J$ because α and β are continuous. We will now state some other characteristics of this set.

Lemma 9.9. *The following assertions relative to the set X defined above are true.*

> **A:** *X is a graph over a union of open subintervals of I and also a graph over a union of subintervals of J which consists of segments with slope $+1$ or slope -1.*

> **B:** *Each segment of X leans its end points on non-parallel sides of $I \times J$.*

Proof. In fact, assertion **A** will be proved if we observe that $(s_1, s_2) \in X \subset I \times J$ is equivalent to $s_2 = (\beta^{-1} \circ \alpha)(s_1) = f(s_1)$, that is, X is the graph of the bijective map f. On the other hand, let σ be one of the segments with slope ± 1 of X. Since X is a closed subset of $I \times J$, if some of the end points of σ were not on a side of the rectangle $I \times J$, this end point would be in σ and, so, its projection on I would not be an open interval. This contradiction implies that the end points of σ lie on the sides of $I \times J$. Finally, if the two end points of the segment σ were on parallel sides of $I \times J$, the projection of σ on one of the two intervals I or J would be surjective. Therefore $U = \alpha(I)$ would be contained in $V = \beta(J)$ or vice-versa, in contraction to our hypotheses. This proves **B**.   □

These two properties imply that X is formed by at most two segments and, in this case, the two segments have the same slope. That is, the only two possible cases, up to a sign for the slope of the segments and a possible interchange of I and J, can be seen in Figure 9.5. The fact that there are only two alternatives for two parametrizations b.a.l. of a simple curve whose images intersect without containing one another leads to the following classification result that we already used in Chapter 4 to prove the Jordan-Brower Theorem 4.16; see the proof of Lemma 4.17.

Theorem 9.10 (Topological classification of simple curves). *If a simple curve of $\mathbb{R}^2$ or of $\mathbb{R}^3$ is connected, then either it is homeomorphic to $\mathbb{R}$ and, in this case, it can be entirely covered by a parametrization, or it is homeomorphic to a circle and, in this case, it is the trace of a periodic regular parametrized curve.*

Proof. Let $\alpha : I \to C$ and $\beta : J \to C$ be two parametrizations b.a.l. of the connected simple curve C, defined on two open intervals of $\mathbb{R}$ and let U and V be their images, which are open in C. Suppose that U and V intersect but that either of them is included in the other. Then, we are in a situation geometrically equivalent to one of the two cases **a** or **b** represented in Figure 9.5. In case **a**, the set $\alpha^{-1}(U \cap V)$ has two connected components I_1 and I_2 and $\beta^{-1}(U \cap V)$ also has two connected components J_1 and J_2, such that $f : I_1 \cup I_2 \to J_1 \cup J_2$ coincides, restricted, for example, to I_2 with a translation $T : \mathbb{R} \to \mathbb{R}$, in such a way that $T(I_2) = J_1$, and restricted to I_1 with another translation R such that $R(I_1) = J_2$. Then

$$K = I \cup T^{-1}(J)$$

is an open interval of $\mathbb{R}$ which strictly contains the interval I and we may define on it a surjective curve

$$\gamma : K \longrightarrow U \cup V \subset C$$

by the expression

$$\gamma(s) = \begin{cases} \alpha(s) & \text{if } s \in I, \\ (\beta \circ T)(s) & \text{if } s \in T^{-1}(J), \end{cases}$$

which is well defined because $\alpha = \beta \circ f = \beta \circ T$ on $I \cap T^{-1}(J) = I_2$ and, so, is differentiable and p.b.a.l. Moreover, the interval K includes two disjoint open subintervals $I_1 \subset I \subset K$ and $T^{-1}(J_2) \subset T^{-1}(J) \subset K$ such that

$$\gamma_{|I_1} = (\gamma \circ S)_{|I_1} \quad \text{and} \quad S(I_1) = T^{-1}(J_2),$$

for the translation $S = T^{-1} \circ R$. That is, $\gamma : K \to U \cup V \subset C$ extends uniquely to a periodic curve whose image must be C. Therefore, in this case, either directly or using Proposition 9.7, we have that C is homeomorphic to a circle. So, we have shown that, if there are two parametrizations of C in situation **a**, then C is a Jordan curve, and that it is the trace of a periodic parametrized curve. Similar reasoning for case **b** would only have given that there is an interval K of $\mathbb{R}$ strictly containing I and a parametrization b.a.l. $\gamma : K \to C$ whose image is $U \cup V$.

Suppose that C is not a Jordan curve, and let $\alpha : I \to C$ be a parametrization b.a.l. of C which is maximal, in the sense that it cannot be extended to another parametrization defined on an interval containing I properly. Suppose also that $\alpha(I)$ is a proper open subset of C. Then, there exists at least a point p in the boundary of C which is not in C. Since C is a simple curve, there is a parametrization b.a.l. $\beta : J \to C$ covering p. Thus, $\beta(J) = V$ is an open neighbourhood of p cutting $U = \alpha(I)$, because p is adherent to U. Moreover, V is not contained in U because $p \notin U$ and U is not included in V by maximality. Hence, the discussion above can be applied to these two parametrizations. Since C is not a Jordan curve, they have to be in case **b**. So, according to the discussion above, this parametrization would not be maximal. Consequently, we have $\alpha(I) = C$ for such a maximal parametrization. Thus, C is topologically the interval I or equivalently, $\mathbb{R}$. $\square$

Exercise 9.11. ↑ Prove that, if C is a connected simple curve, any two curves α and β p.b.a.l., defined on two intervals I and J, respectively, which are homeomorphisms onto the curve C, have to satisfy either $\beta(s) = \alpha(s+c)$ or $\beta(s) = \alpha(-s + c)$ for some $c \in \mathbb{R}$. Show that we can talk about the orientation of simple curves and that, if C is compact, it makes sense to talk about its length.

Exercise 9.12. Define the concepts of the tangent line at a point, the normal line at a point, and the curvature of a simple curve, the latter depending on the orientation.

Exercise 9.13. ↑ Let C be a non-compact connected simple curve which is closed as a subset of Euclidean space. Show that the domain of any curve p.b.a.l. whose trace is C must be the whole of $\mathbb{R}$.

9.3. Results already shown on surfaces

Once we have established the relation between the one-dimensional objects analogous to surfaces, namely, simple curves in $\mathbb{R}^2$ and $\mathbb{R}^3$ and parametrized curves studied in Chapter 1, we will focus mainly on the geometrical properties of simple curves which are closed as subsets of $\mathbb{R}^2$ or $\mathbb{R}^3$. For these kinds of curves, some of the results, suitably adapted, that we proved for closed and compact surfaces of three-space remain true and, moreover, their proofs work with only slight modifications.

9.3.1. Jordan curve theorem. For example, the Jordan-Brower Theorem 4.16, which reveals the separation property of surfaces closed in $\mathbb{R}^3$, has as an analogue the so-called Jordan curve theorem—in our case, the Jordan *differentiable* curve, to be precise. This result comes historically and perhaps logically earlier than the Jordan-Brower theorem, as we already mentioned in the introduction of this chapter. There are various proofs sufficient for the level of our text, but, since our proof of the Jordan-Brower Theorem 4.16 is valid only by replacing the *surface closed in* $\mathbb{R}^3$ by the *simple curve closed in* $\mathbb{R}^2$, we will consider that the following statement is proved.

Theorem 9.14 (Jordan curve theorem). *Let C be a connected simple plane curve which is closed as a subset of $\mathbb{R}^2$. Then $\mathbb{R}^2 - C$ has exactly two connected components with common boundary C.*

Remark 9.15. The Euclidean space $\mathbb{R}^2$ is locally connected and $\mathbb{R}^2 - C$ is open. Hence, its connected components are open in $\mathbb{R}^2$. Furthermore, since $\mathbb{R}^2$ is locally arc-connected, $\mathbb{R}^2 - C$ is locally arc-connected as well and its connected and arc-connected components coincide.

Remark 9.16 (Inner domain determined by a compact connected simple curve). If C is compact, there is a ball B centred at the origin such that $C \subset B$. Thus, the connected set $\mathbb{R}^2 - B$ is contained in $\mathbb{R}^2 - C$ and, consequently, either $\mathbb{R}^2 - B \subset \Omega^+$ or $\mathbb{R}^2 - B \subset \Omega^-$, where Ω^+, Ω^- are the two domains mentioned in Theorem 9.14. Suppose the latter is true. Then, the connected component Ω^- is not bounded and, furthermore, $\Omega^+ \cup C = \mathbb{R}^2 - \Omega^- \subset B$. Thus, Ω^+ is bounded instead. The connected component which is bounded will be called the *inner domain* determined by C and will usually be denoted by Ω. The non-bounded connected component will be called the *outer domain*. Note that $\overline{\Omega} = \Omega \cup C$ is also bounded and, so, it is compact.

Remark 9.17 (Orientation of a compact simple curve). Given a compact connected simple curve C, let Ω be the inner domain determined by C. We know, by Theorem 9.10, that there exists a periodic curve p.b.a.l. $\alpha : \mathbb{R} \to \mathbb{R}^2$, of period $L > 0$, such that $\alpha(\mathbb{R}) = C$. One can show that there are only two possibilities: either the normal vector $N(s)$ of the curve α at $\alpha(s)$, as defined in Section 1.4, points to the inner domain Ω—in the sense of Remark 4.22—for all $s \in \mathbb{R}$ or $N(s)$ points to the outer domain for all

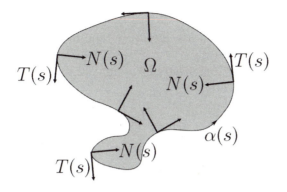

Figure 9.6. *Positive orientation*

$s \in \mathbb{R}$. If the first alternative occurs, we will say that α supplies a positive orientation of C. If not, we will consider the new curve $\beta : \mathbb{R} \to \mathbb{R}^2$ given by $\beta(s) = \alpha(-s)$. Then β is parametrized by arc length, is L-periodic, and $\beta(\mathbb{R}) = C$. Since the Frenet dihedra of the two curves are related as

$$T_\beta(s) = -T_\alpha(-s) \quad \text{and} \quad N_\beta(s) = -N_\alpha(-s),$$

we have that β is instead positively oriented. Consequently, a compact connected simple curve can always be reparametrized so that it is positively oriented. From now on, we will always consider that simple curves homeomorphic to a circle are oriented in this way, unless otherwise specified.

Exercise 9.18. Following the guidelines of Section 4.5, define tubular neighbourhoods and show their existence for compact plane simple curves. After that, define the notion of a *parallel curve*, by analogy to that of the parallel surface, as seen in Remark 4.29.

Exercise 9.19. ↑ If $\alpha : \mathbb{R} \to \mathbb{R}^2$ is a periodic curve p.b.a.l. with period $L > 0$ and whose trace is a simple curve C, compute the curvature and the length of its parallel curves in terms of those of C.

Exercise 9.20. ↑ If C is a compact connected simple plane curve, show that there are points with positive curvature with respect to the positive orientation.

9.3.2. The Sard theorem, the length formula, and consequences.
As in the case of surfaces, in order to prove the Jordan curve Theorem 9.14, we need a version of the Sard theorem for differentiable maps of $\mathbb{R}$ to $\mathbb{R}$. Of course, this one-dimensional version exists and an analogue to the proof made in Chapter 4 remains valid for this case. As a consequence, we will also dispose of a corresponding area formula—the length formula, in this case. A direct consequence of this formula is the following analogue to the Chern-Lashoff Theorem 5.29 proved in Chapter 5.

Proposition 9.21. *Let* $\alpha : \mathbb{R} \to \mathbb{R}^2$ *be a periodic plane curve p.b.a.l. with length* $L > 0$. *If* $k : \mathbb{R} \to \mathbb{R}$ *is its curvature function, then the following hold.*

A: $\int_0^L k^+(s)\, ds \geq 2\pi$, *where* $k^+ = \max\{0, k\}$.

B: $\int_0^L |k(s)|\, ds \geq 2\pi$ *and, if equality holds, then the curvature of the curve* α *is non-negative.*

With regard to this statement, it is worth noting that there are periodic plane curves with non-negative curvature everywhere with respect to the positive orientation for which equality in **B** does not happen, for example, the curve γ in Exercise 9.48.

9.3.3. The divergence theorem. We may also obtain, as in the case of compact simple curves, as a consequence of the length formula, the corresponding divergence theorem, this time referring to integrals over the plane domains in $\mathbb{R}^2$ determined by this class of curves. To do this, let us consider, as in the three-dimensional case, the following definition (compare with Definition 5.30):

Definition 9.22 (Divergence of a field). Let $A \subset \mathbb{R}^2$ and $X : A \to \mathbb{R}^2$ be a vector field defined on A. The *divergence* of X is the function div $X : A \to \mathbb{R}$ given by

$$(\text{div } X)(p) = \text{trace}\,(dX)_p.$$

Then, we have

$$(\text{div } X)(p) = \sum_{i=1}^{2} \langle (dX)_p(e_i), e_i \rangle,$$

where $\{e_1, e_2\}$ is any orthonormal basis of $\mathbb{R}^2$. If we choose $e_1 = (1,0), e_2 = (0,1)$, then

$$(dX)_p(e_1) = \left(\frac{\partial X^1}{\partial x}(p), \frac{\partial X^2}{\partial x}(p) \right) \quad \text{and} \quad (dX)_p(e_2) = \left(\frac{\partial X^1}{\partial y}(p), \frac{\partial X^2}{\partial y}(p) \right),$$

provided that $X = (X^1, X^2)$. Therefore,

$$\text{div } X = \frac{\partial X^1}{\partial x} + \frac{\partial X^2}{\partial y}$$

and, so, div X is a differentiable function.

Theorem 9.23 (Divergence theorem). *Let* C *be a compact connected simple curve in* $\mathbb{R}^2$. *Suppose that* Ω *is the inner domain determined by* C *and that* X *is a differentiable vector field defined on* $\overline{\Omega}$. *Then*

$$\int_{\Omega} \text{div } X = - \int_0^L \langle X(\alpha(s)), N(s) \rangle\, ds,$$

where $\alpha : \mathbb{R} \to \mathbb{R}^2$ is a positively oriented periodic curve p.b.a.l. with period $L > 0$ whose trace is the simple curve C.

Remark 9.24. We have given the divergence theorem for domains Ω determined by a compact connected simple curve in the plane. In fact, the proof is also valid for those bounded connected open subsets of the plane whose boundary consists of finitely many disjoint simple curves of this type. That is (compare with Remark 5.33), we have the following statement.

> *If Ω is a bounded connected open subset of $\mathbb{R}^2$ such that Bdry $\Omega = \bigcup_{i=1}^n C_i$, where each C_i is a compact connected simple curve, and X is a vector field defined on $\overline{\Omega}$, then*
>
> $$\int_\Omega \operatorname{div} X = -\sum_{i=1}^n \int_0^{L_i} \langle X(\alpha_i(s)), N_i(s) \rangle \, ds,$$
>
> *where each curve α_i p.b.a.l. orientates C_i in such a way that its normal vector $N_i(s)$ points to Ω for all $s \in [0, L_i]$.*

Remark 9.25. Let $\alpha : \mathbb{R} \to \mathbb{R}^2$ be a periodic parametrized curve with period $A > 0$, not necessarily p.b.a.l., whose trace is a simple curve. We know that, if $\phi : \mathbb{R} \to \mathbb{R}$ is the inverse diffeomorphism of the arc length function S of α, the curve $\beta = \alpha \circ \phi$ is a curve p.b.a.l. with length $L = S(A)$. If Ω is the inner domain determined by $\beta(\mathbb{R})$—or by $\alpha(\mathbb{R})$—and X is a field defined on $\overline{\Omega}$, one has, by the divergence theorem,

$$\int_\Omega \operatorname{div} X = -\int_0^L \langle X(\beta(s)), N_\beta(s) \rangle \, ds.$$

Now, taking into account that $\beta(s) = \alpha(\phi(s))$ and that $N_\beta(s) = J\beta'(s) = \phi'(s) J\alpha'(\phi(s))$ and applying the change of variables theorem to the integral on the right-hand side for the diffeomorphism $\phi : (0, L) \to (0, A)$, we infer that

$$\int_\Omega \operatorname{div} X = -\int_0^A \langle X(\alpha(t)), J\alpha'(t) \rangle \, dt,$$

which is the expression of the divergence theorem when we use an arbitrary parametrization for the given simple curve.

Example 9.26. Let $\alpha : \mathbb{R} \to \mathbb{R}^2$ be a periodic curve p.b.a.l. with length $L > 0$, whose trace is a simple curve, and let Ω be the inner domain determined by $\alpha(\mathbb{R})$. Consider the vector field given by $X(p) = p$, for all $p \in \overline{\Omega}$, which is clearly differentiable. Since $(dX)_p = I_{|\mathbb{R}^2}$ for all p, one has that div X is always 2. Applying the divergence theorem, we obtain that

$$A(\Omega) = -\frac{1}{2} \int_0^L \langle \alpha(s), N(s) \rangle \, ds,$$

where $A(\Omega)$ is the area of Ω.

Example 9.27. If α is a curve p.b.a.l. with length $L > 0$ which orientates positively a compact connected simple curve C and Ω is the inner domain of C, the vector field

$$X(p) = \frac{p - a}{|p - a|^2}$$

is differentiable on $\overline{\Omega}$ if $a \notin \overline{\Omega}$. One easily checks that

$$(dX)_p(v) = \frac{1}{|p - a|^2} v - 2 \frac{\langle p - a, v \rangle}{|p - a|^4} (p - a)$$

for all $v \in \mathbb{R}^2$. Hence, X has null divergence everywhere. As a consequence

$$\int_0^L \frac{\langle \alpha(s) - a, N(s) \rangle}{|\alpha(s) - a|^2} \, ds = 0, \qquad \forall a \notin \overline{\Omega}.$$

Now suppose that $a \in \Omega$. Since Ω is open, there exists $\varepsilon > 0$ such that the closed ball $\overline{B_\varepsilon(a)}$ with centre a and radius ε is contained in Ω. The set $\Omega' = \Omega - \overline{B_\varepsilon(a)}$ is a bounded connected open subset of the plane. Moreover, its boundary is $\alpha(\mathbb{R}) \cup \mathbb{S}^1_\varepsilon(a)$. Thus, Ω' satisfies the requirements of Remark 9.24. Since the vector field X defined earlier is differentiable on $\overline{\Omega'}$ and has divergence zero, we have

$$0 = \int_0^L \frac{\langle \alpha(s) - a, N(s) \rangle}{|\alpha(s) - a|^2} \, ds + \int_0^{2\pi\varepsilon} \frac{\langle \beta(s) - a, N_\beta(s) \rangle}{|\beta(s) - a|^2} \, ds,$$

where $\beta(s) = a + \varepsilon \left(\sin(s/\varepsilon), \cos(s/\varepsilon) \right)$ is the circle centred at a with radius ε oriented so that the normal vector $N_\beta(s) = (1/\varepsilon)(\beta(s) - a)$ points to Ω'. Computing the second integral explicitly, we get

$$\int_0^L \frac{\langle a - \alpha(s), N(s) \rangle}{|a - \alpha(s)|^2} \, ds = 2\pi, \qquad \forall a \in \Omega.$$

Notice that this equality is a two-dimensional version of the well-known *Gauss theorem* of electrostatics that we saw in Remark 5.34.

Exercise 9.28. ↑ Let α be an L-periodic plane curve p.b.a.l. whose trace is a simple curve, and let Ω be its inner domain. Apply the divergence theorem to prove that

$$\int_0^L \langle N(s), a \rangle \, ds = 0, \qquad \forall a \in \mathbb{R}^2.$$

Exercise 9.29. Let α be a regular periodic plane curve with period $A > 0$, not necessarily parametrized by the arc length, whose trace is a simple curve, and let Ω be its inner domain. If $a \notin \overline{\Omega}$ and $b \in \mathbb{R}^2$, let X be the differentiable field defined on $\overline{\Omega}$ by $X(p) = \langle p - a, b \rangle (p - a)/|p - a|^2$. Prove that

$$(\text{div } X)(p) = \frac{\langle p - a, b \rangle}{|p - a|^2}$$

for each $p \in \overline{\Omega}$. Show that

$$\int_\Omega \frac{\langle p - a, b \rangle}{|p - a|^2} \, dp = -\int_0^A \frac{\langle \alpha(t) - a, b \rangle \, \langle \alpha(t) - a, J\alpha'(t) \rangle}{|\alpha(t) - a|^2} \, dt,$$

using the divergence theorem and Remark 9.25.

Exercise 9.30. ↑ Let $A \subset \mathbb{R}^2$, and let $f, g : A \to \mathbb{R}$ be differentiable functions. Prove that div $J\nabla f = 0$ and that div $gJ\nabla f = \det(\nabla f, \nabla g)$.

Exercise 9.31. ↑ Let C be a compact connected simple plane curve and Ω its inner domain. Let $F = (F^1, F^2) : \overline{\Omega} \to \mathbb{R}^2$ be a differentiable map such that $F_{|C} = I_C$. Show that

$$\int_\Omega \operatorname{Jac} F = A(\Omega).$$

Exercise 9.32. Show from the previous exercise that, if Ω is the inner domain determined by a compact connected simple plane curve C, there exist no differentiable maps $F : \overline{\Omega} \to \mathbb{R}^2$ with $F_{|C} = I_C$ and $F(\Omega) \subset C$. That is, there are no differentiable retractions from $\overline{\Omega}$ to its boundary.

Exercises 9.30, 9.31, and 9.32 above, all derived from the divergence Theorem 9.23, provide a proof of the two-dimensional version of the Brouwer fix point theorem, following the guidelines of the three-dimensional case contained in Theorem 5.41.

9.3.4. The isoperimetric inequality. In the case of three-dimensional Euclidean space, the solution of the isoperimetric problem was postponed until the Alexandrov Theorem 6.17 was proved. The reason was that the equality in the Brunn-Minkowski inequality of Theorem 6.22 was not significant and we saw that surfaces attaining the minimum of the isoperimetric quotient have constant mean curvature, because they are critical points of the area with respect to preserving volume variations. The corresponding question to be solved now is: among all compact connected simple plane curves with prescribed length $L > 0$, which determines an inner domain with the greatest area? The solution to this problem—and to others of an analogous nature—is given by the isoperimetric inequality of the plane. As a first step to stating it, we point out that the same proof of Chapter 6, with suitable minor modifications, supplies us with the plane version of the Brunn-Minkowski inequality.

Theorem 9.33 (Brunn-Minkowski's inequality in the plane). *Let A and B be two bounded open subsets of the Euclidean plane $\mathbb{R}^2$. Then*

$$(\operatorname{area} A)^{1/2} + (\operatorname{area} B)^{1/2} \le (\operatorname{area}(A + B))^{1/2},$$

where the sum of subsets of $\mathbb{R}^2$ must be understood in the sense of Section 6.5.

This theorem, as mentioned before, can be utilized to answer the isoperimetric question for the plane. In this way, we obtain the following result.

Theorem 9.34 (Plane isoperimetric inequality). *Let C be a compact connected simple curve of length L and Ω its inner domain. Then*

$$4\pi A(\Omega) \leq L^2$$

and equality occurs if and only if C is a circle.

Proof. This proof is equal to the proof of Theorem 6.25 unless the following is true. Once the inequality is established for each curve C under our hypotheses, we succeed in concluding that any curve reaching the minimum isoperimetric quotient must have constant curvature. Then, to finish, we use Exercise 1.20 instead of the Alexandrov Theorem 6.17. $\qquad\square$

In the case of the plane that we are dealing with, many intelligent proofs of the isoperimetric inequality have been contrived. This is likely because this problem has been one of the most famous of geometry, due to its ancient beginnings, which even involves Greek and Phoenician myths, and its relatively recent solution. In Exercises (12)–(15) in this chapter, we will outline a proof by Hurwitz which does not utilizes the Brunn-Minkowski inequality. In the following remark, we will sketch another proof having the charm of being a mere consequence of the divergence theorem.

Remark 9.35. We will start by giving an integral expression for the area of the inner domain determined in the plane by a compact connected simple curve.

Lemma 9.36. *Let α be a periodic plane curve p.b.a.l. with length L whose trace is a simple curve, and let Ω be its inner domain. Then*

$$2\pi A(\Omega) = \int_0^L \int_0^L \frac{\langle \alpha(u) - \alpha(s), N(u)\rangle \langle \alpha(s) - \alpha(u), N(s)\rangle}{|\alpha(u) - \alpha(s)|^2} \, du \, ds.$$

Proof. In fact, let $\varepsilon > 0$ be small enough to consider the parallel curves—according to Exercise 9.19—$\alpha_t(s) = \alpha(s) + tN(s)$ with $s \in \mathbb{R}$ and $|t| < \varepsilon$. Then, each α_t is a periodic parametrized curve whose trace is a simple curve and such that $\alpha_t'(s) = \big(1 - tk(s)\big)T(s)$, where k is the curvature function of α. Let Ω_t be the inner domain determined by α_t. Since $\alpha_t(\mathbb{R}) \subset \Omega$, we have $\overline{\Omega}_t \subset \Omega$. Let us consider the continuous function $f : \overline{\Omega}_t \times [0, L] \to \mathbb{R}$ defined by

$$f(p, s) = \frac{\langle p - \alpha(s), N(s)\rangle}{|p - \alpha(s)|^2}.$$

By Example 9.27 after the divergence theorem, one has that

$$\int_0^L f(p, s) \, ds = 2\pi, \qquad \forall p \in \overline{\Omega}_t.$$

Integrating over Ω_t and applying the Fubini theorem of elementary calculus,

$$2\pi A(\Omega_t) = \int_0^L \int_{\Omega_t} f(p, s) \, dp \, ds, \qquad 0 < t < \varepsilon.$$

Utilizing Exercise 9.29 for the curve α_t with $a = \alpha(s)$ and $b = N(s)$, this equality becomes

$$2\pi A(\Omega_t) = -\int_0^L \int_0^L \frac{\langle \alpha_t(u) - \alpha(s), N(s) \rangle \, \langle \alpha_t(u) - \alpha(s), J\alpha_t'(u) \rangle}{|\alpha_t(u) - \alpha(s)|^2} \, du \, ds.$$

Notice that, using the Schwarz inequality, the integrand on the right-hand side has absolute value bounded by $1 + \varepsilon \max |k|$. Taking limits as $t \to 0$, the required equality follows from the dominated convergence theorem applied to the two sides of the equality. $\qquad \square$

The following exercises will be useful for completing the proof of the isoperimetric inequality in the plane as a consequence, only, of the divergence theorem.

Exercise 9.37. Let $u, v, w \in \mathbb{R}^2$. Prove that

$$-4 \langle u, v \rangle \langle u, w \rangle \le \langle u, v - w \rangle^2$$

and that equality holds if and only if u is perpendicular to $v + w$.

Exercise 9.38. ↑ Let $\alpha : I \to \mathbb{R}^2$ be a plane curve p.b.a.l. Suppose that, for each $s, u \in I$, we have $\langle \alpha(s) - \alpha(u), N(s) + N(u) \rangle = 0$. Demonstrate that α is a segment of a straight line or an arc of a circle. Interpret the above hypothesis geometrically.

From the integral formula obtained in Lemma 9.36 for the area of the domain Ω, along with Exercise 9.37 above, the inequality

$$2\pi A(\Omega) \le \frac{1}{4} \int_0^L \int_0^L \frac{\langle \alpha(u) - \alpha(s), N(u) - N(s) \rangle^2}{|\alpha(u) - \alpha(s)|^2} \, du \, ds$$

follows and equality is attained, according to Exercise 9.38, if and only if α is a circle. Using the Schwarz inequality, we have

$$2\pi A(\Omega) \le \frac{1}{4} \int_0^L \int_0^L |N(u) - N(s)|^2 \, du \, ds$$

$$= \frac{1}{2} \int_0^L \int_0^L \left(1 - \langle N(u), N(s) \rangle \right) du \, ds = \frac{1}{2} L^2,$$

where we have also taken Exercise 9.28 into account to get the last equality.

9.3.5. Positively curved simple curves. In Chapter 6, we studied some of the features of closed or compact surfaces with positive Gauss curvature and their connection to the theory of convex sets. With these precedents, it is natural to use the term *ovals* for the compact connected simple plane curves whose curvature does not vanish anywhere or, equivalently, that has positive curvature everywhere; recall that we are considering positive orientations and see Exercise 9.20. The concept of a *convex* curve is in relation to the above notion of an oval.

Definition 9.39 (Convex plane curve). A simple plane curve will be called convex when it lies entirely on the closed half-plane determined by the affine tangent line at each of its points. Also, it will be called *strictly convex* when it is convex and touches each of its affine tangent lines at only a point. In this sense, Exercise (8) of Chapter 1 says that an oval is always locally strictly convex and that each convex curve has non-negative curvature, with respect to the positive orientation. Note that it also makes sense to talk about convex curves in the context of parametrized curves, with this same definition.

As in the case of surfaces, we will be able to obtain results, not only on compact curves, but on curves closed in Euclidean space. If C is a connected simple curve closed in $\mathbb{R}^2$ with positive curvature everywhere, we will denote by Ω the domain of $\mathbb{R}^2 - C$ that the normal vector points to, and we will call it its *inner domain*. Some of the results that we showed about closed positively curved surfaces in the Hadamard-Stoker Theorems 6.1 and 6.5 and in the Bonnet Theorem 6.8 can also be proved, using the same methods, for simple plane curves with positive curvature. We summarize them in the following statement.

Theorem 9.40. *Let C be a connected simple plane curve with positive curvature everywhere which is closed as a subset of $\mathbb{R}^2$. Then the following hold.*

A: *Its inner domain Ω is convex.*

B: *C is strictly convex.*

C: *If C is compact with length $L > 0$ and $\alpha : \mathbb{R} \to \mathbb{R}^2$ is a parametrization b.a.l. of C, the normal vector $N : \mathbb{R} \to \mathbb{S}^1$ defines an L-periodic map injective on $[0, L)$.*

D: *If C is not compact and $\alpha : \mathbb{R} \to \mathbb{R}^2$ is a parametrization b.a.l. of C, then the normal vector $N : \mathbb{R} \to \mathbb{S}^1$ is injective and C is a graph over an open interval of a straight line of $\mathbb{R}^2$.*

E: *If $\inf_{p \in C} k(p) > 0$, then C is an oval.*

Exercise 9.41. ↑ Prove that the ellipse

$$C = \{(x,y) \in \mathbb{R}^2 \mid \frac{x^2}{a^2} + \frac{y^2}{b^2} = 1\}$$

with half-axes $a \geq b > 0$ is an oval.

Exercise 9.42. ↑ Show that the parabola

$$C = \{(x,y) \in \mathbb{R}^2 \mid y^2 = 2px\}$$

with $p > 0$ is a simple curve with positive curvature everywhere and is closed in $\mathbb{R}^2$.

Exercise 9.43. Show that, if C is an oval and R a straight line, we have $R \cap C = \emptyset$, or $R \cap C = \{p\}$ and R is tangent to C at p, or $R \cap C$ consists of exactly two points; see Exercise (6) of Chapter 6 and Remark 6.3.

Exercise 9.44. ↑ *The four-vertices theorem for ovals.* Let $\alpha : \mathbb{R} \to \mathbb{R}^2$ be a periodic curve p.b.a.l. of length $L > 0$ and curvature function k. Show, using the Frenet formulas, that

$$\int_0^L k'(s)\langle \alpha(s) - a, v \rangle \, ds = 0$$

for all $a, v \in \mathbb{R}^2$. Use this and the previous exercise to prove that, if the trace of α is an oval, then there exist at least four points in $[0, L)$ where the derivative of k vanishes.

9.4. Rotation index of plane curves

Represent by $\alpha : \mathbb{R} \to \mathbb{R}^2$ a periodic regular parametrized plane curve with period $A > 0$. Then the map $T : \mathbb{R} \to \mathbb{S}^1 \subset \mathbb{R}^2$ taking each $t \in \mathbb{R}$ to the unit tangent vector of the curve α at the time t, that is,

$$T(t) = \frac{\alpha'(t)}{|\alpha'(t)|},$$

provides us a differentiable map which is also A-periodic. We may, then, talk about its degree—see Appendix 9.7 and compare with Definition 8.1— $\deg T \in \mathbb{Z}$ which is, roughly speaking, the oriented number of times that the tangent vector of the curve α turns around the circle as we go along the trace of the curve exactly once. For this reason, we will call this integer the *rotation index* of the curve α and we will denote it by

$$i(\alpha) = \deg T = \deg \frac{\alpha'}{|\alpha'|}.$$

Since T is differentiable, this degree defining the rotation index can be calculated, by Proposition 9.74, by using

$$\operatorname{Jac} T = \det(T, T') = k|\alpha'|,$$

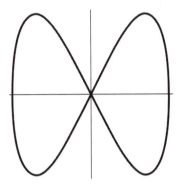

Figure 9.7. *Eight curve*

where k is the curvature of the regular curve α, as defined in Remark 1.15. Therefore

(9.1) $$2\pi\, i(\alpha) = \int_0^A k(t)|\alpha'(t)|\,dt = \int_0^L k(s)\,ds.$$

This formula closely resembles the Gauss-Bonnet formula proved in part **C** of Theorem 8.38 for compact surfaces in $\mathbb{R}^3$. As in the cases of surfaces, the right-hand side of this equality is usually called the *total curvature* of the curve α and we will see in the text that follows that the left-hand side also has a topological meaning. This equation could have been inferred, in a heuristic way, from Exercise 1.20.

In the following exercises, which one can easily solve from the results of Appendix 9.7, we will list most of the remarkable properties of the rotation index of a periodic regular plane curve.

Exercise 9.45. ↑ *(Invariance through increasing reparametrizations.)* If $f : \mathbb{R} \to \mathbb{R}$ is an increasing diffeomorphism such that $f(s + B) = f(s) + A$ for each $s \in \mathbb{R}$ and $\alpha : \mathbb{R} \to \mathbb{R}^2$ is an A-periodic regular curve, show that $\beta = \alpha \circ f$ is a B-periodic regular curve and that $i(\beta) = i(\alpha)$. If f is decreasing, then $i(\beta) = -i(\alpha)$.

Exercise 9.46. ↑ Two A-periodic, $A > 0$, regular curves $\alpha_0, \alpha_1 : \mathbb{R} \to \mathbb{R}^2$ are said to be *regularly homotopic* if there exists a continuous homotopy between them through periodic regular curves with the same period A. Prove that, if α_0 and α_1 are regularly homotopic, their rotation indices coincide.

Exercise 9.47. ↑ Prove that the rotation index of periodic regular plane curves is invariant under direct rigid motions and homotheties of $\mathbb{R}^2$ and that it changes sign under inverse rigid motions.

Exercise 9.48. ↑ Show that the rotation index of the circle

$$\alpha(t) = (r\cos t, r\sin t), \qquad \forall t \in \mathbb{R},$$

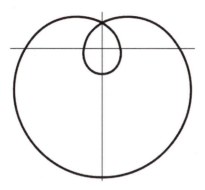

Figure 9.8. *Index two curve*

when considered as a 2π-periodic curve, is 1 and that the rotation indices of the curves

$$\beta(t) = (\cos t, \sin 2t), \qquad \forall t \in \mathbb{R},$$
$$\gamma(t) = ((1 - 2\sin t)\cos t, (1 - 2\sin t)\sin t + \tfrac{1}{2}), \qquad t \in \mathbb{R},$$

whose traces appear in Figures 9.7 and 9.8, are 0 and 2, respectively.

The topological contents of the rotation index of periodic regular plane curves will be plain in the following theorem.

Theorem 9.49 (Whitney-Grauenstein's theorem). *Let $\alpha_0, \alpha_1 : \mathbb{R} \to \mathbb{R}^2$ be two periodic regular curves. Then, their rotation indices coincide, i.e., $i(\alpha_0) = i(\alpha_1)$ if and only if, up to increasing reparametrizations, α_0 and α_1 are regularly homotopic, that is, there exists a homotopy between the periodic curves, suitably reparametrized, through regular curves.*

Proof. The equality of rotation indices is necessary because of Exercises 9.45 and 9.46, since the existence of a regular homotopy between the curves implies the existence of a homotopy between the respective tangent maps. Conversely, suppose that $i(\alpha_0) = i(\alpha_1)$. If we define $\beta : \mathbb{R} \to \mathbb{R}^2$ by

$$\beta(s) = \alpha_1\left(\frac{A_1}{A_0}s\right),$$

where A_0 and A_1 are the periods of α_0 and α_1, respectively, we have that β is an increasing reparametrization of α_1 which is A_0-periodic. By Exercise 9.45, its rotation index is the same as that of α_1. This shows that, without lost of generality, we may assume that the two periods coincide, i.e., that $A_0 = A_1$. We may also suppose that both curves have the same length $L > 0$ thanks to Exercise 9.47, which asserts that homothetic curves have the same rotation indices, and to the fact that two homothetic curves are always homotopic. Therefore, we may suppose that both are p.b.a.l., since the reparametrization by the arc length is an increasing reparametrization arising from another periodic curve. Thus, by the definition of the rotation index, the tangent maps $T_0 = \alpha_0'$ and $T_1 = \alpha_1'$ are L-periodic and have the same degree. Using Exercise 9.71 in Appendix 9.7, we check that α_0' and α_1' are homotopic as L-periodic curves taking values in the circle. Let $F = p \circ \tilde{F} : \mathbb{R} \times [0,1] \to \mathbb{S}^1$ be the homotopy that we have found, where $\tilde{F} = (1-t)\tilde{\alpha_0'} + t\tilde{\alpha_1'}$ and $\tilde{\alpha_0'}$ and $\tilde{\alpha_1'}$ are arbitrary lifts of α_0' and α_1', respectively. Let us define $H : \mathbb{R} \times [0,1] \to \mathbb{R}^2$ by

$$H(s,t) = \int_0^s F(u,t)\,du - \frac{s}{L}\int_0^L F(u,t)\,du, \qquad \forall (s,t) \in \mathbb{R} \times [0,1].$$

This map is continuous and satisfies

$$H(s,0) = \int_0^s F(u,0)\,du - \frac{s}{L}\int_0^L F(u,0)\,du = \alpha_0(s) - \alpha_0(0),$$

for all $s \in \mathbb{R}$, and, analogously,

$$H(s,1) = \alpha_1(s) - \alpha_1(0), \qquad \forall s \in \mathbb{R}.$$

Moreover, since F is L-periodic in its first variable,

$$H(s+L,t) - H(s,t) = \int_0^{s+L} F(u,t)\,du - \int_0^s F(u,t)\,du - \int_0^L F(u,t)\,du$$

$$= \int_s^{s+L} F(u,t)\,du - \int_0^L F(u,t)\,du = 0.$$

This shows that $\alpha_0 - \alpha_0(0)$ and $\alpha_1 - \alpha_1(0)$ are homotopic as L-periodic curves. From this, it is not difficult to see that α_0 and α_1 are homotopic as well. It remains to show that each L-periodic curve of the homotopy is regular. In fact,

$$\frac{d}{ds}H(s,t) = F(s,t) - \frac{1}{L}\int_0^L F(u,t)\,du,$$

for each $s \in \mathbb{R}$ and each $t \in [0,1]$. If, for some pair (s,t), the above derivative vanishes, since F takes its values in $\mathbb{S}^1$, the Schwarz inequality would force $u \mapsto F(u,t)$ to be constant, that is, one of the homotopy curves would be constant. From the uniqueness shown in Proposition 9.66, the curve $u \mapsto \tilde{F}(u,t)$ would also be constant and furthermore, by the invariance of degree under homotopies, one would have $\deg \alpha_0' = \deg \alpha_1' = 0$. Thus, $\tilde{\alpha_0'}$ and $\tilde{\alpha_1'}$ would be L-periodic functions and, so, would attain their minima. Suppose that, reparametrizing one of them if necessary by means of a translation, the two curves reach their minima at $s = 0$. Then, since $t \in [0,1]$, we have

$$\tilde{F}(s,t) = (1-t)\tilde{\alpha_0'}(s) + t\tilde{\alpha_1'}(s) \geq \tilde{F}(0,t).$$

Now then, since $\tilde{F}(s,t)$ is constant in s, it follows that both $\tilde{\alpha_0'}$ and $\tilde{\alpha_1'}$ are constant functions. As a consequence, the tangent maps α_0' and α_1' of the

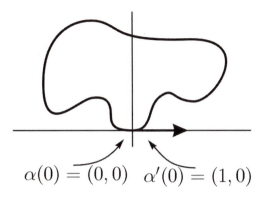

$$\alpha(0) = (0,0) \quad \alpha'(0) = (1,0)$$

Figure 9.9. *The curve α*

two curves are constant. Integrating the tangent maps, we would deduce that these two curves are straight lines and this contradicts the fact that they are periodic. Therefore, the homotopy that we found is through regular curves. ☐

The following theorem computes the rotation index for those regular parametrized plane curves whose trace is a simple curve, that is, whose trace has no self-intersections.

Theorem 9.50 (Umlaufsatz: turning tangent theorem). *Let $\alpha : \mathbb{R} \to \mathbb{R}^2$ be an L-periodic, $L > 0$, curve p.b.a.l. which is injective on $[0, L)$, i.e., whose trace is a simple curve. Then, the rotation index of α is either $+1$ or -1. Therefore*

$$\int_0^L k(s)\, ds = 2\pi$$

if k is the curvature of α and α is positively oriented.

Proof. We may assume, after applying a rigid motion of $\mathbb{R}^2$ to the curve α, that its trace $C = \alpha(\mathbb{R})$ belongs to the closed half-plane $\{(x, y) \in \mathbb{R}^2 \mid y \geq 0\}$, that the origin $(0,0)$ lies on C, and that the straight line $y = 0$ is the affine tangent line of the curve α at this point; see Figure 9.9. This process does not modify the rotation index, according to Exercise 9.47. Also using Exercise 9.45, we can consider that α is p.b.a.l. and that $\alpha(0) = (0,0)$ and $\alpha'(0) = (1,0)$. The reparametrizations that, if necessary, we should carry out would change, at most, the sign of $i(\alpha)$.

Once we have accepted these assumptions on the curve, we define a map $F : T \to \mathbb{S}^1$, where T is the triangle

$$T = \{(s_1, s_2) \in \mathbb{R}^2 \mid 0 \leq s_1 \leq s_2 \leq L\},$$

by means of the expression

$$F(s_1, s_2) = \begin{cases} \dfrac{\alpha(s_2) - \alpha(s_1)}{|\alpha(s_2) - \alpha(s_1)|} & \text{if } s_1 < s_2 \quad \text{and } (s_1, s_2) \neq (0, L), \\[2ex] \alpha'(t) & \text{if } s_1 = s_2 = s, \\[2ex] -\alpha'(0) & \text{if } (s_1, s_2) = (0, L), \end{cases}$$

which provides a well-defined map because α is injective on $[0, L)$. Let us check that this map F is continuous. It suffices to do this along the hypotenuse of T and at the opposite vertex $(0, L)$. In fact, if $(s_1, s_2) \in [0, L] \times [0, L]$,

(9.2)
$$\alpha(s_2) - \alpha(s_1) = \int_0^1 \frac{d}{dr} \alpha(s_1 + r(s_2 - s_1))\, dr = (s_2 - s_1) \int_0^1 \alpha'(s_1 + r(s_2 - s_1))\, dr.$$

Hence, if $s_1 \neq s_2$, we obtain that

$$F(s_1, s_2) = \frac{\displaystyle\int_0^1 \alpha'(s_1 + r(s_2 - s_1))\, dr}{\left| \displaystyle\int_0^1 \alpha'(s_1 + r(s_2 - s_1))\, dr \right|}.$$

Since the integral appearing in this expression depends continuously on s_1 and s_2, one has

$$\lim_{(s_1, s_2) \to (s,s)} F(s_1, s_2) = \frac{\displaystyle\int_0^1 \alpha'(s)\, dr}{\left| \displaystyle\int_0^1 \alpha'(s)\, dr \right|} = \alpha'(s),$$

for each $s \in [0, L]$, and, consequently, F is continuous along the hypotenuse of T. On the other hand, since α is L-periodic,

$$F(s_1, s_2) = \frac{\alpha(s_2 - L) - \alpha(s_1)}{|\alpha(s_2 - L) - \alpha(s_1)|}$$

when $(s_1, s_2) \in T$ and $s_1 < s_2$. Applying equality (9.2) here and observing that $s_2 - L - s_1 < 0$ if $(s_1, s_2) \neq (0, L)$, we deduce that

$$F(s_1, s_2) = -\frac{\displaystyle\int_0^1 \alpha'(s_1 + r(s_2 - L - s_1))\, dr}{\left| \displaystyle\int_0^1 \alpha'(s_1 + r(s_2 - L - s_1))\, dr \right|},$$

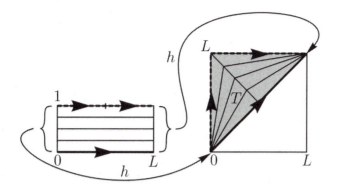

Figure 9.10. *The map h*

and consequently

$$\lim_{(s_1,s_2)\to(0,L)} F(s_1, s_2) = -\frac{\displaystyle\int_0^1 \alpha'(0)\,dr}{\left|\displaystyle\int_0^1 \alpha'(0)\,dr\right|} = -\alpha'(0).$$

Hence, F is also continuous at the vertex $(0, L)$ of the triangle T.

Now consider a continuous map $h : [0, L] \times [0, 1] \to T$ such that, as indicated in Figure 9.10,

$$h(s, 0) = (s, s) \quad \text{if } s \in [0, L],$$

$$h(s, 1) = \begin{cases} (0, 2s) & \text{if } s \in [0, L/2], \\ \\ (2s - L, L) & \text{if } s \in [L/2, L] \end{cases}$$

and that $h(0, t) = (0, 0)$ and $h(L, t) = (L, L)$ for each $t \in [0, 1]$. From this, the continuous map $H = F \circ h : [0, L] \times [0, 1] \to \mathbb{S}^1$ is a homotopy between the lap γ given by

$$s \in [0, L] \longmapsto \gamma(s) = H(s, 0) = \alpha'(s) \in \mathbb{S}^1$$

and the lap β, also taking values on the unit circle, given by

$$s \in [0, L] \longmapsto \beta(s) = H(s, 1) = \begin{cases} \alpha'(0) & \text{if } s = 0, \\[2mm] \dfrac{\alpha(2s)}{|\alpha(2s)|} & \text{if } s \in (0, L/2), \\[2mm] -\alpha'(0) & \text{if } s = L/2, \\[2mm] -\dfrac{\alpha(2s)}{|\alpha(2s)|} & \text{if } s \in (L/2, L), \\[2mm] \alpha'(0) & \text{if } s = L. \end{cases}$$

Moreover $H(0, t) = F(0, 0) = \alpha'(0)$ and $H(L, t) = F(L, L) = \alpha'(L) = \alpha'(0)$ for each $t \in [0, 1]$. Then, the definitions of the rotation index, degree of a periodic map to $\mathbb{S}^1$, and the invariance of the degree under homotopies (see Appendix 9.7) supply us with the equality

$$i(\alpha) = \deg \alpha' = \deg \alpha'_{|[0,L]} = \deg \gamma = \deg \beta.$$

Then, the proof will be finished if we show that the degree of the lap β is 1. To show this, let us call $f : [0, L] \to \mathbb{R}$ the lift of β such that $f(0) = 0$; recall that $\beta(0) = \alpha'(0) = (1, 0)$. So, we obtain that $f(s) \leq \pi$ for $0 < s < L/2$ since on this interval $p(f(s)) = \beta(s) = \alpha(2s)/|\alpha(2s)|$ and α has the trace over the x-axis, that is, its second component is always greater than or equal to zero. This, together with the fact that $p(f(L/2)) = \beta(L/2) = -\alpha'(0) = (-1, 0)$, implies that $f(L/2) = \pi$. Reasoning in a similar way on the interval $(L/2, L)$ and, taking into account that β, on this interval, takes exactly the opposite values, one gets that $f(L) = 2\pi$. Hence

$$i(\alpha) = \deg \beta = \frac{1}{2\pi}[f(L) - f(0)] = 1,$$

as we expected.   □

An obvious consequence of the turning tangent theorem, combined with the Whitney-Grauenstein Theorem 9.49, is that each compact simple plane curve is regularly homotopic to a circle. Another result is an improvement of Proposition 9.21, where we succeed in characterizing the equality in the integral inequality that we got there.

Corollary 9.51. *Let $\alpha : \mathbb{R} \to \mathbb{R}^2$ be a periodic curve p.b.a.l. with length $L > 0$ whose trace is a simple curve. If k is its curvature function, then*

$$\int_0^L |k(s)| \, ds \geq 2\pi$$

and equality is attained if and only if $k(s) \geq 0$ for all $s \in \mathbb{R}$; recall that we presuppose the positive orientation.

A third consequence is the characterization of convex simple plane curves in terms of its curvature.

Corollary 9.52. *Let $\alpha : \mathbb{R} \to \mathbb{R}^2$ be a periodic curve p.b.a.l. whose trace $C = \alpha(\mathbb{R})$ is a simple curve, and let k be its curvature. Then C is convex if and only if $k \geq 0$.*

Proof. The condition about the curvature is necessary, as we already pointed out in Definition 9.39, because of Exercise (8) of Chapter 1. Conversely, suppose that $k(s) \geq 0$ for all $s \in \mathbb{R}$. We will see in (9.4) of Appendix 9.7 that

$$\tilde{\alpha}'(s) = \int_{s_0}^{s} k(t)\, dt$$

is a lift of α', for every $s_0 \in \mathbb{R}$. Hence, since $k \geq 0$, this lift is non-decreasing. Moreover, by the turning tangent Theorem 9.50,

$$1 = \deg \alpha' = \frac{1}{2\pi}[\tilde{\alpha}'(s_0 + L) - \tilde{\alpha}'(s_0)] = \frac{1}{2\pi}\tilde{\alpha}'(s_0 + L),$$

that is, $\tilde{\alpha}'$ takes the interval $[s_0, s_0 + L)$ to the interval $[0, 2\pi)$, on which $p : \mathbb{R} \to \mathbb{S}^1$ is injective. Thus, our hypotheses on the curve imply that, if there exist $s_1, s_2 \in [0, L)$ such that $s_1 < s_2$ with $\alpha'(s_1) = \alpha'(s_2)$, then α' is constant on the interval $[s_1, s_2]$. On the other hand, if C is not convex, then there is a point $\alpha(s_0) \in C$ with $s_0 \in [0, L)$ with the property that its affine tangent line separates the points of C. As a consequence, the height function $f : \mathbb{R} \to \mathbb{R}$ given by

$$f(s) = \langle \alpha(s) - \alpha(s_0), N(s_0) \rangle$$

reaches its maximum and its minimum at different points and at points different from $\alpha(s_0)$, namely, at $\alpha(s_1)$ and at $\alpha(s_2)$, respectively, where $s_1, s_2 \in [0, L)$ and $s_0 \neq s_1$, $s_0 \neq s_2$, and $s_1 \neq s_2$. Since s_1 and s_2 are critical points for f, we have $f'(s_1) = f'(s_2) = 0$, that is, the vectors $\alpha'(s_1)$ and $\alpha'(s_2)$ are either perpendicular to $N(s_0)$ or they are proportional to $\alpha'(s_0)$. Therefore, two of the three tangent vectors of α at s_0, s_1, and s_2 coincide. From the property that we proved at the beginning for the curve α, α' should be constant on a closed interval whose end points are two of three numbers s_0, s_1, and s_2. This would lead, on this interval, to the height function f being constant, and this contradicts the choice that we made of the three points of the curve. So, the curve C must be convex. □

Remark 9.53. If we go over the above proof again, we can observe the fact that the trace of the curve α is a simple curve, that is, the fact that α has no self-intersections, has been invoked only when we used that the rotation index of α—positively oriented—is 1. For this reason, we have really proved the following statement.

> *A periodic regular parametrized plane curve with rotation index ± 1 is convex if and only if its curvature does not change sign.*

9.5. Periodic space curves

In this section, we will continue, in some sense, the study that we carried out on the total curvature of periodic plane curves in the previous section. Now, we want to consider the total curvature of skew curves, that is, of curves whose trace is not, in general, contained in any plane of $\mathbb{R}^3$. We will see that, also in this case, this global geometrical quantity associated to the curve includes some topological information. For plane curves, this total curvature controlled the regular homotopy class. Here, it is also related to the manner in which the curve embeds in three-space. We will start with a property of periodic curves whose trace is not too long and which is contained in the unit sphere, which will be applied later to the tangent curve of any periodic regular curve.

Proposition 9.54. *Let $\gamma : \mathbb{R} \to \mathbb{S}^2$ be a periodic parametrized—not necessarily regular—spherical curve with length L. The following assertions are true.*

> **A:** *If $L < 2\pi$, then the trace of γ is included in some open hemisphere of the sphere $\mathbb{S}^2$.*
>
> **B:** *If $L = 2\pi$, then either the trace of γ is contained in some open hemisphere or it is the union of two great half-circles with the same end points.*

Proof. Part **A**. We may consider, of course, that $L > 0$. If $A > 0$ is the period of γ and $S : [0, A] \to \mathbb{R}$ is its arc length function with $S(0) = 0$, we have that $S(A) = L$. So, since S is continuous, there must exist some $x \in (0, A)$ such that

$$S(x) = L_0^x(\gamma) = \frac{L}{2} = L_x^A(\gamma).$$

Thus, if d is the intrinsic distance (consult Exercise (5) of Chapter 7) on the unit sphere, one has

$$d(\gamma(0), \gamma(x)) \leq L_0^x(\gamma) < \pi,$$

and, consequently (see Exercise (22) of Chapter 7) the points $\gamma(0)$ and $\gamma(x)$ are not antipodal. Then, we may, according to Exercise (23) of Chapter 7, dispose of its middle point $m \in \mathbb{S}^2$. Now let $s \in [0, A)$ be such that the corresponding point of the curve satisfies $d(m, \gamma(s)) < \pi/2$. Then, if we represent by f the 180 degrees rotation around the point m of $\mathbb{S}^2$, the point m is also the middle point of $\gamma(s)$ and $f(\gamma(s))$. Hence

$$d(\gamma(s), f(\gamma(s))) = 2d(m, \gamma(s))$$

and, so, we have

$$2d(m, \gamma(s)) \leq d(\gamma(s), \gamma(x)) + d(\gamma(x), f(\gamma(s))).$$

But f is an involutive isometry, and so

$$d(\gamma(x), f(\gamma(s))) = d(f(\gamma(x)), \gamma(s)) = d(\gamma(0), \gamma(s)),$$

since m is the middle point of $\gamma(0)$ and $\gamma(x)$. Combining the previous inequalities, we deduce that

$$2d(m, \gamma(s)) \leq d(\gamma(0), \gamma(s)) + d(\gamma(s), \gamma(x)) \leq L_0^s(\gamma) + L_s^x(\gamma) = L_0^x(\gamma) = \frac{L}{2}.$$

We have proved, then, that

$$d(m, \gamma(s)) < \frac{\pi}{2} \implies d(m, \gamma(s)) \leq \frac{L}{4}, \qquad \forall s \in \mathbb{R}.$$

Since, by hypothesis, $L/4 < \pi/2$, it follows that, for each $s \in \mathbb{R}$,

$$\text{either} \quad d(m, \gamma(s)) \leq \frac{L}{4} < \frac{\pi}{2} \quad \text{or} \quad d(m, \gamma(s)) \geq \frac{\pi}{2}.$$

Since the function $s \mapsto d(m, \gamma(s))$ is continuous and

$$d(m, \gamma(0)) = \frac{1}{2}d(\gamma(0), \gamma(x)) \leq \frac{1}{2}L_0^x(\gamma) = \frac{L}{4},$$

only the first alternative is admissible. Therefore

$$d(m, \gamma(s)) \leq \frac{L}{4} < \frac{\pi}{2}, \qquad \forall s \in \mathbb{R},$$

that is, the trace of γ is contained in the open hemisphere whose pole is the point m, as we wanted to prove.

Part **B**. Now, we modify our assumption and assume that $L = L_0^A(\gamma) = 2\pi$. Suppose, first, that, in the trace of γ, there are two antipodal points of $\mathbb{S}^2$. For instance, imagine that $\gamma(0)$ and $\gamma(x)$ are antipodal for some $x \in (0, A)$. Then, utilizing Exercise (22) of Chapter 7 again, we obtain

$$\pi = d(\gamma(0), \gamma(x)) \leq L_0^x(\gamma) \quad \text{and} \quad \pi = d(\gamma(x), \gamma(A)) \leq L_x^A(\gamma).$$

Adding these two inequalities, we have

$$2\pi \leq L_0^x(\gamma) + L_x^A(\gamma) = L_0^A(\gamma) = 2\pi,$$

and so the curve γ, restricted to the subintervals $[0, x]$ and $[x, A]$, minimizes the length between its end points. We apply Theorem 7.42 and deduce that the trace of γ on each of these intervals coincides with the trace of a geodesic of length π, that is, with a great half-circle. Then, in this case, the trace of

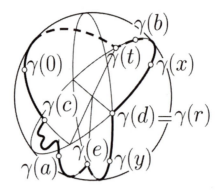

Figure 9.11. *The trace of* γ

γ is the union of two great half-circles with the same end points, which is one of the two possibilities expected in **B**.

Suppose, instead, that γ does not pass through any pair of antipodal points. We will prove that, in this case, it is possible to find two numbers a and b such that $a < b < a + A$, $L_a^b(\gamma) = L_b^{a+A}(\gamma) = \pi$, and

$$d(\gamma(a), \gamma(s)) + d(\gamma(s), \gamma(b)) < \pi, \qquad \forall s \in \mathbb{R}.$$

To show this, let us start by taking, as in the previous case, some $x \in (0, A)$ such that

$$L_0^x(\gamma) = L_x^A(\gamma) = \frac{L}{2} = \pi.$$

If the pair $0, x$ does not serve as the required pair a, b, then there exists $t \in [0, A]$ such that

$$\pi \le d(\gamma(0), \gamma(t)) + d(\gamma(t), \gamma(x)).$$

Notice that $t \in (0, x) \cup (x, A)$ because $\gamma(0)$ and $\gamma(x)$ are not antipodal. We will assume, without lost of generality, that $t \in (0, x)$. On the other hand, since

$$d(\gamma(0), \gamma(t)) + d(\gamma(t), \gamma(x)) \le L_0^t(\gamma) + L_t^x(\gamma) = L_0^x(\gamma) = \pi,$$

we see that γ, restricted to $[0, t]$ and to $[t, x]$, minimizes the length between its end points. Again Theorem 7.42 tells us that the trace of γ, on these intervals, is contained in two arcs of great circles which must be different because, otherwise, since the sum of its lengths does not exceed π, we would have, from Exercise (11) of Chapter 7, that

$$d(\gamma(0), \gamma(x)) = d(\gamma(0), \gamma(t)) + d(\gamma(t), \gamma(x)) = \pi,$$

which is impossible. Thus, the arc of a great circle joining $\gamma(0)$ and $\gamma(t)$ and belonging to the trace of γ finishes at $\gamma(t)$, but it could likely start before $\gamma(0)$. Suppose that, in fact, it starts at $\gamma(c)$ for some $c \le 0$. Hence, $\gamma_{|[c,t]}$ is an arc of a great circle which cannot be extended inside of the trace of

γ and which must have length less than π because $\gamma(c)$ and $\gamma(t)$ cannot be antipodal. In the same way, the arc of the great circle joining $\gamma(t)$ and $\gamma(x)$ inside of the trace of γ could not finish at the point $\gamma(x)$. Assume that this goes on until a point $\gamma(d)$, with $d \in [x, A)$. Then, $\gamma_{|[t,d]}$ is also an arc of a great circle with length less than π; see Figure 9.11. Now, we repeat the same argument replacing $\gamma(0) = \gamma(A)$ by $\gamma(t) = \gamma(t + A)$ and we find $y \in (t, t + A)$ such that

$$L_t^y(\gamma) = L_y^{t+A}(\gamma) = \frac{L}{2} = \pi.$$

Notice that it must necessarily occur that $y > d$. If the pair t, y is not the required pair a, b either, then there exists $r \in (t, y) \cup (y, t + A)$ playing the role that t played before. Suppose, without lost again of generality, that r is in (t, y). Consequently, the traces of $\gamma_{|[t,r]}$ and $\gamma_{|[r,y]}$ are arcs of different great circles. So, we have no alternative but $\gamma(r) = \gamma(d)$. As before, the arc of the great circle joining $\gamma(r) = \gamma(d)$ and $\gamma(y)$ can be extended, inside of the trace of γ, until a last point $\gamma(e)$ for some $e \in [y, c + A]$, in such a manner that the arc of the great circle joining $\gamma(d)$ and $\gamma(e)$ also has length less than π. If $e = c + A$, we would have a spherical triangle $\gamma(c), \gamma(t), \gamma(d)$ with perimeter 2π and this would force its three sides to be parts of the same great circle. Therefore, $e < c + A$. Then, we may choose a number $a \in (e, c + A)$ and repeat the same above discussion a third time this time replacing $\gamma(0)$ by $\gamma(a)$. So, we find $b \in (a, a + A)$ such that

$$L_a^b(\gamma) = L_b^{a+A}(\gamma) = \frac{L}{2} = \pi.$$

It is necessary that $b \geq t + A$ because, otherwise,

$$L_b^{a+A}(\gamma) > L_{t+A}^{y+A}(\gamma) = L_t^y(\gamma) = \pi.$$

For the same reason $b \leq x + A$. From what we already know about the trace of γ, neither the trace of $\gamma_{|[a,b]}$ nor the trace of $\gamma_{|[b,a+A]}$ can be the union of two arcs of great circles. Therefore

$$d(\gamma(a), \gamma(s)) + d(\gamma(s), \gamma(b)) < \pi$$

for all $s \in \mathbb{R}$, as required.

We also know that the points $\gamma(a)$ and $\gamma(b)$ are not antipodal and, then, as in the proof of part **A**, we may consider its middle point $m \in \mathbb{S}^2$. Suppose that a point $\gamma(s)$, $s \in \mathbb{R}$, of the trace of γ is on the great circle with pole m, that is, $d(m, \gamma(s)) = \pi/2$, and let f be the 180 degree rotation around the point m. Taking Exercise (11) of Chapter 7 into account, we may write

$$\pi = 2d(m, \gamma(s)) = d(m, \gamma(s)) + d(m, f(\gamma(s))) = d(\gamma(s), f(\gamma(s))).$$

But, since m is the middle point of $\gamma(a)$ and $\gamma(b)$,

$$\pi \leq d(\gamma(s), \gamma(b)) + d(\gamma(b), f(\gamma(s))) = d(\gamma(s), \gamma(b)) + d(\gamma(a), \gamma(s)) < \pi,$$

where we have used the particular choice of a and b. This contradiction implies that the continuous function $s \mapsto d(m, \gamma(s))$ never takes the value $\pi/2$. Since, moreover, $d(m, \gamma(a)) < \pi/2$ because $\gamma(a)$ and $\gamma(b)$ are not antipodal, we conclude that

$$d(m, \gamma(s)) < \frac{\pi}{2}, \qquad \forall s \in \mathbb{R},$$

that is, the trace of γ is contained in the open hemisphere whose pole is the point m.   □

Remark 9.55. Proposition 9.54 is still valid for Lipschitz curves because we only required this type of regularity in Theorem 7.42. In fact, for regular curves, we could have done a quite easier proof and the second alternative of part **B** would not be admissible, unless the two arcs were in the same great circle. The important point is that it can be applied to curves that are *not necessarily regular*. This will permit us to prove the Fenchel theorem for any regular curve in $\mathbb{R}^3$, without the usual assumption that the curve has positive curvature everywhere.

Now suppose that $\alpha : \mathbb{R} \to \mathbb{R}^3$ is an A-periodic, $A > 0$, regular curve. Then, the tangent map $T : \mathbb{R} \to \mathbb{R}^3$ which takes each $s \in \mathbb{R}$ to the unit tangent vector of α at s, that is,

$$T(s) = \frac{\alpha'(s)}{|\alpha'(s)|}, \qquad \forall s \in \mathbb{R},$$

is a differentiable map of $\mathbb{R}$ to $\mathbb{S}^2$ which is A-periodic as well. Moreover, note that, for each $a \in \mathbb{S}^2$,

$$\int_0^A \langle \alpha'(s), a \rangle \, ds = \int_0^A \langle \alpha(s), a \rangle' \, ds = 0$$

and, hence, the trace of T cannot be contained in the open hemisphere with pole a and, if it were contained in the corresponding closed hemisphere, the function $\langle T, a \rangle = \langle \alpha', a \rangle / |\alpha'|$ would not change its sign and would be, in this way, identically zero, and, consequently, α would be a plane curve. That is, if α is a periodic regular curve, the trace of its unit tangent map T on the unit sphere is not contained in any open hemisphere and, if it is not a plane curve, in any closed hemisphere. From this information, we may state the following result concerning the total curvature of periodic regular curves in $\mathbb{R}^3$.

Theorem 9.56 (Fenchel's theorem). *If $\alpha : \mathbb{R} \to \mathbb{R}^3$ is an A-periodic regular curve, its total curvature satisfies*

$$\int_0^A k(t) \, dt \geq 2\pi$$

and equality is reached if and only if α is a convex plane curve.

Proof. The discussion before the statement implies that we may apply Proposition 9.54 to the curve T of the unit tangent vectors of α. Thus, since we have seen that the trace of T is not included in any open hemisphere of $\mathbb{S}^2$, the length of T has to be greater than or equal to 2π, that is,

$$2\pi \le L_0^A(T) = \int_0^A |T'(s)|\,ds = \int_0^A k(s)\,ds$$

and, if equality occurs, then the trace of T is the union of two great half-circles. In this case, it will be contained in a closed hemisphere and so α is a plane curve whose curvature, as a plane curve, does not change its sign, according to part **B** of Proposition 9.21. Moreover, the rotation index of α is ± 1, because it has total curvature 2π. Hence, we may apply Remark 9.53 and conclude that α is convex. $\qquad\square$

We will proceed by stating and proving an integral formula by Crofton, which measures the total number of intersections that a periodic regular curve whose trace is in the unit sphere has with its great circles. After stating this, one can guess that the corresponding proof must be closely related to Proposition 9.54.

Theorem 9.57 (Crofton's formula). *Let $\gamma : \mathbb{R} \to \mathbb{S}^2$ be an A-periodic regular curve. Then, the function $n_\gamma : \mathbb{S}^2 \to \mathbb{N} \cup \{\infty\}$ given by*

$$n_\gamma(a) = \#\{s \in [0, A) \mid \langle \gamma(s), a \rangle = 0\}, \qquad \forall a \in \mathbb{S}^2,$$

which measures the number of points—with their multiplicity—intersecting each great circle of the sphere, is integrable and moreover

$$\int_{\mathbb{S}^2} n_\gamma(a)\,da = 4\,L_0^A(\gamma).$$

Proof. Let us represent by O the open rectangle of $\mathbb{R}^2$ given by $(0, A) \times (0, 2\pi)$, and let $\phi : O \to \mathbb{S}^2$ be the differentiable map given by

$$\phi(s, t) = \cos t\, T(s) + \sin t\, T(s) \wedge \gamma(s),$$

where $(s, t) \in O$ and $T(s)$ is the unit tangent vector of the curve γ at time s, that is, $T(s) = \gamma'(s)/|\gamma'(s)|$, which is a well-defined map because γ is regular. Now, we apply the area formula of Theorem 5.28 to that map ϕ and to the function $f : O \to \mathbb{R}$ constantly $+1$ and deduce that the function $n(\phi, 1)$ is integrable on O and that

$$\int_O |\mathrm{Jac}\,\phi| = \int_{\mathbb{S}^2} n(\phi, 1).$$

Bearing in mind the definition of the absolute value of the Jacobian given in Section 5.2 or in the exercises cited there, we have

$$|\mathrm{Jac}\,\phi|(s, t) = |\det(\phi_s, \phi_t, \phi)|(s, t) = |\cos t|\,|\gamma'(s)|.$$

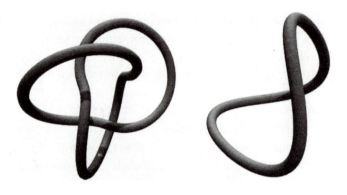

Figure 9.12. *Knotted and unknotted curves*

Therefore, using the Fubini theorem of calculus,

$$\int_O |\mathrm{Jac}\,\phi| = \int_0^A \int_0^{2\pi} |\cos t|\,|\gamma'(s)|\,dt\,ds = 4\,L_0^A(\gamma).$$

Thus, it is enough to show that the functions $n(\phi, 1)$, defined in Chapter 5 to give the area formula, and n_γ, defined in the statement that we want to prove, coincide except on a subset of measure zero in $\mathbb{S}^2$. To do this, it suffices to show that

$$n(\phi, 1)(a) = n_\gamma(a)$$

for each $a \in \mathbb{S}^2$ which is a regular value for ϕ (see the Sard Theorem 4.33) and such that $\langle \gamma(0), a \rangle \neq 0$—because, with this restriction, we eliminate only a great circle. Then take a point $a \in \mathbb{S}^2$ with these characteristics. We know that, in this situation, the set $\phi^{-1}(\{a\})$ is finite and that, taking the definition of ϕ and the fact that $\langle \gamma(0), a \rangle \neq 0$ into account,

$$n(\phi, 1)(a) = \#\phi^{-1}(\{a\}) = \#\{(s,t) \in [0, A] \times (0, 2\pi) \mid \langle \gamma(s), a \rangle = 0\}.$$

It only remains to show that, for the last set on the right-hand side whose cardinal we want to compute, the projection on the first component is bijective. This is true because, for each $s \in \mathbb{R}$, the map

$$t \in [0, 2\pi) \longmapsto \phi(s, t) = \cos t\, T(s) + \sin t\, T(s) \wedge \gamma(s)$$

is injective.   □

Remark 9.58. From the Crofton formula, we can deduce a proof of the inequality provided in the Fenchel Theorem 9.56, valid only for curves with positive curvature everywhere. In fact, it is enough to realize that, if the curve is plane, the desired inequality is already shown in Proposition 9.21 and, if not, then $n_T \geq 2$ because T is not contained in any open hemisphere and this implies, by the periodicity of T, that, for each $a \in \mathbb{S}^2$, we have $\langle T, a \rangle = 0$ at least two times in $[0, A]$.

The Crofton formula will allow us to show that, for a periodic curve, whose trace is a simple curve, to be *knotted* in $\mathbb{R}^3$, it is necessary that the curve bend at least twice the amount that we would expect for a general curve; see Figure 9.12. Before stating the corresponding result, proved first by Fary and Milnor, we need to make clear what we mean when we say that a curves is knotted. Let C be a compact connected simple curve in $\mathbb{R}^3$. We will say that C is knotted or that it is a non-trivial *knot* when we cannot find any injective continuous map from the closed unit disc of $\mathbb{R}^2$ to $\mathbb{R}^3$ whose restriction to the frontier circle is a homeomorphism onto C. Otherwise, we will say that the curve C is unknotted.

Theorem 9.59 (Fary-Milnor's theorem). *Let $\alpha : \mathbb{R} \to \mathbb{R}^3$ be an A-periodic regular curve with positive curvature everywhere such that its trace is a simple curve C. Then, if C is a knot, we have*

$$\int_0^A k(s)\,ds \geq 4\pi,$$

where k is the curvature function of α.

Proof. Assume without lost of generality that α is p.b.a.l. Since the curvature k of α does not vanish, the tangent curve $T : \mathbb{R} \to \mathbb{S}^2$ is regular and A-periodic. We may, then, apply to it the above Crofton formula in Theorem 9.57 and obtain

$$\int_0^A k(s)\,ds = L_0^A(T) = \frac{1}{4}\int_{\mathbb{S}^2} n_T(a)\,da,$$

with n_T defined as in the statement of the Crofton theorem. Suppose that the conclusion of the theorem were not true. In this case, we would find a point $a \in \mathbb{S}^2$ such that $n_T(a) < 4$. That is, the function

$$s \in [0, A) \longmapsto \langle T(s), a \rangle = \langle \gamma, a \rangle'(s)$$

has at most three zeroes, or, in other words, the function

$$s \in [0, A) \longmapsto h(s) = \langle \gamma, a \rangle(s)$$

has at most three critical points. Since h is continuous and A-periodic, it attains its maximum and its minimum in $[0, A)$. If they were reached at the same time, then they would coincide and, so, α would be a plane curve and, thus, since its curvature does not vanish, it would be an oval. In this case, its trace C could not be a knot, by part **A** of Theorem 9.40 about ovals. Hence, the maximum and the minimum are attained at different times and take different values $r_2 > r_1$. Then, for each $r \in (r_1, r_2)$, the function $h - r$ has exactly two zeroes, because, otherwise, the Cauchy mean value theorem would supply two different critical points of h also different from the absolute maximum and minimum. Since the curve has a simple trace, each plane of

$\mathbb{R}^3$ whose height is between r_1 and r_2 cuts C at exactly two points. Let σ_r be the closed segment of $\mathbb{R}^3$ joining these two points at height r. We may construct an injective continuous map from a closed disc of $\mathbb{R}^2$ with diameter $r_2 - r_1$ to $\mathbb{R}^3$ taking each horizontal segment of the disc to each of those of σ_r. This contradicts the fact that C was assumed to be a knot. $\square$

9.6. The four-vertices theorem

The four-vertices theorem asserts that a compact connected simple plane curve C has at least four vertices. We use the term *vertex* of C for any point $\alpha(s) \in C$ for which the curvature satisfies $k'(s) = 0$, where $\alpha : \mathbb{R} \to \mathbb{R}^2$ is any periodic regular parametrization of C. It is easy to be convinced, according to the notion of curvature in Remark 1.15, that this definition of a vertex does not depend upon the utilized parametrization. Thus, for instance, it is not difficult to check that an ellipse which is not a circle has exactly four vertices at its intersections with its axes. A first version of this theorem appeared as early as in 1909—as we mentioned in the introduction of this chapter—for convex curves; see Exercise (4) of this chapter. In this context, the theorem provides a sufficient condition for a positive periodic function to be the curvature of a simple curve. We will prove here the theorem for simple curves not necessarily convex following an idea due to Osserman. Consider the circle circumscribed to the curve and count the points of contact with the curve. Between each two contact points, there are necessarily local extremes of the curvature and, so, vertices, in such a way that a curve with exactly four vertices can have only two contact points with the circumscribed circle. We will start by studying the circle circumscribed to any compact subset of the plane.

Proposition 9.60. *Let E be a compact subset of $\mathbb{R}^2$ with more than one point. Let $\mathcal{C}_E$ be the family of all the circles of the plane such that the corresponding discs contain E. There exists in $\mathcal{C}_E$ a unique circle C with the least radius which will be called the* circumscribed circle *of E.*

Proof. Let $F : \mathbb{R}^2 \to \mathbb{R}^+$ be the function given by

$$F(p) = \max_{x \in E} |x - p|, \qquad \forall p \in \mathbb{R}^2,$$

which is a well-defined function because E is compact. If, for some p, we had that $F(p) = 0$, then $E \subset \{p\}$, which is impossible by the hypothesis. On the other hand, if $p, q \in \mathbb{R}^2$, then

$$|x - q| \le F(p) + |p - q| \quad \text{and} \quad |x - p| \le F(q) + |p - q|$$

for each $x \in E$. Taking maxima, we obtain

$$|F(p) - F(q)| \le |p - q|, \qquad \forall p, q \in \mathbb{R}^2.$$

In particular, F is continuous. Furthermore, if $C(p, F(p))$ is the circle centred at p with radius $F(p)$, we have that, since

$$|x - p| \leq F(p), \qquad \forall x \in E,$$

$E \subset C(p, F(p))$—we will deliberately confuse circles and discs—and so

$$C(p, F(p)) \in \mathcal{C}_E \qquad \forall p \in \mathbb{R}^2.$$

Also, if $C \in \mathcal{C}_E$ has radius $R > 0$, we can find a point $x_0 \in E$ with $|x_0 - p| = F(p)$ and so

$$F(p) = |x_0 - p| \leq R.$$

Thus, $C(p, F(p))$ is the least circle centred at $p \in \mathbb{R}^2$ containing E.

Now let $x_0 \in E$ such that $F(0) = \max_{x \in E} |x| = |x_0|$, and let $p \in \mathbb{R}^2$ such that $|p| \geq 2|x_0|$. Then, for each $x \in E$ we have

$$|p - x| \geq |p| - |x| \geq 2|x_0| - |x_0| = F(0).$$

Taking maximum relative to x, we get

$$F(p) \geq F(0) \quad \text{if } |p| \geq R_0 = 2|x_0| > 0.$$

By continuity, F attains its minimum in the closed ball $B_{R_0}(0)$ at some point p_0. Therefore

$$F(p_0) \leq \begin{cases} F(p) & \text{if } |p| \leq R_0, \\ \\ F(0) \leq F(p) & \text{if } |p| \geq R_0. \end{cases}$$

Hence, F has an absolute minimum at the point p_0. That is, the circle $C = C(p_0, F(p_0))$ is a circle of the plane containing E and it has minimum radius among all the circles including it.

If there were two different circles C_1 and C_2 with minimum radius $F(p_0)$ containing E, then we would have $E \subset C_1 \cap C_2$. But the intersection of two circles of the same radius and different centres is always contained in another circle C_3 with smaller radius. So, we would deduce that $E \subset C_3$, which is impossible. $\qquad\square$

Another fundamental property of circles circumscribed to compact sets is, besides its uniqueness, the way in which they touch the set.

Proposition 9.61. *Let E be a compact subset of the plane with more than one point, and let C be its circumscribed circle. Then, there are points of E on each closed half-circle of C.*

Proof. Suppose that C is centred at the origin and that its radius is $R > 0$. Consider the two closed half-circles S_1 and S_2 determined by the y-axis and admit that $E \cap S_1 = \emptyset$; see Figure 9.13. Since E and S_1 are disjoint compact subsets, we have that $\text{dist}\,(E, S_1) = \varepsilon > 0$. Then E is contained in the

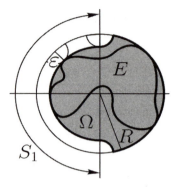

Figure 9.13. $E \cap S_1 = \emptyset$

closed region Ω which is the complement in the disc of a neighbourhood of S_1 with radius ε. One can see that this region is included in a circle with radius less than R—with centre on the x-axis to the right of the origin. This contradicts the fact that R is the radius of the circle circumscribed to the set E. $\square$

Corollary 9.62. *Let E be a compact subset of $\mathbb{R}^2$ with more than one point, and let C be its circumscribed circle. The intersection $E \cap C$ has at least two points. Moreover, if it has exactly two, they are antipodal.*

Finally, we may give the four—or more—vertices theorem in its strongest version due to Osserman.

Theorem 9.63 (Four-vertices theorem). *Let C be a compact connected simple curve of the plane $\mathbb{R}^2$. Then C has at least four vertices. Moreover, if it has exactly four, then C touches its circumscribed circle at exactly two antipodal points.*

Proof. We know that C is a compact subset of $\mathbb{R}^2$ with more than one point and, so, by Proposition 9.60, it has a circumscribed circle S whose radius will be denoted by $R > 0$. By Corollary 9.62, we know that $C \cap S$ has at least two points. Suppose that $p \in C \cap S$ is an arbitrary point in this intersection. Exercise (1) of Chapter 1 tells us that the curves C and S are tangent at p and, since the inner domain determined by C (see the Jordan curve Theorem 9.14) is contained in the circle S, the normal vectors corresponding to the positive orientations of C and S coincide at p. We may, then, apply the result of the comparison of plane curves of Exercise (16) of Chapter 1 and deduce that $k(p) \geq 1/R$, where k is the curvature of the curve C; see Figure 9.14.

Suppose now that p and q are two different points in $C \cap S$, and let Γ be any of the two connected components of $C - \{p, q\}$. Let us show, and for

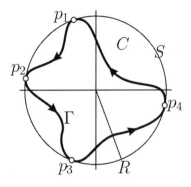

Figure 9.14. *Circle circumscribed to C*

this it is fundamental that C is simple, that either $\Gamma \subset S$ or there exists a point $r \in \Gamma$ such that $k(r) < 1/R$. In fact, assume without lost of generality that S is centred at the origin and the points p and q are on some vertical straight line V; see Figure 9.15. Since this line V cuts S transversely and so C at the points p and q, there must be points of C on the right-hand and on the left-hand sides of V; see, for instance, Exercise (17) at the end of Chapter 1.

Now, we will pay attention to the component Γ of C with end points p and q which, in each neighbourhood of p, has points on the right-hand side of V. Since Γ, together with the two disjoint half-lines of V starting from p and q, forms a curve without self-intersections separating the plane into two components—which we will naturally call right and left regions—we have that the inner normal of C restricted to Γ points towards the left region. We move the circle S to the left up to the last position S' in which it intersects Γ. There are only two alternative cases: either Γ is contained in S and $S' = S$ or Γ is not contained in S, S' touches Γ at a non-end point and the two curves are tangent at this point. In both cases, Γ and S' are tangent at a non-end point, of Γ and the inner normal of Γ at this point is the inner normal of S'. We deduce, by comparison, that the curvature of Γ at this contact point is less than or equal to $1/R$. Thus, the curvature of Γ attains its minimum at a non-end point; recall that the curvature of C at p and q was greater than or equal to $1/R$. The same argument supplies another local minimum of the curvature of C in the interior of the other arc determined by p and q. Since in each arc of C determined by two local minima there has to be at least a local maximum, we infer that the curvature of C has at least four different local extremes. This finishes the proof of the theorem. □

Exercise 9.64. ↑ Find an example of a periodic regular parametrized plane curve whose trace is not a simple curve and for which the four-vertices theorem is false.

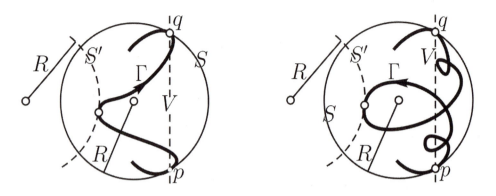

Figure 9.15. *Cases of a simple curve and a curve with self-intersections*

Exercise 9.65. Consider the curve $\alpha : \mathbb{R} \to \mathbb{R}^2$ given by

$$\alpha(t) = (t, t^2), \qquad \forall t \in \mathbb{R}.$$

Show that the trace of α is a simple curve and, however, α has a unique vertex.

Exercises

(1) ↑ Let $\Omega \subset \mathbb{R}^2$ be a bounded open set and X a vector field defined on $\overline{\Omega}$. Suppose that there exists $\varepsilon > 0$ such that, for $|t| < \varepsilon$, the map $\phi_t : \Omega \to \phi_t(\Omega)$ given by $\phi_t(p) = p + tX(p)$ is a diffeomorphism. Prove that

$$\frac{d}{dt}\bigg|_{t=0} A\big(\phi_t(\Omega)\big) = \int_\Omega \operatorname{div} X.$$

Taking this equality into account and interpreting X as the field of velocities of a fluid with constant density, give a heuristic *proof* of the divergence theorem for the plane.

(2) ↑ Let C be a compact connected simple plane curve with length L and Ω its inner domain. If C is contained in a closed disc of radius $r > 0$, prove that

$$A(\Omega) \le \frac{r}{2} L$$

and equality holds if and only if C is a circle.

(3) ↑ Given a periodic curve α p.b.a.l. with length L, whose trace is a simple curve, if Ω is its inner domain, show that

$$2A(\Omega) \le \left(\int_0^L |\alpha(s)|^2 \, ds \right)^{\frac{1}{2}} L^{\frac{1}{2}}$$

and that equality occurs if and only if α is a circle centred at the origin.

(4) ↑ Let α be a periodic curve p.b.a.l. with length L, whose trace is an oval and Ω its inner domain. Suppose that Ω contains the disc $B_r = \{p \in \mathbb{R}^2 \mid |p| < r\}$. Prove the following statements.

- $- \langle N(s), \alpha(s) \rangle \geq r$ for each $s \in \mathbb{R}$.
- $A(\Omega) \geq (r/2)L$ and equality is attained if and only if α is the circle centred at the origin with radius r.

(5) Let $A \subset \mathbb{R}^2$, X, Y be vector fields defined on A, $f : A \to \mathbb{R}$ be a differentiable function, and $a, b \in \mathbb{R}$. Prove the following equalities.

$$\operatorname{div}(aX + bY) = a \operatorname{div} X + b \operatorname{div} Y \quad \text{and} \quad \operatorname{div}(fX) = f \operatorname{div} X + \langle \nabla f, X \rangle.$$

(6) ↑ Let X be a vector field defined on the whole of the plane. Show that X is a conservative field if and only if $\operatorname{div} JX = 0$.

(7) Let $A \subset \mathbb{R}^2$, and let $f, g : A \to \mathbb{R}$ be two differentiable functions. We define the *Laplacian* of f by

$$\Delta f = \frac{\partial^2 f}{\partial x^2} + \frac{\partial^2 f}{\partial y^2}.$$

Prove that $\Delta f = \operatorname{div} \nabla f$ and $\Delta(fg) = g\Delta f + f\Delta g + 2 \langle \nabla f, \nabla g \rangle$.

(8) Let α be an L-periodic curve p.b.a.l. whose trace is a simple curve and let Ω be its inner domain. If $f, g : \overline{\Omega} \to \mathbb{R}$ are differentiable functions, show that

$$\int_\Omega f\Delta g + \int_\Omega \langle \nabla f, \nabla g \rangle = -\int_0^L f\big(\alpha(s)\big)(dg)_{\alpha(s)}\big(N(s)\big)\, ds,$$

$$\int_\Omega f\Delta g - \int_\Omega g\Delta f$$
$$= -\int_0^L f\big(\alpha(s)\big)(dg)_{\alpha(s)}\big(N(s)\big)\, ds + \int_0^L g\big(\alpha(s)\big)(df)_{\alpha(s)}\big(N(s)\big)\, ds.$$

These identities are known as *Green's formulas.*

(9) If $\Omega \subset \mathbb{R}^2$ is an open set and $f : \Omega \to \mathbb{R}$ is a differentiable function, we say that f is *harmonic* when $\Delta f = 0$ on Ω. Prove that the functions

$$g : \mathbb{R}^2 \longrightarrow \mathbb{R}, \qquad g(p) = \langle a, p \rangle + b \text{ with } a \in \mathbb{R}^2, b \in \mathbb{R},$$

$$h : \mathbb{R}^2 - \{0\} \longrightarrow \mathbb{R}, \qquad h(p) = \log |p|^2$$

are harmonic.

(10) ↑ Let Ω be the inner domain determined by a compact connected simple plane curve C, and let $f : \overline{\Omega} \to \mathbb{R}$ be a differentiable function such that $\Delta f = 0$ on Ω and $f_{|C} = 0$. Prove that $f = 0$ on $\overline{\Omega}$.

(11) Let $\Omega \subset \mathbb{R}^2$ be an open set and $f : \Omega \to \mathbb{R}$ a differentiable function. Show that the following assertions are equivalent.

- f is harmonic.
- f satisfies the *mean value* property, that is, for each $p_0 \in \Omega$ and each ball $B_r(p_0)$ centred at p_0 with radius $r > 0$ contained in Ω, one has

$$f(p_0) = \frac{1}{\pi r^2} \int_{B_r(p_0)} f(p) \, dp.$$

- For each $p_0 \in \Omega$ and each circle α centred at p_0 with radius $r > 0$ contained in Ω

$$\int_0^{2\pi r} (df)_{\alpha(s)} \big(\alpha(s) - p_0\big) \, ds = 0.$$

(12) ↑ Let $A < B$ be two real numbers and $f : [A, B] \to \mathbb{R}$ a function of class C^2 such that $f(A) = f(B) = 0$. Show that there exists a function $h : [A, B] \to \mathbb{R}$ of class C^1 such that

$$f(t) = h(t) \sin \frac{\pi}{B - A}(t - A)$$

for each $t \in [A, B]$.

(13) ↑ *Wirtinger's inequality.* Let $f : [A, B] \to \mathbb{R}$ be a function of class C^2 such that $f(A) = f(B) = 0$. Then

$$\int_A^B \big(f'(t)\big)^2 \, dt \geq \frac{\pi^2}{(B - A)^2} \int_A^B \big(f(t)\big)^2 \, dt$$

and equality occurs if and only if $f(t) = a \sin \frac{\pi}{B-A}(t - A)$ for some $a \in \mathbb{R}$.

(14) ↑ Let $\alpha : \mathbb{R} \to \mathbb{R}^2$ be an L-periodic curve p.b.a.l. with simple trace. Prove that there exists $s_0 \in (0, L/2)$ and two perpendicular straight lines R_1 and R_2 of the plane, in such a way that $\alpha(0), \alpha(L/2) \in R_1$ and such that $\alpha(s_0), \alpha(s_0 + L/2) \in R_2$.

(15) ↑ Under the hypotheses of the previous exercise, choosing an affine frame of the plane

$$O = R_1 \cap R_2, \qquad e_1 = \frac{\alpha(\frac{L}{2}) - \alpha(0)}{|\alpha(\frac{L}{2}) - \alpha(0)|}, \qquad e_2 = \frac{\alpha(s_0 + \frac{L}{2}) - \alpha(s_0)}{|\alpha(s_0 + \frac{L}{2}) - \alpha(s_0)|},$$

one has

$$\int_0^L |\alpha(s)|^2 \, ds \;=\; \int_0^L \langle \alpha(s), e_1 \rangle^2 \, ds + \int_0^L \langle \alpha(s), e_2 \rangle^2 \, ds$$

$$= \int_{s_0}^{s_0 + \frac{L}{2}} \langle \alpha(s), e_1 \rangle^2 \, ds + \int_{s_0 + \frac{L}{2}}^{s_0 + L} \langle \alpha(s), e_1 \rangle^2 \, ds$$

$$+ \int_0^{\frac{L}{2}} \langle \alpha(s), e_2 \rangle^2 \, ds + \int_{\frac{L}{2}}^L \langle \alpha(s), e_2 \rangle^2 \, ds.$$

Using this equality and the Wirtinger inequality (Exercise (13) above) prove that

$$\int_0^L |\alpha(s)|^2 \, ds \leq \frac{L^3}{4\pi^2}.$$

This inequality, together with Exercise (3), shows the isoperimetric inequality. (This is a simplification of the Hurwitz proof.)

(16) If α is a periodic curve with simple trace and length L, Ω its inner domain and if $f : \Omega \to \mathbb{R}$ is a differentiable function, prove that

$$\int_\Omega \Delta f = - \int_0^L (df)_{\alpha(s)} \big(N(s) \big) \, ds.$$

(17) ↑ What can you say about the length of a periodic parametrized plane curve whose curvature k satisfies $0 \leq k \leq 1$?

Hints for solving the exercises

Exercise 9.1: Let R be the straight line passing through $p \in \mathbb{R}^3$ with direction $v \in \mathbb{S}^2$. Then, the affine map $\alpha : \mathbb{R} \to R$ given by $\alpha(t) = p + tv$ is a regular parametrization covering the whole of the line R. Let C be the circle centred at $a \in \mathbb{R}^2$ with radius $r > 0$. Then, for each $\theta \in \mathbb{R}$, the map $\gamma : (\theta, \theta + 2\pi) \to C$ given by $\gamma(t) = a + (r \cos t, r \sin t)$ is a regular parametrization of C covering $C - \{(r \cos \theta, r \sin \theta)\}$. We move θ in $\mathbb{R}$ and cover the whole of C.

Exercise 9.2: Let $\mathcal{R} = \alpha(\mathbb{R})$ be the four-petal rose. We know that the origin is in $\mathcal{R}$ because $\alpha(0) = (0,0)$. If $\mathcal{R}$ were a simple curve, the origin would have a neighbourhood in $\mathcal{R}$ homeomorphic to $\mathbb{R}$. This is impossible because we are going to show that each punctured neighbourhood of the origin in $\mathcal{R}$ has at least four connected components. In fact, α takes the interval $(0, \pi/2)$ to the first quadrant because

$$t \in (0, \frac{\pi}{2}) \Longrightarrow \begin{cases} \sin 2t > 0, \\ (\cos t, \sin t) & \text{is in the first quadrant.} \end{cases}$$

On the other hand, since

$$t \in (\frac{\pi}{2}, \pi) \Longrightarrow \begin{cases} \sin 2t < 0, \\ (\cos t, \sin t) & \text{is in the second quadrant,} \end{cases}$$

the map α takes the interval $(\pi/2, \pi)$ to the fourth quadrant. Reasoning in this way, one sees that α maps $(\pi, 3\pi/2)$ to the third quadrant and $(3\pi/2, 2\pi)$ to the second quadrant. Since α is clearly 2π-periodic, one has that $\mathcal{R} - \{(0,0)\}$ has four connected components, which are the images through α of these four intervals. Moreover

$$\lim_{t \to 0} \alpha(t) = \lim_{t \to \pi/2} \alpha(t) = \lim_{t \to 3\pi/2} \alpha(t) = (0,0),$$

and so there are points of C as close to the origin as we want in each of the four components of $\mathcal{R} - \{(0,0)\}$.

Exercise 9.3: In fact, for each $t \in (-1, +\infty)$, we have

$$\alpha'(t) = \left(\frac{3(1 - 2t^3)}{(1 + t^3)^2}, \frac{3t(2 - t^3)}{(1 + t^3)^2} \right),$$

which, obviously, can never vanish. Furthermore, if $t, t' \in (-1, +\infty)$ and $\alpha(t) = \alpha(t')$, it is clear that $t = 0$ if and only if $t' = 0$. Suppose that t and t' are non-null. Then $x(t) = x(t')$ and $y(t) = y(t')$, where x and y are the components of α. Hence

$$t = \frac{y(t)}{x(t)} = \frac{y(t')}{x(t')} = t',$$

and, so, α is injective. However, the folium $\mathcal{F} = \alpha((-1, +\infty))$ cannot be a simple curve because $(0, 0) = \alpha(0) \in \mathcal{F}$ has a basis of disconnected neighbourhoods. Indeed, since α is continuous and $\alpha(0) = (0, 0)$ and, moreover, since $\lim_{t \to +\infty} \alpha(t) = (0, 0)$, if B is a ball of $\mathbb{R}^2$ centred at the origin to which, for example, the point $\alpha(1) \neq (0, 0)$ does not belong, then there exist $\varepsilon > 0$ and $r > 0$ such that $U = \alpha((-\varepsilon, \varepsilon)) \subset B$ and $V = \alpha((r, +\infty)) \subset B$. Notice that U and V are connected and disjoint, since α is injective and $(-\varepsilon, \varepsilon)$ and $(r, +\infty)$ are disjoint. If not, the union of these two intervals would give $(-\varepsilon, +\infty)$ and $\alpha(1) \in B$. Since this occurs for each ball with radius less than the radius of B with centre at the origin, we conclude.

Exercise 9.11: Since α and β are bijective, we may consider the composition $f = \alpha^{-1} \circ \beta$ which is a bijective map of J to I. Hence $\beta = \alpha \circ f$. Suppose that we have already seen that f is differentiable. Then $f'(s)\alpha'(f(s)) =$

$\beta'(s)$, for all $s \in J$, and, since α and β are p.b.a.l., $f'(s) = \pm 1$, and so, since f' is continuous, f is of the form $f(s) = s + c$ or $f(s) = -s + c$, for all $s \in J$ and some $c \in \mathbb{R}$, as we wanted. Therefore, it is a question of showing that f is differentiable. We can do it exactly as in Theorem 2.26, by proving that each point of C has a neighbourhood on which one of the projections p to the coordinate axes is bijective. Thus, if we restrict f to this neighbourhood, $f = (p \circ \alpha)^{-1} \circ (p \circ \beta)$ is a composition of differentiable functions.

After the previous discussion, on a non-compact connected simple curve C, an orientation will be a choice of a curve p.b.a.l. $\alpha : I \to \mathbb{R}^3$ which is a homeomorphism onto C. Two of these curves will determine the same orientation when they differ by a translation and they will determine the opposite orientation if, moreover, they differ by a symmetry relative to the origin. If C is compact, by Theorem 9.10, there exists a curve p.b.a.l. $\alpha : \mathbb{R} \to \mathbb{R}^3$ with period, say $L > 0$. If β is another curve of this type with another period $M > 0$, since α restricted to $(0, L)$ and β restricted to $(0, M)$ are homeomorphisms onto C minus a point, by the above, these two intervals must have the same length. Hence $L = M$, and so it makes sense to talk about the length of C, and moreover either $\beta(s) = \alpha(s)$ or $\beta(s) = \alpha(-s + L)$ for each $s \in (0, L)$. In the first case, we say that α and β determine the same orientation of C, and in the second case, the opposite orientation.

Exercise 9.13: We know that there exists a curve p.b.a.l. $\alpha : I \subset \mathbb{R} \to \mathbb{R}^3$ which is a homeomorphism between I and C, according to Theorem 9.10. If I were not the whole of $\mathbb{R}$, then there would be some $s \in \mathrm{Bdry}\, I - I$. Let $\{s_n\}_{n \in \mathbb{N}}$ be a sequence of points of I converging to s. We have

$$|\alpha(s_n) - \alpha(s_m)| = \left| \int_{s_n}^{s_m} \alpha'(s)\, ds \right| \le |s_n - s_m|,$$

and, so, $\{\alpha(s_n)\}_{n \in \mathbb{N}}$ is a Cauchy sequence. Hence, it converges to a point of $\mathbb{R}^3$, which must belong to $C = \alpha(I)$, by closedness. Thus $\lim_{n \to \infty} \alpha(s_n) = \alpha(x)$ for some $x \in I$. Since α is a homeomorphism, $s = \lim_{n \to \infty} s_n = x \in I$ and we have a contradiction. Then, the homeomorphism α is defined on $\mathbb{R}$.

Now let $\beta : J \subset \mathbb{R} \to \mathbb{R}^3$ be any curve p.b.a.l. whose image is C. Suppose that, after a translation and a symmetry relative to the origin if necessary, $0 \in J$, $\alpha(0) = \beta(0)$, and $\alpha'(0) = \beta'(0)$. Then, the set

$$A = \{s \in J \mid \alpha(s) = \beta(s) \text{ and } \alpha'(s) = \beta'(s)\}$$

is a non-empty closed subset of J. Since, by the inverse function theorem applied to the composition of β with some of the projections of $\mathbb{R}^3$, the curve β is also a local homeomorphism, if we use Exercise 9.1, we have that A is also open. Therefore $A = J$ and so $\beta = \alpha_{|J}$. Hence $C = \beta(J) = \alpha(J)$ and, so, $J = \mathbb{R}$.

Exercise 9.19: The parallel curve at distance t is the image of $\alpha_t : \mathbb{R} \to \mathbb{R}^2$ defined by $\alpha_t(s) = \alpha(s) + tN(s)$, for all $s \in \mathbb{R}$ and where N is the inner normal of $\alpha(\mathbb{R})$. By the Frenet equations,

$$\alpha_t'(s) = (1 - tk(s))\, N(s), \qquad \forall s \in \mathbb{R},$$

and, moreover, $1 - tk(s) > 0$ for all $s \in \mathbb{R}$, with t small enough. Consequently,

$$L(\alpha_t) = \int_0^L (1 - tk(s))\, ds = L - t \int_0^L k(s)\, ds = L - 2\pi t,$$

where the last equality follows from the turning tangent Theorem 9.50. Furthermore, by Remark 1.15,

$$k_t(s) = \frac{1}{|\alpha_t'(s)|} \det(\alpha_t'(s), \alpha_t''(s)) = \frac{k(s)}{1 - tk(s)},$$

for $s \in \mathbb{R}$ and t small.

Exercise 9.20: Let $a \in \mathbb{R}^2$ be arbitrary. If $\alpha : \mathbb{R} \to \mathbb{R}^2$ is an L-periodic curve p.b.a.l. whose trace is C, the periodic function $f : \mathbb{R} \to \mathbb{R}$ given by $f(s) = |\alpha(s) - a|^2$, $s \in \mathbb{R}$, must attain its maximum at some $s_0 \in [0, L)$. Let $r = f(s_0) > 0$, and let $\mathcal{C}$ be the circle centred at a with radius r. Then $C \subset D$, where D is the closed disc determined by the circle. Hence, the non-bounded connected set $\mathbb{R}^2 - D$ is contained in $\mathbb{R}^2 - C$. Then, the domain Ω determined by C is contained in D. It suffices now to apply Exercise (16) of Chapter 1 to the curves C and $\mathcal{C}$, taking into account that the inner normals of $\mathcal{C}$ and C coincide at $\alpha(s_0)$.

Exercise 9.28: If Ω is the inner domain determined by $\alpha(\mathbb{R})$, we consider on $\overline{\Omega}$ the constant vector field with value $a \in \mathbb{R}^3$, which has vanishing divergence, and we apply the divergence Theorem 9.23. Otherwise, notice that

$$N(s) = JT(s) = J\alpha'(s) = (J\alpha(s))', \qquad \forall s \in \mathbb{R},$$

because J is linear. Now, it is enough to apply the fundamental theorem of calculus on $[0, L]$, bearing in mind that α is an L-periodic function.

Exercise 9.30: The components of the field ∇f are the partial derivatives f_x and f_y. Then $J\nabla f = (-f_y, f_x)$ and, so,

$$\operatorname{div} J\nabla f = \frac{\partial}{\partial x}(-f_y) + \frac{\partial}{\partial y} f_x = 0,$$

by the Schwarz theorem of calculus. On the other hand,

$$\operatorname{div} g J \nabla f = \frac{\partial}{\partial x}(-g f_y) + \frac{\partial}{\partial y}(g f_x) = \det \begin{pmatrix} f_x & f_y \\ g_x & g_y \end{pmatrix} = \det(\nabla f, \nabla g).$$

Exercise 9.31: It suffices to realize that $\operatorname{Jac} F = \det(\nabla F^1, \nabla F^2)$. From Exercise 9.30 and the divergence Theorem 9.23,

$$\int_\Omega \operatorname{Jac} F = -\int_0^L F^2(\alpha(s)) \langle (J \nabla F^1)(\alpha(s)), N(s) \rangle \, ds,$$

where $\alpha : \mathbb{R} \to \mathbb{R}^2$ is an L-periodic curve p.b.a.l. whose trace is C. Since $F_{|C} = I_C$, if we denote by x and y the components of α, we have

$$\int_\Omega \operatorname{Jac} F = -\int_0^L y(s) x'(s) \, ds = -\frac{1}{2} \int_0^L (y(s) x'(s) - x(s) y'(s)) \, ds,$$

where the last equality follows after applying the fundamental theorem of calculus. But

$$y(s) x'(s) - x(s) y'(s) = \langle \alpha(s), N(s) \rangle, \qquad \forall s \in \mathbb{R}.$$

Hence

$$\int_\Omega \operatorname{Jac} F = -\frac{1}{2} \int_0^L \langle \alpha(s), N(s) \rangle \, ds.$$

On the other hand, let us consider the identity field $X : \overline{\Omega} \to \mathbb{R}^2$, taking each $p \in \overline{\Omega}$ to its own position $X(p) = p$. Then $(dX)_p = I_{\mathbb{R}^2}$ for all $p \in \Omega$ and, so, $\operatorname{div} X = 2$. Applying the divergence Theorem 9.23 to this field, we obtain

$$2A(\Omega) = -\int_0^L \langle \alpha(s), N(s) \rangle \, ds$$

and we conclude.

Exercise 9.38: Differentiating alternatively with respect to s and u in the equality $\langle \alpha(s) - \alpha(u), N(s) + N(u) \rangle = 0$ and using the Frenet equations, we have

$$(k(s) - k(u)) \langle T(s), T(u) \rangle = 0, \qquad \forall s, u \in I.$$

Differentiating again with respect to s, we obtain

$$k'(s) \langle T(s), T(u) \rangle + (k(s) - k(u)) k(s) \langle N(s), T(u) \rangle = 0,$$

for all $s, u \in I$. Putting $s = u \in I$ in the previous equality, we deduce that k' is identically zero, that is, α has constant curvature. We finish by Exercise 1.19.

As for the geometrical interpretation of the condition, since we may rewrite it as

$$\langle \alpha(s) - \alpha(u), N(u) \rangle = \langle \alpha(u) - \alpha(s), N(s) \rangle,$$

for each $s, u \in I$, this means that the distance from $\alpha(s)$ to the tangent line of α at u coincides with the distance from $\alpha(u)$ to the tangent line of α at s.

Exercise 9.41: The curve $\alpha : \mathbb{R} \to \mathbb{R}^2$ given by

$$\alpha(t) = (a \cos t, b \sin t), \qquad \forall t \in \mathbb{R},$$

is a regular 2π-periodic curve whose trace is the ellipse C. Hence, using Remark 1.15,

$$k(t) = \frac{ab}{(a^2 \cos^2 t + b^2 \sin^2 t)^{3/2}} > 0, \qquad \forall t \in \mathbb{R}.$$

Since C is diffeomorphic to a circle, it is a compact connected simple curve with positive curvature.

Exercise 9.42: Since it is the graph of a differentiable function defined on the real line $\mathbb{R}$, the parabola C is a connected simple curve closed in $\mathbb{R}^2$. Moreover, $\alpha : \mathbb{R} \to \mathbb{R}^2$ given by $\alpha(t) = (t^2/2p, t)$ is clearly a regular parametrization for it. Using the same formula as in Exercise 9.1, it follows that

$$k(t) = -\frac{\sqrt{p}}{\sqrt{p^2 + t^2}^3} \qquad \forall t \in \mathbb{R},$$

and the parabola has non-vanishing curvature everywhere, and thus, it is positive for the opposite orientation.

Exercise 9.44: The function $f : \mathbb{R} \to \mathbb{R}$ given by

$$f(s) = k(s)\langle \alpha(s) - a, v \rangle, \qquad \forall s \in \mathbb{R},$$

is L-periodic and, using the Frenet formulas,

$$f'(s) = k'(s)\langle \alpha(s) - a, v \rangle + \langle N(s), v \rangle, \qquad \forall s \in \mathbb{R}.$$

Integrating between 0 and L, taking into account the fundamental theorem of calculus for the integral on the left-hand side and Exercise 9.28 for the second summand in the right-hand side, we find the required equality, for any $a, v \in \mathbb{R}^3$.

To prove the assertion of the four-vertices theorem for ovals, we may suppose that the curvature k of α is not constant on any subinterval of $[0, L]$. Then, there exist two numbers $s_1 < s_2$ in $[0, L)$ where k attains its extreme values, since k is continuous and L-periodic. Suppose that, on the interval (s_1, s_2), the function k' does not vanish. In this case, if this function k' changed sign on the other interval $(s_2, s_1 + L)$, it would have at least two zeroes on it, since it has the same sign in a neighbourhood on the right of s_2 and in a neighbourhood on the left of $s_1 + L$. Then, we may consider that

k' is, for example, strictly positive on (s_1, s_2) and that it is not positive on $(s_2, s_1 + L)$. Thus, if a is a point of the line joining $\alpha(s_1)$ and $\alpha(s_2)$ and v is a direction vector of this line, the function $k' \langle \alpha - a, v \rangle$, using Exercise 9.3, does not change sign and is non-zero on a non-empty open set. Thus, the previous integral cannot vanish. This contradiction is solved because k' has to vanish on each of the intervals (s_1, s_2) and $(s_2, s_1 + L)$.

Exercise 9.45: It is clear that, if $s \in \mathbb{R}$,

$$\beta(s + B) = \alpha(f(s + B)) = \alpha(f(s) + A) = \alpha(f(s)) = \beta(s)$$

and so β is B-periodic, because f is a diffeomorphism and it is impossible to find other periods smaller than B. On the other hand, from the relation $\beta = \alpha \circ f$, one obtains by differentiation that $\beta' = f'(\alpha' \circ f)$. Consequently, the tangent map of β is

$$T_\beta = \frac{\beta'}{|\beta'|} = \frac{\alpha' \circ f}{|\alpha' \circ f|} = T_\alpha \circ f.$$

Thus $T_{\beta|[0,B]} = T_\alpha \circ f_{|[0,B]}$. Since $f : [0, B] \to [0, A]$ is a diffeomorphism, one has

$$i(\beta) = \deg T_\beta = \deg T_{\beta|[0,B]} = \deg T_{\alpha|[0,A]} = \deg T_\alpha = i(\alpha),$$

where we have used Exercise 9.68.

Exercise 9.46: Let $F : \mathbb{R} \times [0, 1] \to \mathbb{R}^2$ be the homotopy between α_0 and α_1. We know that, if we fix $t \in [0, 1]$, there exists the derivative of $s \in \mathbb{R} \mapsto F(s, t)$ and it is non-zero. We define, then, $H : \mathbb{R} \times [0, 1] \to \mathbb{S}^1$ by

$$H(s, t) = \frac{\dfrac{d}{ds} F(s, t)}{\left| \dfrac{d}{ds} F(s, t) \right|}, \qquad s \in \mathbb{R}, \quad t \in [0, 1],$$

which is continuous and A-periodic in the s-variable, since F is. Now, applying Exercise 9.70, one has that

$$i(\alpha_0) = \deg \frac{\alpha_0'}{|\alpha_0'|} = \deg H_0 = \deg H_1 = \deg \frac{\alpha_1'}{|\alpha_1'|} = i(\alpha_1)$$

and we conclude.

Exercise 9.47: Let $\phi : \mathbb{R}^2 \to \mathbb{R}^2$ be a rigid motion or a homothety. Then, if $\alpha : \mathbb{R} \to \mathbb{R}^2$ is a periodic regular curve, we have

$$i(\phi \circ \alpha) = \deg \frac{(\phi \circ \alpha)'}{|(\phi \circ \alpha)'|} = \deg A \frac{\alpha'}{|\alpha'|},$$

where A is the linear part of ϕ, if ϕ is a motion, or the identity map, if it is a homothety. We finish, then, using Exercise 9.72.

Exercise 9.48: Equality (9.1) allows us to compute the rotation index of α, β, and γ integrating their curvatures multiplied by the length of the velocity vector. To compute these integrands, we take Remark 1.15 into account and get $k_\alpha(t)|\alpha'(t)| = 1$ and

$$k_\beta(t)|\beta'(t)| = -2\frac{2\sin t \sin 2t + \cos t \cos 2t}{\sin^2 t + 4\cos^2 2t}, \quad k_\gamma(t)|\gamma'(t)| = 3\frac{3 - 2\sin t}{5 - 4\sin t},$$

for each $t \in \mathbb{R}$. Integrating these three functions between 0 and 2π, we obtain the required rotation indices. It is clear, from the first equality, that $i(\alpha) = 1$. The second function to integrate clearly satisfies

$$k_\beta(t + \pi)|\beta'(t + \pi)| = -k_\beta(t)|\beta'(t)|, \qquad \forall t \in \mathbb{R}.$$

Thus, since

$$i(\beta) = \frac{1}{2\pi}\int_\theta^{\theta+2\pi} k_\beta(t)|\beta'(t)|\,dt,$$

for each $\theta \in \mathbb{R}$, making the change of variables given by $t \mapsto t + \pi$, we have

$$i(\beta) = \frac{1}{2\pi}\int_{\theta-\pi}^{\theta+\pi} k_\beta(t + \pi)|\beta'(t + \pi)|\,dt = -\frac{1}{2\pi}\int_{\theta-\pi}^{\theta+\pi} k_\beta(t)|\beta'(t)|\,dt = -i(\beta),$$

and, so, $i(\beta) = 0$. As for the third curve, it is not difficult to see that the function $k_\gamma|\gamma'|$ is increasing in $\sin t$. Thus, its extremes are $5/3$, which is attained by $t = 3\pi/2$, and 3, which is attained by $t = \pi/2$. Therefore

$$\frac{5}{3} \le \frac{1}{2\pi}\int_0^{2\pi} k_\gamma(t)|\gamma'(t)|\,dt = i(\gamma) \le 3.$$

Equality on the right-hand side would force $k_\gamma|\gamma'|$ to be always equal to 3, which is not true. Hence, $i(\gamma)$ is an integer in $[5/3, 3)$. That is, $i(\gamma) = 2$.

Exercise 9.64: The curve γ of Exercise 9.48 is regular, 2π-periodic, does not have a simple trace, and its curvature satisfies

$$k_\gamma(t) = 3\frac{3 - 2\sin t}{(5 - 4\sin t)^{3/2}}, \qquad \forall t \in \mathbb{R},$$

according to the resolution of this exercise. Then

$$k_\gamma'(t) = 3\frac{2 - \sin t}{(5 - 4\sin t)^{5/2}}\cos t, \qquad \forall t \in \mathbb{R},$$

which vanishes only when $t = \pi/2$ and $t = 3\pi/2$.

Exercise (1): By the change of variables theorem for the Lebesgue integral on $\mathbb{R}^2$, we have

$$A(\phi_t(\Omega)) = \int_\Omega \text{Jac } \phi_t = \int_\Omega \det(d\phi_t)_p\,dp.$$

But, if $p \in \Omega$ and $v \in \mathbb{R}^2$, by definition

$$(d\phi_t)_p(v) = v + t(dX)_p(v).$$

From this, a simple calculation gives

$$\det(d\phi_t)_p = 1 + t(\operatorname{div} X)(p) + t^2 \det(dX)_p,$$

and, now, integrating over Ω, we get

$$A(\phi_t(\Omega)) = A(\Omega) + t \int_\Omega \operatorname{div} X + t^2 \int_\Omega \operatorname{Jac} X.$$

Taking the derivative with respect to t at $t = 0$, we are finished.

Exercise (2): We put the origin of the plane at the centre of the disc where C is included. Thus, we have $|\alpha(s)| \leq r$ for all $s \in \mathbb{R}$. On the other hand, we showed, in the solution to Exercise 9.31, that

$$A(\Omega) = -\frac{1}{2} \int_0^L \langle \alpha(s), N(s) \rangle \, ds.$$

From the Schwarz inequality, we obtain

$$A(\Omega) \leq \frac{1}{2} \int_0^L |\alpha(s)| \, ds \leq \frac{r}{2} L.$$

If equality occurred, among other things, we would have that $|\alpha(s)| = r$ for all $s \in \mathbb{R}$. That is, C would be contained in the circle of radius r and centre at the origin. Since they are both compact simple curves, C and this circle must coincide.

Exercise (3): The same expression that we gave in the solution to Exercise (2) for the area of Ω allows us to write

$$A(\Omega) \leq \frac{1}{2} \int_0^L |\alpha(s)| \, ds \leq \frac{1}{2} \left(\int_0^L |\alpha(s)|^2 \, ds \right)^{1/2} L^{1/2},$$

where the last inequality is nothing more than the Schwarz inequality in $L^2([0, L])$. If equality occurs, $|\alpha|$ is constant and we conclude by Exercise (2) above.

Exercise (4): The function $f(s) = -\langle \alpha(s), N(s) \rangle$ measures the distance from the origin to the tangent line of the curve α at the time s. Let us fix $s \in \mathbb{R}$, and let σ be the segment of the straight line joining the origin with the tangent line of α at s and being perpendicular to it. As mentioned before, the length of σ is the value $f(s)$. If x is the end point of σ different from the origin, by Theorem 9.40 and since $\alpha(\mathbb{R})$ is an oval, $x \notin \Omega$ and, so, $x \notin B_r$. Hence, the length of σ has to be greater than or equal to r. That is, $f(s) \geq r$ for all $s \in \mathbb{R}$. Now utilizing the expression for the area of Ω that we wrote in previous exercises, we have

$$A(\Omega) = -\frac{1}{2} \int_0^L \langle \alpha(s), N(s) \rangle \, ds \geq \frac{r}{2} L.$$

If equality occurs, the function $f(s)$ takes the constant value r. Differentiating, using the Frenet formulas, and recalling that the trace of α is an oval, we obtain that $\langle \alpha(s), T(s) \rangle$ vanishes identically. From this, all the normal lines of α pass through the origin. We may conclude by using Exercise (5) at the end of Chapter 1.

Exercise (6): The necessary condition can be easily checked. In fact, in the case where X is conservative, that is, $X = \nabla f$ for a certain differentiable function f defined on $\mathbb{R}^2$, it is enough to use Exercise 9.30. Now suppose that the vector field X defined on the whole of the plane satisfies $\operatorname{div} JX = 0$. By definition, this means that

$$\frac{\partial X^2}{\partial x} = \frac{\partial X^1}{\partial y}$$

on $\mathbb{R}^2$, and this is equivalent to

$$\langle (dX)_p(v), w \rangle = \langle v, (dX)_p(w) \rangle$$

for all $v, w \in \mathbb{R}^2$. We define a function $f : \mathbb{R}^2 \to \mathbb{R}$ by

$$f(p) = \int_0^1 \langle X(tp), p \rangle \, dt$$

for each $p \in \mathbb{R}^2$. Since the map $(t, p) \mapsto \langle X(tp), p \rangle$ is differentiable, the function f that we have just defined is differentiable as well. Now, if $p, v \in \mathbb{R}^2$, we have

$$\langle (\nabla f)(p), v \rangle = \frac{d}{ds}\bigg|_{s=0} f(p + sv) = \frac{d}{ds}\bigg|_{s=0} \int_0^1 \langle X(t(p + sv)), p + sv \rangle \, dt$$

$$= \int_0^1 \{ \langle (dX)_{tp}(tv), p \rangle + \langle X(tp), v \rangle \} \, dt.$$

Using the hypothesis, we obtain

$$\langle (\nabla f)(p), v \rangle = \int_0^1 \langle t(dX)_{tp}(p) + X(tp), v \rangle \, dt.$$

On the other hand,

$$\frac{d}{dt} \langle tX(tp), v \rangle = \langle X(tp), v \rangle + t \langle (dX)_{tp}(p), v \rangle,$$

and, so, by the fundamental theorem of calculus

$$\langle (\nabla f)(p), v \rangle = \int_0^1 \frac{d}{dt} \langle tX(tp), v \rangle \; dt = \langle X(p), v \rangle$$

for all $v \in \mathbb{R}^2$. Therefore $X = \nabla f$.

Exercise (10): The first equality of Exercise (8) of this chapter, applied to the case $f = g$, and bearing in mind that f vanishes along C and is harmonic, gives us

$$\int_\Omega |\nabla f|^2 = 0.$$

Thus, ∇f vanishes on $\overline{\Omega}$, and so f must be constant. Since it vanishes along C, it has to be identically zero on $\overline{\Omega}$.

Exercise (12): We define a function h on the interval $[A, B]$ by

$$h(t) = \begin{cases} f(t)/\sin \dfrac{\pi}{B-A}(t-A) & \text{if } t \in (A, B), \\[2mm] \dfrac{B-A}{\pi} f'(A) & \text{if } t = A, \\[2mm] -\dfrac{B-A}{\pi} f'(B) & \text{if } t = B. \end{cases}$$

We see that h is continuous at the end points by using that $f(A) = f(B) = 0$ and the L'Hôpital rule. Moreover, by definition, we have

$$f(t) = h(t) \sin \frac{\pi}{B-A}(t-A), \qquad \forall t \in [A, B].$$

On the other hand, if $t \in (A, B)$, we have

$$h'(t) = \frac{f'(t) \sin \dfrac{\pi}{B-A}(t-A) - \dfrac{\pi}{B-A} f(t) \cos \dfrac{\pi}{B-A}(t-A)}{\sin^2 \dfrac{\pi}{B-A}(t-A)}.$$

Consequently, using the L'Hôpital rule again,

$$\lim_{t \to A} h'(t) = \frac{B-A}{2\pi} f''(A) \quad \text{and} \lim_{t \to B} h'(t) = -\frac{B-A}{2\pi} f''(B),$$

and so h is of class C^1.

Exercise (13): Set

$$f(t) = h(t) \sin \frac{\pi}{B-A}(t-A), \qquad t \in [A, B],$$

for some function h of class C^1, according to Exercise (12). A direct calculation from this expression of f leads to the equation

$$f'(t)^2 - \frac{\pi^2}{(B-A)^2} f(t)^2 = h'(t)^2 \sin^2 \frac{\pi}{B-A}(t-A)$$

$$+ \frac{d}{dt} \left(\frac{\pi}{B-A} h(t)^2 \sin \frac{\pi}{B-A}(t-A) \cos \frac{\pi}{B-A}(t-A) \right).$$

Integrating between A and B and using the fundamental theorem of calculus, we get

$$\int_A^B f'(t)^2 \, dt - \frac{\pi^2}{(B-A)^2} \int_A^B f(t)^2 \, dt = \int_A^B h'(t)^2 \sin^2 \frac{\pi}{B-A}(t-A) \, dt \geq 0,$$

and equality holds if and only if $h'(t) = 0$ for each $t \in [A, B]$, that is, if $h(t) = a$ for some $a \in \mathbb{R}$.

Exercise (14): Let R_1 be the straight line determined by the points $\alpha(0)$ and $\alpha(L/2)$, which are different because α is injective on $[0, L)$. Hence, it is clear that $\alpha(0)$ and $\alpha(L/2)$ lie on R_1. Now, set

$$e_1 = \frac{\alpha(L/2) - \alpha(0)}{|\alpha(L/2) - \alpha(0)|}.$$

The function $f : [0, L/2] \to \mathbb{R}$ given by

$$f(s) = \langle \alpha(s + L/2) - \alpha(0), e_1 \rangle, \qquad \forall s \in [0, L/2],$$

is continuous and satisfies

$$f(0) = |\alpha(L/2) - \alpha(0)| > 0 \quad \text{and} \quad f(L/2) = -|\alpha(L/2) - \alpha(0)| < 0.$$

Thus, there exists $s_0 \in (0, L/2)$ such that the straight line R_2 passing through $\alpha(s_0)$ and $\alpha(s_0 + L/2)$ is perpendicular to R_1.

Exercise (15): By our choice of the frame of the plane, we may apply the Wirtinger inequality of Exercise (13) to each of the four integrals in which we have decomposed

$$\int_0^L |\alpha(s)|^2 \, ds.$$

Each of them is less than or equal to $L^2/4\pi^2$ times the corresponding integral of the derivative. Grouping the four resulting integrals, we obtain

$$\int_0^L |\alpha(s)|^2 \, ds \leq \frac{L^2}{4\pi^2} \int_0^L |\alpha'(s)| \, ds = \frac{L^3}{4\pi^2},$$

as required.

Exercise (17): The length of such a curve must be greater than or equal to that of the unit circle, that is, 2π and equality holds only for this circle. In fact, if $L \leq 2\pi$, integrating the inequality and taking the definition of rotation index into account,

$$0 \leq i(\alpha) = \frac{1}{2\pi} \int_0^L k(s)\, ds \leq \frac{L}{2\pi} \leq 1$$

and so $i(\alpha)$ is 0 or $+1$ and either of the two equalities forces the curvature to be constantly 0 or $+1$.

9.7. Appendix: The one-dimensional degree theory

In the study of the fundamental group of the circle $\mathbb{S}^1$, it is usual to utilize the map $p : \mathbb{R} \to \mathbb{S}^1$ given by the expression

$$p(s) = (\cos s, \sin s), \qquad \forall s \in \mathbb{R},$$

which is differentiable as a map from $\mathbb{R}$ to $\mathbb{R}^2$ and has the following fundamental property, characteristic of covering maps.

Proposition 9.66. *Let $\alpha : [a, b] \to \mathbb{S}^1$ be a continuous curve in the circle, and let $s_0 \in \mathbb{R}$ such that $p(s_0) = \alpha(a)$. Then, there is a unique continuous curve $\tilde{\alpha} : [a, b] \to \mathbb{R}$ such that*

$$\tilde{\alpha}(a) = s_0 \quad and \quad p \circ \tilde{\alpha} = \alpha.$$

We say that $\tilde{\alpha}$ is a lift of α and that s_0 is its initial condition at a. Moreover, the same result is true for continuous maps defined on a rectangle $[a, b] \times [c, d] \subset \mathbb{R}^2$, where the initial condition is now taken at the point (a, c).

Proof. It is clear that, for each $q = (\cos \theta, \sin \theta) \in \mathbb{S}^1$, the open neighbourhood $V^q = p((\theta - \pi, \theta + \pi))$ of q satisfies the property that $p^{-1}(V^q)$ is the disjoint union of the open subsets $U^{q,k} = (\theta + (2k - 1)\pi, \theta + (2k + 1)\pi)$, with $k \in \mathbb{Z}$, of $\mathbb{R}$. Furthermore, the map p is a homeomorphism onto its image restricted to each of these intervals. Thus, the family $\{V^q\}$, with q varying along $\mathbb{S}^1$, covers $\mathbb{S}^1$. Hence, if $\alpha : [a, b] \to \mathbb{S}^1$ is a continuous curve with $\alpha(a) = q_0$, we deduce that $\{\alpha^{-1}(V^q)\}$, with $q \in \mathbb{S}^1$, is a cover by open subsets of the interval $[a, b]$. Let $\delta > 0$ be the Lebesgue number of this cover. Then

$$\left.\begin{array}{l} A \subset [a, b], \\ \operatorname{diam} A < \delta \end{array}\right\} \Longrightarrow \alpha(A) \subset V^q \text{ for some } q \in \mathbb{S}^1.$$

Let us consider a partition $\{t_0 = a < t_1 < \cdots < t_k = b\}$ with norm less than this δ, and let us lift the curve α on each subinterval of the partition. We want to assemble the diverse lifts, taking care of preserving the continuity of the final curve. By the choice of the partition, for each $i = 0, \ldots, k-1$, there exists a point $r_i \in \mathbb{S}^1$ such that $\alpha([t_i, t_{i+1}]) \subset V^{r_i}$. Then $\alpha(t_0) = \alpha(a) = q_0 \in V^{r_0}$. Since $s_0 \in \mathbb{R}$ is a point such that $p(s_0) = q_0$, we have

$$s_0 \in p^{-1}(V^{r_0}) = \bigcup_{j \in \mathbb{N}} U^{r_0, j},$$

and so s_0 belongs to a unique U^{r_0, j_0}. We define $\tilde{\alpha}_0 : [t_0, t_1] \to \mathbb{R}$ by

$$\tilde{\alpha}_0 = (p_{|U^{r_0, j_0}})^{-1} \circ \alpha,$$

which is continuous and satisfies $\tilde{\alpha}_0(a) = s_0$ and $p \circ \tilde{\alpha}_0 = \alpha$ on $[t_0, t_1]$. The other end of the lifted curve satisfies $p(\tilde{\alpha}_0(t_1)) = \alpha(t_1) \in \alpha([t_1, t_2]) \subset V^{r_1}$. Therefore

$$\tilde{\alpha}_0(t_1) \in p^{-1}(V^{r_1}) = \bigcup_{j \in \mathbb{N}} U^{r_1, j}.$$

So, $\tilde{\alpha}_0(t_1)$ belongs to a unique term of this union, which will be denoted by U^{j_1}. We define $\tilde{\alpha}_1 : [t_1, t_2] \to S$ by

$$\tilde{\alpha}_1 = (p_{|U^{j_1}})^{-1} \circ \alpha,$$

which is continuous as well and satisfies $p \circ \tilde{\alpha}_1 = \alpha$ and, thus, $p(\tilde{\alpha}_1(t_1)) = \alpha(t_1) = p(\tilde{\alpha}_0(t_1)) \in V^{r_1}$, which is the homeomorphic image of U^{j_1} through p. Then

$$\tilde{\alpha}_0(t_1) = \tilde{\alpha}_1(t_1).$$

We repeat this process finitely many times to obtain k continuous curves on $\mathbb{R}$, $\tilde{\alpha}_0, \tilde{\alpha}_1, \ldots, \tilde{\alpha}_{k-1}$ defined, respectively, on the successive subintervals of the partition and such that

$$p \circ \tilde{\alpha}_i = \alpha_{|[t_i, t_{i+1}]} \quad \text{and} \quad \tilde{\alpha}_i(t_i) = \tilde{\alpha}_{i-1}(t_i),$$

for $i = 0, \ldots, k - 1$. Finally, we define $\tilde{\alpha} : [a, b] \to \mathbb{R}$ by

$$\tilde{\alpha}(t) = \tilde{\alpha}_i(t) \text{ if } t \in [t_i, t_{i+1}].$$

This $\tilde{\alpha}$ is clearly continuous by construction and satisfies $p \circ \tilde{\alpha} = \alpha$ and $\tilde{\alpha}(a) = s_0$.

As for uniqueness, suppose that there are two continuous curves $\beta, \gamma : [a, b] \to \mathbb{R}$ with the required conditions. Let us denote by A the subset of the interval $[a, b]$ formed by those t such that $\beta(t) = \gamma(t)$. It is trivial that A is closed and it is non-empty because $\beta(a) = \gamma(a) = s_0$. Let us show now that it is also open. In fact, let $t_0 \in A$. Then $\beta(t_0) = \gamma(t_0)$ and $p(\beta(t_0)) = p(\gamma(t_0)) = \alpha(t_0) \in \mathbb{S}^1$. Since p is a local homeomorphism, there exists an open neighbourhood U of $\beta(t_0) = \gamma(t_0)$ in $\mathbb{R}$ and another neighbourhood V of $\alpha(t_0)$ in $\mathbb{S}^1$ such that p is a homeomorphism of U to V. Since β and γ are continuous, there is a neighbourhood I of t_0 such that $\beta(I), \gamma(I) \subset U$. Then $\beta_{|I} = \gamma_{|I}$. Hence $I \subset A$ and A is open. For maps defined on a rectangle, a completely similar proof works. □

The property shown in Proposition 9.66 will permit us to define the *degree* of a continuous curve in the unit sphere $\phi : [a, b] \to \mathbb{S}^1$, provided that it is *closed* in the sense of having the same end points. Let $\tilde{\phi} : [a, b] \to \mathbb{R}$ be any lift of ϕ. Then, we have

$$p(\tilde{\phi}(a)) = \phi(a) = \phi(b) = p(\tilde{\phi}(b))$$

and, so, $\tilde{\phi}(a)$ and $\tilde{\phi}(b)$ differ by an integer number of turns. The integer

$$\deg \phi = \frac{1}{2\pi}[\tilde{\phi}(b) - \tilde{\phi}(a)] \in \mathbb{Z}$$

is called the degree of ϕ. Perhaps the most remarkable property of this degree is the *invariance under homotopies relative to* $\phi(a) = \phi(b)$, which is also a consequence of Proposition 9.66 above. In fact, if $F : [a, b] \times [0, 1] \to \mathbb{S}^1$ is a continuous homotopy between two closed curves ϕ_0 and ϕ_1 defined on $[a, b]$ with the same origin $\phi_0(a) = \phi_1(a) \in \mathbb{S}^1$, let us consider its lift $\tilde{F} : [a, b] \times [0, 1] \to \mathbb{R}$ with an arbitrary initial condition. Then, $\tilde{F}(s, 0)$ and $\tilde{F}(s, 1)$ are lifts for the curves ϕ_0 and ϕ_1, respectively, and moreover $\tilde{F}(a, t)$ and $\tilde{F}(b, t)$ are constant curves. Hence

$$\deg \phi_0 = \frac{1}{2\pi}[\tilde{F}(b, 0) - \tilde{F}(a, 0)] = \frac{1}{2\pi}[\tilde{F}(b, 1) - \tilde{F}(a, 1)] = \deg \phi_1.$$

Exercise 9.67. Let $\phi : [a, b] \to \mathbb{S}^1$ be a closed continuous curve in the circle, and let $G \in SO(2)$ be a rotation of angle $\theta \in \mathbb{R}$. Prove that, if $\tilde{\phi}$ is a lift of ϕ, then $\tilde{\phi} + \theta$ is a lift of $G \circ \phi$. Deduce that $\deg(G \circ \phi) = \deg \phi$. Moreover, prove that, if $S \in O(2)$ is a symmetry, then $\deg(S \circ \phi) = -\deg \phi$.

Exercise 9.68. Let $h : [c, d] \to [a, b]$ and $\phi : [a, b] \to \mathbb{S}^1$ be a continuous function and a continuous curve in the unit circle. Show that, if $\tilde{\phi}$ is a lift of ϕ, then $\tilde{\phi} \circ h$ is a lift of $\phi \circ h$. Deduce from this that, if h is a homeomorphism, then $\deg(\phi \circ h) = \pm \deg \phi$, depending upon whether $h(c) = a$ and $h(d) = b$ or $h(c) = b$ and $h(d) = a$.

Exercise 9.69. Let $\phi, \psi : [a, b] \to \mathbb{S}^1$ be two continuous curves in the circle such that $\phi(a) = \psi(a) = \phi(b) = \psi(b)$ and such that $\deg \phi = \deg \psi$. Show that, if $\tilde{\phi}$ and $\tilde{\psi}$ are lifts of ϕ and ψ, respectively, with the same initial condition, then $F = p \circ \tilde{F}$ is a homotopy between them, where $\tilde{F}(s, t) = (1 - t)\tilde{\phi}(s) + t\tilde{\psi}(s)$ for all $(s, t) \in [a, b] \times [0, 1]$. This homotopy is clearly differentiable, provided that the original curves are differentiable.

From the above, it is not difficult to define the *degree* of a periodic continuous map $\phi : \mathbb{R} \to \mathbb{S}^1$ with period $A > 0$ as follows. For each $s \in \mathbb{R}$, we obtain, by restriction, a closed curve $\phi_s : [s, s + A] \to \mathbb{S}^1$ in the circle, to which we may associate, according to the discussion above, a degree $\deg \phi_s \in \mathbb{Z}$. Now then, it is not difficult to see, using Proposition 9.66 and an induction process, that there exists a unique continuous map $\tilde{\phi} : \mathbb{R} \to \mathbb{R}$

lifting ϕ, that is, such that $p \circ \tilde{\phi} = \phi$ with an initial condition given, for example, at $0 \in \mathbb{R}$. Thus, each restriction $\tilde{\phi}_s : [s, s+A] \to \mathbb{R}$ is clearly a lift of the curve ϕ_s. Hence

$$\deg \phi_s = \frac{1}{2\pi}[\tilde{\phi}_s(s+A) - \tilde{\phi}_s(s)] = \frac{1}{2\pi}[\tilde{\phi}(s+A) - \tilde{\phi}(s)],$$

for each $s \in \mathbb{R}$. On the other hand, since ϕ is A-periodic,

$$p(\tilde{\phi}(s+A)) = \phi(s+A) = \phi(s) = p(\tilde{\phi}(s))$$

and, so, $\tilde{\phi}(s+A) - \tilde{\phi}(s)$ is a continuous function in s taking only integer values. Therefore, it is constant. Consequently, $\deg \phi_s$ is independent of s. Then, we may define the degree of the A-periodic map $\phi : \mathbb{R} \to \mathbb{S}^1$ by

$$\deg \phi = \deg \phi_s = \frac{1}{2\pi}[\tilde{\phi}(s+A) - \tilde{\phi}(s)]$$

for any $s \in \mathbb{R}$. Of course, the properties of this degree of periodic maps are similar to those of closed curves. We list them in the following exercises.

Exercise 9.70. Let $\phi, \psi : \mathbb{R} \to \mathbb{S}^1$ be two A-periodic continuous maps. A homotopy between them is, by definition, a continuous map $F : \mathbb{R} \times [0, 1] \to \mathbb{S}^1$ such that $F(s, 0) = \phi(s)$, $F(s, 1) = \psi(s)$ for all $s \in \mathbb{R}$ and $s \mapsto F(s, t)$ is an A-periodic map for all $t \in [0, 1]$. Using a lift to $\mathbb{R}$ of the homotopy F, show that $\deg \phi = \deg \psi$.

Exercise 9.71. Suppose that $\phi, \psi : \mathbb{R} \to \mathbb{S}^1$ are two A-periodic continuous maps with the same degree. Prove that $F = p \circ \tilde{F}$ is a homotopy between them, where $\tilde{F}$ is the map defined by $\tilde{F}(s, t) = (1 - t)\tilde{\phi}(s) + t\tilde{\psi}(s)$ for all $(s, t) \in \mathbb{R} \times [0, 1]$. Then, the homotopy is differentiable if and only if the original maps are differentiable.

Exercise 9.72. Demonstrate that the degree of periodic continuous maps to the circle remains unchanged through rotations of $\mathbb{R}^2$ and changes sign when we compose with plane symmetries.

Now suppose that the curve $\phi : [a, b] \to \mathbb{S}^1 \subset \mathbb{R}^2$ is differentiable; it is enough to assume piecewise C^1. We define the *Jacobian* of ϕ, in agreement with Exercise (13) of Chapter 2, Exercise 3.15 and Section 5.2, as the function $\operatorname{Jac} \phi : [a, b] \to \mathbb{R}$ given by

$$\operatorname{Jac} \phi = \det(\phi, \phi') = \langle J\phi, \phi' \rangle,$$

where $J : \mathbb{R}^2 \to \mathbb{R}^2$ is, as throughout Chapter 1, the 90 degree counterclockwise rotation of the plane.

Lemma 9.73. *Let $\phi, \psi : [a, b] \to \mathbb{S}^1$ be two differentiable curves. Then ϕ and ψ differ by a rotation of $\mathbb{R}^2$ if and only if $\operatorname{Jac} \phi = \operatorname{Jac} \psi$.*

Proof. The necessary condition is clear because, if $A \in SO(2)$ and $\psi = A\phi$, one has

$$\text{Jac}\,\psi = \det(\psi, \psi') = \det(A\phi, A\phi') = \det(\phi, \phi') = \text{Jac}\,\phi.$$

Conversely, if this last equality holds,

$$(9.3) \qquad \frac{d}{ds}|\phi - \psi|^2 = 2\frac{d}{ds}[1 - \langle \phi, \psi \rangle] = -2[\langle \phi', \psi \rangle + \langle \phi, \psi' \rangle].$$

Now then, since $|\phi|^2 = |\psi|^2 = 1$, taking derivatives, one obtains

$$\langle \phi, \phi' \rangle = \langle \psi, \psi' \rangle = 0$$

and, so, ϕ' and ψ' are proportional to $J\phi$ and $J\psi$, respectively. Then

$$\phi' = \det(\phi, \phi')J\phi \quad \text{and} \quad \psi' = \det(\psi, \psi')J\psi.$$

Substituting these equalities in (9.3), since the two Jacobians are equal, we get

$$\frac{d}{ds}|\phi - \psi|^2 = 0,$$

and, so, the distance $|\phi - \psi|$ is constant. If the two curves ϕ and ψ coincided at a point, then they would be equal. But, in any case, we can find a rotation $A \in SO(2)$ such that ϕ and ψ coincide at a given point of $[a, b]$. We conclude by applying the above to the curves $A\phi$ and ψ. $\qquad\square$

Let us proceed by supposing that $\phi : [a, b] \to \mathbb{S}^1$ is a differentiable curve. Define another map $\tilde{\phi} : [a, b] \to \mathbb{R}$ by the equality

$$(9.4) \qquad \tilde{\phi}(s) = c + \int_{s_0}^s (\text{Jac}\,\phi)(t)\,dt,$$

where $s_0 \in [a, b]$ is arbitrary and $c \in \mathbb{R}$ will be determined later. So $\psi = p \circ \tilde{\phi}$ is a differentiable to the circle defined on $[a, b]$, satisfying

$$\psi' = (\cos\tilde{\phi}, \sin\tilde{\phi})' = \tilde{\phi}'(-\sin\tilde{\phi}, \cos\tilde{\phi}).$$

Hence, since $\tilde{\phi}' = \text{Jac}\,\phi$, one has that

$$\text{Jac}\,\psi = \text{Jac}\,\phi.$$

By Lemma 9.73, $\psi = p \circ \tilde{\phi}$ and ϕ differ by a rotation. But $\psi(s_0) = p(c) = (\cos c, \sin c)$. Thus, if we choose $c \in \mathbb{R}$ so that $p(c) = \phi(s_0)$, we arrive at the equality. That is, (9.4) is an explicit expression for any lift of a differentiable curve to the circle. From this expression, it follows that the degree of a closed differentiable curve ϕ will be given by

$$\deg\phi = \frac{1}{2\pi}[\int_{s_0}^b \text{Jac}\,\phi - \int_{s_0}^a \text{Jac}\,\phi] = \frac{1}{2\pi}\int_a^b \text{Jac}\,\phi.$$

Using this, we may prove the following result.

Proposition 9.74. *Let* $\phi : \mathbb{R} \to \mathbb{S}^1$ *be a differentiable map. Then* ϕ *is periodic with period* $A > 0$ *if and only if there exists an integer* $n \in \mathbb{Z}$ *such that*

$$2\pi n = \int_s^{s+A} (\mathrm{Jac}\, \phi)(t)\, dt$$

for each $s \in \mathbb{R}$. *In such a case, the number* n *is the degree of the* A-*periodic map* ϕ.

Proof. If ϕ is A-periodic, it is enough to take $n \in \mathbb{Z}$ as its degree as defined before and to bear in mind the last equality in the statement. Conversely, suppose that there exists $n \in \mathbb{Z}$ such that

$$\int_s^{s+A} \mathrm{Jac}\, \phi = 2\pi n, \qquad \forall s \in \mathbb{R}.$$

Then, defining a map $\tilde{\phi} : \mathbb{R} \to \mathbb{R}$ as in (9.4), one has that

$$\tilde{\phi}(s + A) - \tilde{\phi}(s) = 2\pi n$$

for all $s \in \mathbb{R}$. Thus, $\phi(s + A) = p(\tilde{\phi}(s + A)) = p(\tilde{\phi}(s)) = \phi(s)$, that is, ϕ is A-periodic. $\qquad\square$

Bibliography

[1] M. Berger, B. Gostiaux, *Differential Geometry: Manifolds, Curves and Surfaces*, Springer-Verlag, 1988.

[2] M. P. do Carmo, *Differential Geometry of Curves and Surfaces*, Prentice-Hall, 1976.

[3] L. A. Cordero, M. Fernández, A. Gray, *Geometría Diferencial de Curvas y Superficies*, Addison-Wesley Iberoamericana, 1995.

[4] A. Goetz, *Introduction to Differential Geometry*, Addison-Wesley, 1970.

[5] H. Hadwiger, *Vorlesungen über Inhalt, Oberfläche und Isoperimetrie*, Springer-Verlag, 1957.

[6] H. Hadwiger, D. Ohmann, *Brunn-Minkowskischer Satz und Isoperimetrie*, Math. Z., 66 (1956), 1–8.

[7] C. C. Hsiung, *A First Course in Differential Geometry*, Wiley-Interscience, 1981.

[8] W. Klingenberg, *Curso de Geometría Diferencial*, Alhambra, 1978.

[9] D. Lehmann, C. Sacré, *Géométrie et Topologie des Surfaces*, Presses Universitaires de France, 1982.

[10] R. S. Millman, G. D. Parker, *Elements of Differential Geometry*, Prentice-Hall, 1977.

[11] J. W. Milnor, *Topology from the Differentiable Viewpoint*, University of Virginia Press, 1965.

[12] S. Montiel, A. Ros, *Curvas y Superficies*, Proyecto Sur de Ediciones, 1997, 1998.

[13] B. O'Neill, *Elementary Differential Geometry*, Academic Press, 1966.

[14] R. Osserman, *The four-or-more vertex theorem*, Amer. Math. Monthly, 92 (1985), no. 5, 332–337.

[15] A. V. Pogorelov, *Geometría Diferencial*, Mir, 1977.

[16] N. Prakash, *Differential Geometry*, Tata McGraw-Hill, 1981.

[17] M. Spivak, *A Comprehensive Introduction to Differential Geometry*, vols. 3 and 5, Publish or Perish, 1979.

[18] J. J. Stoker, *Differential Geometry*, Wiley-Interscience, 1969.

[19] D. J. Struik, *Geometría Diferencial Clásica*, Aguilar, 1970.

[20] J. A. Thorpe, *Elementary Topics in Differential Geometry*, Springer-Verlag, 1979.

[21] E. Vidal Abascal, *Introducción a la Geometría Diferencial*, Dossat, 1956.

[22] A. Wallace, *Differential Topology*, Benjamin, 1968.

Index

Alexandrov, 172, 187
Ampère, 2
Arc length, 1
Archimedes, 161
Area, 141

Bartels, 2
Bernouilli, D., 1
Bernouilli, John, 1, 206
Bernouilli, Joseph, 2
Bertrand, 68
Bianchi, 31
Binormal, 2, 14
  line, 22
Blaschke, 172
Bonnet, 182, 204, 276, 296, 302
Brower, 108, 116, 118, 161, 288, 310
Brunn, 173, 189, 311, 325
Brunn-Minkowski inequality, 189, 325

Catenoid, 95, 239
Cauchy, 2, 32, 205
Chain rule, 48
Chasles, 68
Chern, 156
Circumscribed circle, 346
Clairaut, 2, 31, 240
Cohn-Vossen, 205, 218
Comparison
  of curves, 22
  of lines and surfaces, 111, 112
  of planes and surfaces, 113
  of surfaces, 96
Cone, 38, 39
Congruence, 207
Connected component, 36

Crofton, 312, 343
Curvature
  centre, 21
  Gauss, 68, 77
  mean, 68, 77
  of a curve, 1
    in $\mathbb{R}^3$, 14
    plane, 10
  radius, 1
  total, 330
Curve, 3
  convex, 328
  coordinate, 35
  integral, 258
  knotted, 345
  on a surface, 42
  p.b.a.l., 9
  parallel, 321
  parametrized, 313
  periodic, 314
  regular, 8
  simple, 38, 313, 318
Cylinder, 37, 44, 71, 80, 95, 225

Darboux, 2, 68, 240
  trihedron, 240
Degree
  between curves, 365
  between surfaces, 277
  local, 281
Dependence of parameters, 149
Descartes folium, 314
Diffeomorphism, 43
  local, 54
Differential, 46, 48
Diquet, 241

373

Distance, 239
Divergence, 157, 322
  on surfaces, 184
Domain
  inner, 118, 174
  inner of a curve, 320
  outer, 118
  outer of a curve, 320
  regular, 159
Dominated convergence, 148
Dupin, 2

Efimov, 173
Ellipsoid, 36, 37, 80, 174
Equations
  Darboux, 240
  Frenet, 10
Euler, 1, 31, 67, 206, 312
  characteristic, 295, 302, 304
Evolute, 21
Evolvent, 21
Existence and uniqueness
  theorem of, 17, 227
Exponential map, 230

Fary, 312, 345
Fenchel, 312, 342
Field
  conformal on a sphere, 292
  conservative, 351
  non-degenerate, 289, 292
  normal, 69, 119
  on $\mathbb{R}^2$, 322
  on $\mathbb{R}^3$, 157
  on a surface, 69
  tangent, 69, 257
First fundamental form, 68, 207
First integral, 259
Flow of a field, 258
Formula
  area, 152
  Bertrand-Puiseux, 241
  change of variables, 144
  Crofton, 343
  Diquet, 241
  Gauss-Bonnet, 296, 302
  Herglotz, 217
  integration in polar coordinates, 147
  length, 321
  Poincaré-Hopf, 302
  variation of area, 193
  variation of length, 222
  variation of volume, 194
Formulas
  Frenet, 16
  Green, 351
  Minkowski, 184

Four-petal rose, 313
Fourier, 2
Frenet, 2, 16
  dihedron, 9
  trihedron, 14
Fubini, 146
Function
  differentiable, 40
  distance, 41, 87
  height, 41, 85, 87, 301
  implicit, 36, 49
  integrable, 139
  inverse, 58
  on a sphere, 150
  support, 94, 177

Gauss, 2, 31, 68, 157, 172, 203, 210, 232,
  275, 296, 302
  lemma, 232
  map, 68, 76
Geodesic, 1, 223
  ball, 232
  existence, 228
  homogeneity, 229
  minimizes locally, 233
Germain, 68
Gluck, 313
Gradient, 161
  on surfaces, 197
Graph, 35, 53, 71
Grauenstein, 312, 331
Green, 157
Gromov, 311
Group of isometries, 209

Hélein, 311
Hadamard, 172, 175, 178, 311
Hartman, 173, 273
Heintze, 173, 186
Heintze-Karcher's inequality, 186
Helicoid, 91, 239
Helix, 4, 16, 19, 22
Herglotz, 205, 217, 312
Hessian, 85, 87
Hilbert, 91, 172
Hopf, 206, 238, 312
Hurwitz, 311
Hyperboloid, 37, 44

Index
  of a field, 289
  rotation, 329
Integral
  on $S \times \mathbb{R}$, 137
  on a surface, 139
Isometry, 208
  local, 207

Isoperimetric inequality
  in the plane, 326
  in the space, 195

Jackson, 312
Jacobi, 206
Jacobian, 76, 277
  absolute value, 57, 136
  on curves, 368
Jellett, 91, 172, 185
Jordan, 107, 116, 310, 320

Karcher, 173, 186
Kelvin, 135
Kneser, 312
Krämer, 31

Lagrange, 205
Lancret, 2, 23
Laplace, 2
Laplacian, 351
  on $\mathbb{R}^3$, 164
  on surfaces, 197
Lashoff, 156
Legendre, 275
Leibniz, 1
Length
  arc, 5, 8
  of a curve, 7
  of a periodic curve, 315
Liebmann, 91, 172, 184, 205
Liouville, 206
Logarithmic spiral, 9

Malus, 2
Map
  constant, 41
  differentiable, 40
  identity, 41
  inclusion, 41
Massey, 173, 273
Maxwell, 135, 311
Measure zero, 126, 131
  on a surface, 142
Meusnier, 2, 68
Milnor, 312, 345
Minding, 172, 205, 241
Minkowski, 172, 189, 311, 325
Moebius strip, 71, 74
Monge, 1, 31, 67
Monodromy, 196
Monotone convergence, 148
Mukhopadhyaya, 312
Myers, 182

Neighbourhood
  coordinate, 35

normal, 236
tubular, 122
  of a curve, 321
Newton, 1
Nirenberg, 173, 273
Normal
  line, 8, 319
  of a curve
    in the plane, 9
    in the space, 14
  of a surface, 46, 51
  sections, 81

Oresme, 1
Orientation, 119
  of curves, 319, 320
  of surfaces, 70
  preserving, 76
Oriented distance, 125
Osculating
  circle, 21
  plane, 23
Osserman, 313, 346
Oval, 328, 329
Ovaloid, 172, 174
  intersection with lines, 177

Pappus, 311
Paraboloid, 37, 90, 174, 256
  hyperbolic, 80
Parametrization, 35
  by curvature lines, 261
  change of, 39, 40
  orthogonal, 95, 261
Partition of unity, 143
Period
  of a periodic curve, 314
Poincaré, 276, 295, 302
Point
  critical, 48
  elliptic, 83, 87
  hyperbolic, 83, 87
  of double intersection, 115
  of first contact, 115
  parabolic, 83
  planar, 83
  umbilical, 83
Poisson, 2
Poncelet, 2
Principal curvatures, 78
Principal directions, 79, 84
Projection, 125
Ptolemy, 31
Puiseux, 241

Quadric, 37, 80

Region of a surface, 136
Regular value, 36, 111
Reilly, 173
Reparametrization, 8
  by arc length, 9
Riemann, 32, 239
Rigid motion, 7, 80, 144, 207
Rinow, 206, 238
Rodrigues, 2, 68
Rustihauser, 312

Saint-Simon, 68
Saint-Venant, 2
Samelson, 108, 118, 312
Sard, 112, 132, 311, 321
Schmidt, 311
Schwarz, 173
Second fundamental form, 68, 76, 77, 207
Second variation of the length, 253
Segre, 312
Senff, 2
Separation
  global, 116
  local, 108
Serret, 2
Sphere, 36, 45, 71, 79, 143, 145, 213, 225,
      292
Steiner, 173
Stoker, 172, 175, 178, 273, 311
Stokes, 135, 157, 311
Surface, 32, 34
  convex, 174
  of revolution, 55, 163
  orientable, 70
  oriented, 70
  parallel, 123, 145, 147
  star-shaped, 57, 95, 177
  strictly convex, 175
  triply periodic, 56
  umbilical, 84
  with parallel s.f.f., 263
Surfaces
  congruent, 207
  diffeomorphic, 43
  isometric, 208

Tangent
  line, 8, 319
  plane, 44
  to a curve, 3
Theorem
  Alexandrov, 187
  Bonnet, 181, 254
  Brower, 288
  Brower fix point, 163
  Brower-Samelson, 118
  Chern-Lashoff, 156

Cohn-Vossen, 218
divergence, 158, 159, 184, 322
Euler, 82
Fary-Milnor, 345
Fenchel, 342
four-vertices, 329, 348
Fubini, 146
fundamental, 11, 18, 209
Gauss, 160
Hadamard-Stoker, 175, 178
Hilbert, 91
Hilbert-Liebmann, 94, 184
Hopf-Rinow, 238
inverse function, 52
Jellett-Liebmann, 93, 185
Jordan curve, 320
Jordan-Brower, 116
Minding, 241
Pappus, 163
Rolle, 20
Sard, 112, 132, 321
sphere rigidity, 213
turning tangent, 333
Whitney-Grauenstein, 331
Theorema Egregium, 68, 205, 210
Torsion, 2, 15
Torus, 37, 142, 284
Transversality
  of curves, 22
  of curves and surfaces, 109
  of planes and surfaces, 113
  of surfaces, 50, 51

Umbilical, 84, 85
Umlaufsatz, 333

Variation
  of a curve, 220
  of a surface, 193
Vertex of a curve, 346
Volume, 147, 159

Weierstrass, 311
Whitney, 312, 331
Wirtinger, 352
Wirtinger's inequality, 352

Titles in This Series

69 **Sebastián Montiel and Antonio Ros,** Curves and surfaces, 2005

68 **Luis Caffarelli and Sandro Salsa,** A geometric approach to free boundary problems, 2005

67 **T.Y. Lam,** Introduction to quadratic forms over fields, 2004

66 **Yuli Eidelman, Vitali Milman, and Antonis Tsolomitis,** Functional analysis, An introduction, 2004

65 **S. Ramanan,** Global calculus, 2004

64 **A. A. Kirillov,** Lectures on the orbit method, 2004

63 **Steven Dale Cutkosky,** Resolution of singularities, 2004

62 **T. W. Körner,** A companion to analysis: A second first and first second course in analysis, 2004

61 **Thomas A. Ivey and J. M. Landsberg,** Cartan for beginners: Differential geometry via moving frames and exterior differential systems, 2003

60 **Alberto Candel and Lawrence Conlon,** Foliations II, 2003

59 **Steven H. Weintraub,** Representation theory of finite groups: algebra and arithmetic, 2003

58 **Cédric Villani,** Topics in optimal transportation, 2003

57 **Robert Plato,** Concise numerical mathematics, 2003

56 **E. B. Vinberg,** A course in algebra, 2003

55 **C. Herbert Clemens,** A scrapbook of complex curve theory, second edition, 2003

54 **Alexander Barvinok,** A course in convexity, 2002

53 **Henryk Iwaniec,** Spectral methods of automorphic forms, 2002

52 **Ilka Agricola and Thomas Friedrich,** Global analysis: Differential forms in analysis, geometry and physics, 2002

51 **Y. A. Abramovich and C. D. Aliprantis,** Problems in operator theory, 2002

50 **Y. A. Abramovich and C. D. Aliprantis,** An invitation to operator theory, 2002

49 **John R. Harper,** Secondary cohomology operations, 2002

48 **Y. Eliashberg and N. Mishachev,** Introduction to the h-principle, 2002

47 **A. Yu. Kitaev, A. H. Shen, and M. N. Vyalyi,** Classical and quantum computation, 2002

46 **Joseph L. Taylor,** Several complex variables with connections to algebraic geometry and Lie groups, 2002

45 **Inder K. Rana,** An introduction to measure and integration, second edition, 2002

44 **Jim Agler and John E. McCarthy,** Pick interpolation and Hilbert function spaces, 2002

43 **N. V. Krylov,** Introduction to the theory of random processes, 2002

42 **Jin Hong and Seok-Jin Kang,** Introduction to quantum groups and crystal bases, 2002

41 **Georgi V. Smirnov,** Introduction to the theory of differential inclusions, 2002

40 **Robert E. Greene and Steven G. Krantz,** Function theory of one complex variable, 2002

39 **Larry C. Grove,** Classical groups and geometric algebra, 2002

38 **Elton P. Hsu,** Stochastic analysis on manifolds, 2002

37 **Hershel M. Farkas and Irwin Kra,** Theta constants, Riemann surfaces and the modular group, 2001

36 **Martin Schechter,** Principles of functional analysis, second edition, 2002

35 **James F. Davis and Paul Kirk,** Lecture notes in algebraic topology, 2001

34 **Sigurdur Helgason,** Differential geometry, Lie groups, and symmetric spaces, 2001

33 **Dmitri Burago, Yuri Burago, and Sergei Ivanov,** A course in metric geometry, 2001

32 **Robert G. Bartle,** A modern theory of integration, 2001

TITLES IN THIS SERIES

31 **Ralf Korn and Elke Korn,** Option pricing and portfolio optimization: Modern methods of financial mathematics, 2001

30 **J. C. McConnell and J. C. Robson,** Noncommutative Noetherian rings, 2001

29 **Javier Duoandikoetxea,** Fourier analysis, 2001

28 **Liviu I. Nicolaescu,** Notes on Seiberg-Witten theory, 2000

27 **Thierry Aubin,** A course in differential geometry, 2001

26 **Rolf Berndt,** An introduction to symplectic geometry, 2001

25 **Thomas Friedrich,** Dirac operators in Riemannian geometry, 2000

24 **Helmut Koch,** Number theory: Algebraic numbers and functions, 2000

23 **Alberto Candel and Lawrence Conlon,** Foliations I, 2000

22 **Günter R. Krause and Thomas H. Lenagan,** Growth of algebras and Gelfand-Kirillov dimension, 2000

21 **John B. Conway,** A course in operator theory, 2000

20 **Robert E. Gompf and András I. Stipsicz,** 4-manifolds and Kirby calculus, 1999

19 **Lawrence C. Evans,** Partial differential equations, 1998

18 **Winfried Just and Martin Weese,** Discovering modern set theory. II: Set-theoretic tools for every mathematician, 1997

17 **Henryk Iwaniec,** Topics in classical automorphic forms, 1997

16 **Richard V. Kadison and John R. Ringrose,** Fundamentals of the theory of operator algebras. Volume II: Advanced theory, 1997

15 **Richard V. Kadison and John R. Ringrose,** Fundamentals of the theory of operator algebras. Volume I: Elementary theory, 1997

14 **Elliott H. Lieb and Michael Loss,** Analysis, 1997

13 **Paul C. Shields,** The ergodic theory of discrete sample paths, 1996

12 **N. V. Krylov,** Lectures on elliptic and parabolic equations in Hölder spaces, 1996

11 **Jacques Dixmier,** Enveloping algebras, 1996 Printing

10 **Barry Simon,** Representations of finite and compact groups, 1996

9   **Dino Lorenzini,** An invitation to arithmetic geometry, 1996

8   **Winfried Just and Martin Weese,** Discovering modern set theory. I: The basics, 1996

7   **Gerald J. Janusz,** Algebraic number fields, second edition, 1996

6   **Jens Carsten Jantzen,** Lectures on quantum groups, 1996

5   **Rick Miranda,** Algebraic curves and Riemann surfaces, 1995

4   **Russell A. Gordon,** The integrals of Lebesgue, Denjoy, Perron, and Henstock, 1994

3   **William W. Adams and Philippe Loustaunau,** An introduction to Gröbner bases, 1994

2   **Jack Graver, Brigitte Servatius, and Herman Servatius,** Combinatorial rigidity, 1993

1   **Ethan Akin,** The general topology of dynamical systems, 1993